Research on Integrated Systems

Proceedings of Previous Conferences

Proceedings of Caltech Conference on Very Large Scale Integration, 1979, Charles L. Seitz, editor. Computer Science Department, California Institute of Technology, 256-80, Pasadena, CA 91125.

Proceedings of Second Caltech Conference on Very Large Scale Integrations, 1981, Charles L. Seitz, editor. Computer Science Department, California Institute of Technology, 256-80, Pasadena, CA 91125.

Proceedings, Conference on Advanced Research in VLSI, MIT, January 1982, Paul Penfield, Jr., editor. Artech House, Inc., 610 Washington St., Dedham, MA 02026.

Third Caltech Conference on Very Large Scale Integration, 1983, Randal Bryant, editor. Computer Science Press, Inc., 1803 Research Boulevard, Rockville, MD 20850.

Proceedings, Conference on Advanced Research in VLSI 1984, Paul Penfield Jr., editor. Artech House, Inc., 610 Washington St., Dedham, MA 02026.

1985 Chapel Hill Conference on Very Large Scale Integration, Henry Fuchs, editor. Computer Science Press, Inc., 1803 Research Boulevard, Rockville, MD 20850.

Advanced Research in VLSI: Proceedings of the Fourth MIT Conference, Charles Leiserson, editor. MIT Press, 55 Hayward Street, Cambridge, MA 02142.

Advanced Research in VLSI: Proceedings of the 1987 Stanford Conference, Paul Losleben, editor. MIT Press, 55 Hayward Street, Cambridge, MA 02142.

Advanced Research in VLSI: Proceedings of the Fifth MIT Conference, March 1988, Jonathan Allen and F. Thomas Leighton, editors. MIT Press, 55 Hayward Street, Cambridge, MA 02142.

Advanced Research in VLSI: Proceedings of the 1989 Decennial Caltech Conference, Charles L. Seitz, editor. The MIT Press, 55 Hayward Street, Cambridge, MA 02142.

Advanced Research in VLSI: Proceedings of the Sixth MIT Conference, April 1990, William J. Dally, editor. The MIT Press, 55 Hayward Street, Cambridge, MA 02142.

Advanced Research in VLSI: Proceedings of the 1991 University of California / Santa Cruz Conference, Carlo H. Séquin, editor. The MIT Press, 55 Hayward Street, Cambridge, MA 02142.

Advanced Research in VLSI and Parallel Systems: Proceedings of the 1992 Brown / MIT Conference, Thomas Knight and John Savage, editors. The MIT Press, 55 Hayward Street, Cambridge, MA 02142.

All of the above volumes are available directly from the publishers.

Research on Integrated Systems

Proceedings of the 1993 Symposium

edited by
Gaetano Borriello and Carl Ebeling

The MIT Press
Cambridge, Massachusetts
London, England

This book was photographed from copy prepared by the authors and was printed and bound in the United States of America.

Library of Congress Cataloging-in-Publication Data

Research on integrated systems : proceedings of the 1993 symposium / edited by Gaetano Borriello and Carl Ebeling.
p. cm.
Includes bibliographical references and index.
ISBN 0-262-02357-1
1. Integrated circuits—Very large scale integration—Congresses. I. Borriello, Gaetano. II. Ebeling, Carl.
TK7874.R47 1993
621.39'5—dc20 93-12230
CIP

Contents

Preface

The range of VLSI research has broadened greatly in the 15 years since the first "VLSI Conference" held at Caltech. Primarily, this change is evident in the emphasis on system solutions that now include a wide range of components and technologies rather than the earlier focus on chip-level solutions. The change in the name of the conference to the "Symposium on Integrated Systems" reflects this trend.

The research papers in this volume represent the 15th edition of a conference with a rich history. The first conference, held in 1979 at the California Institute of Technology, brought together leading researchers in the rapidly emerging field of VLSI. Since then, it has been held three more times at Caltech, six times at MIT, and once each at UNC-Chapel Hill, Stanford University, UC-Santa Cruz, Brown University, and this year at the University of Washington.

The broadened view to integrated systems is a natural direction for what has traditionally been a highly interdisciplinary and application-driven research community. This has been a community that has aggressively applied advances in architecture, circuit design, devices and processes, design automation, and algorithm design to practical problems in the construction of novel high-performance microelectronic systems. It is a very dynamic group that occupies the boundaries between many scientific and engineering disciplines and is in an excellent position to capitalize on advances in any one of the many interconnected fields. Our hope is that this conference will serve to attract a larger group of researchers from both industry and academia to share their experiences. There is no other venue in the field of microelectronics that has as broad and interdisciplinary a mission as this conference and we feel strongly that it serves an important function in enriching the community.

This year's conference includes 4 invited talks in addition to the 21 papers selected by the program committee. Of the invited talks, the keynote address by John Hennessy and the presentation by Chuck Thacker addressed issues in the design of high-performance computers using two of the most advanced processors introduced in 1992. The other two invited talks by Federico Faggin and Mark Weiser focused on the interesting new technologies required for intelligent sensory-motor systems and personal communication devices that promise to change how we live, yet again, before the end of this century.

The 21 regular papers range in scope from theoretical models for programmable devices to the construction of large disk arrays. The full range of "VLSI research" is represented including papers on the design of communication chips and high-speed processor data-paths as well as high-performance interconnection techniques. Design automation research is also represented in papers on optimization, testing, partitioning, and verification. Integrated systems described in this year's conference have applications in computer vision, packet-switching networks, programmable hardware accelerators, highly-parallel computers, and large secondary storage systems.

As program chairs for this conference, we must thank the researchers who submitted papers and the program committee, which did a conscientious job of selecting among these for an excellent program that meets the needs of such a wide-ranging community. Appreciation is due to Mike Foster and Bernard Chern at the National Science Foundation for their continued support for this conference. A special thanks is due to Kay Beck who has single-handedly taken care of the many logistics inherent in organizing such a meeting. Finally, we would like to thank the entire community for providing such interesting work to be showcased.

Gaetano Borriello
Carl Ebeling
Seattle, January 1993

Program Committee

Larry Snyder, University of Washington (general chair)
Gaetano Borriello, University of Washington (program co-chair)
Carl Ebeling, University of Washington (program co-chair)

Anant Agarwal, Massachusetts Institute of Technology
Bob Brodersen, University of California at Berkeley
Paul Chow, University of Toronto
Allan Fisher, Carnegie Mellon University
Ed Frank, Sun Microsystems
Mark Horowitz, Stanford University
Norm Jouppi, Digital Equipment Corporation
Dick Lyon, Apple Computer
Alain Martin, California Institute of Technology
Trevor Mudge, University of Michigan
John Poulton, University of North Carolina at Chapel Hill
Alberto Sangiovanni-Vincentelli, University of California at Berkeley
Martine Schlag, University of California at Santa Cruz
Bob Sproull, Sun Microsystems
Tom Szymanski, AT&T Bell Labs
Jean Vuillemin, Digital Equipment Corporation-Paris Research Lab

Invited Talks

Keynote Address:
Designing a Computer as a Microprocessor: Experience and Lessons from the MIPS 4000
John Hennessy

From Chips to Systems
Charles Thacker

Adaptive Analog Neural Networks
Federico Faggin

System and Technology Challenges in Ubiquitous Computing
Mark Weiser

Research on Integrated Systems

The Design of the Caltech Mosaic C Multicomputer

Charles L. Seitz, Nanette J. Boden, Jakov Seizovic, and Wen-King Su
Computer Science 256-80
California Institute of Technology
Pasadena, CA 91125

Abstract and Introduction

The Caltech Mosaic C is an experimental, fine-grain multicomputer that employs single-chip nodes and advanced packaging technology to demonstrate the performance/cost advantages of the fine-grain-multicomputer architecture. Each Mosaic node includes 64KB of single-clock-cycle dynamic RAM; 2KB of self-test and bootstrap ROM; an 11MIPS processor; a packet interface; and a 60MB/s, two-dimensional, self-timed router. The node is a single, 9.25mm×10.00mm, 1.2μm-feature-size, CMOS chip that, at $V_{dd} = 5$V, operates at 30MHz and dissipates 0.5W. These chips are packaged by tape-automated bonding (TAB) in 8×8 arrays on circuit boards that can, in turn, be composed in two dimensions to construct arbitrarily large arrays of nodes.

In addition to the 8×8 boards, complete Mosaic systems require host-interface boards that allow workstations to send and receive packets on Mosaic channels, and high-bandwidth cables. The Mosaic host-interface boards are built using memoryless Mosaic chips with external SRAM that can also be accessed by the workstation. The host-interface boards were a critical part of the development strategy: They were used both to develop and refine the Mosaic processor, packet interface, and router through MOSIS runs; and also to provide early, operational, Mosaic multicomputers for system-software and test-program development.

The Mosaic programming system consists of a compiler for a source programming notation called C+−, and a distributed runtime system. C+− is an extension of C++ that supports concurrent processes in much the same way that C++ supports program objects. The runtime system, which is written in C+−, provides copyless message handling, highly distributed resource management, automatic process placement, and scheduling.

The Mosaic is intended both for large-scale-computing and for embedded-systems applications. A 16K-node Mosaic system is being constructed at Caltech to explore the programmability and application span of the architecture for large-scale computations. The ATOMIC local-area network [8] is an early example of an embedded-system application of Mosaic components.

1 Multicomputer Scaling Tracks

Multicomputers have already established themselves as a highly scalable, high-performance form of multiple-instruction-multiple-data (MIMD) architecture [1, 3, 4, 15, 16, 17]. As shown in Figure 1, a multicomputer consists of an ensemble of computing *nodes* connected by a message-passing network. Each node of a multicomputer is a computer, including read-write memory, read-only memory for initialization and bootstrap, one or more instruction-interpreting processors, and an interface to the message-passing network. The network connects the nodes in a regular topology such as a hypercube or mesh, so that message packets can be routed algorithmically from any node to any other node.

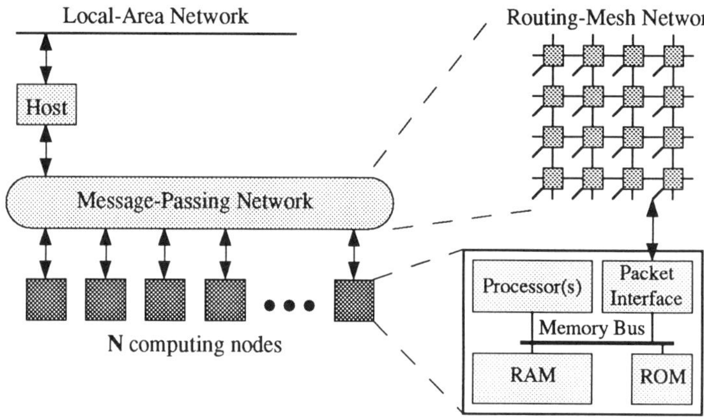

Figure 1: Logical and physical organization of a multicomputer.

Multicomputers can usefully be implemented with nodes that range in size today from a part of a single chip to a large circuit board containing hundreds of chips. Of course, the relationship between node size and complexity evolves with advances in microelectronic technology. In order to appreciate some of the rationale and motivation for the Mosaic experiment, it is helpful to understand the implications of these technology-scaling tracks.

In a January 1985 paper about the Cosmic Cube [14], Charles L. Seitz wrote that: "The Cosmic Cube nodes [implemented as 84 chips on a circuit board, including 128KB of memory] were designed as a simulation in hardware of the nodes we expect to be able to integrate onto one or two chips about five years hence. Future machines of thousands of nodes are feasible ..." The Mosaic C multicomputer closely fulfills the prediction in this 1985 paper. The first Mosaic multicomputers, with processor, packet interface, and router on one chip, and with 128KB of memory on separate chips, were demonstrated in 1990. The first Mosaics based on single-chip nodes, which include 64KB of memory, were demonstrated in 1991. The evolution from the Cosmic Cube to the Mosaic is an example of a nearly *constant-node-*

complexity scaling track (see Figure 2), in which advances in technology are employed to re-implement nodes of a similar logical complexity, but faster, smaller, lower power, and less expensive.

The Mosaic C nodes have only half as much primary memory as the Cosmic Cube nodes; they are, however, $\approx$60 times faster, use $\approx$20 times less power, are $\approx$100 times smaller, and are $\approx$25 times less expensive to manufacture. Such is the remarkable progress in microelectronics over the past decade that the Mosaic C nodes exhibit approximately $60 \times 25 = 1500$ times higher performance/cost than the Cosmic Cube nodes. Thus, the Mosaic C appears in the constant-cost plot of Figure 2 at a point $\log_{10} 1500 \approx 3.2$ decades above the Cosmic Cube in aggregate performance.

We are currently designing the successor to the Mosaic C chip. The Mosaic T is targeted for an 0.8μm-feature-size, three-level-metal, CMOS process; and 50MHz, 3.3V, 0.5W operation. It is being designed to come as close as we can to the following specifications: 128KB of single-clock-cycle internal memory in two interleaved banks of 16K$\times$32, providing 400MB/s memory bandwidth; $\approx$40MIPS, 32-bit, RISC processor; 200MB/s channels; and provision for external, secondary, dRAM memory. If the Mosaic C were taken as the origin of a single-chip-node scaling track, the Mosaic T will be descended from the Mosaic C along a nearly *constant-node-cost scaling track.* It is also possible to use advances in technology to continue the constant-node-complexity track by implementing multiple, faster nodes of complexity similar to the Mosaic C on a single chip.

The Cosmic Cube was also the origin of a different scaling track that has been followed by the commercial, medium-grain multicomputers. The scaling track that, for example, Intel has followed (see Figure 2) has been using advances in technology to scale node performance and memory size. However, Intel has also been slowly but steadily increasing the physical size and cost of the nodes. There are other examples of commercial multicomputers, such as those manufactured by N-CUBE, that would lie closer to a constant-node-cost scaling track from the Cosmic Cube. One of the advantages of the scaling track followed by the commercial, medium-grain multicomputers is that the nodes, because they have been similar in complexity to contemporary personal and workstation computers, have been able to employ the same commodity processor, memory, and software technology. The disadvantage is that the performance/cost scales much more slowly than constant-node-complexity scaling. For example, the 1993 Intel Paragon multicomputer has only about 100 times higher performance/cost than the 1983 Cosmic Cube.

In summary, one motivation for exploring the domain of fine-grain multicomputers is that they provide substantially higher performance/cost than medium-grain multicomputers, and this advantage will continue to grow with future advances in microelectronics. Another motivation is that fine-grain multicomputers can deliver this cost-effective computing in relatively small, low-cost configurations for embedded-system applications, *eg,* they are not aimed exclusively at the "Teraops race." More detailed expositions of our

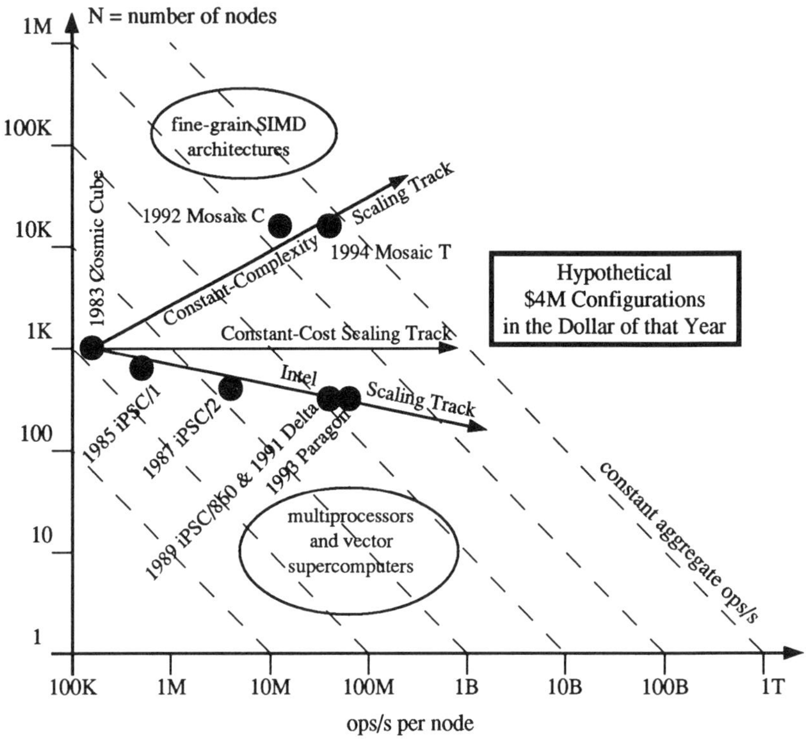

Figure 2: This log-log graph plots node performance on the abscissa, and the reciprocal of node cost on the ordinate by showing how many nodes could be purchased for \$4M. The straight, dashed, diagonal lines are, on this log-log plot, hyperbolas of constant aggregate performance whose values can be read on the abscissa. The price of commercial nodes varied over the product life, and the most attractive price we know of was used to locate the dark circles. For non-commercial machines, the price of a node was estimated to be 3.5 times its leveled-out manufacturing cost. The precise position for these points is arguable, but even a factor of two in cost or performance displaces a dark circle in this log-log plot by only slightly more than its own diameter.

research group's perspectives on multicomputer scaling tracks can be found in [6, 9, 13, 15, 17, 19, 22].

2 The Mosaic C Node

Although medium- and fine-grain multicomputers are structurally similar, a fine-grain multicomputer cannot be designed simply as a variant of a medium-grain multicomputer with scaled-down nodes. The parameters that guided the design of the Mosaic C came from measurements produced from experimental programming systems running fine-grain application programs [2, 5, 9, 22]. Smaller node memories; smaller, more interdependent processes; and a larger number of nodes all imply that the message-passing network must provide relatively higher bandwidth to and from the node and across its bisections. A smaller processor does not readily accommodate features such as address translation for multiprogramming support; instead, code is linked dynamically, and processes execute within the same address space. However, the most critical design consideration is that the smaller processes generate an intense traffic of shorter messages. Medium-grain multicomputers exhibit a typical software overhead for sending or receiving a message of *several hundred* instructions. Such a large software overhead for messages would be completely intolerable for a fine-grain multicomputer. An acceptably small software overhead was achieved for the Mosaic C by a combination of hardware features of the processor and packet interface, and a more streamlined, compilation-based programming system.

2.1 Overview of the Mosaic C Node

The Mosaic C multicomputer node is a single 9.25mm×10.00mm chip fabricated in a 1.2μm-feature-size, two-level-metal, CMOS process. At $V_{dd} = 5$V, the synchronous parts of the chip operate with large margins at a 30MHz clock rate, and the chip dissipates ≈0.5W. The transistor count is approximately 1.2M including the 2^{19} transistors used as storage capacitors in the dynamic RAM. Typical yield using the Hewlett-Packard CMOS34 process (MOSIS SCMOS with $\lambda = 0.6\mu$m (n-well)) in prototype runs is 40%, which results in an unpackaged cost of each tested, working chip of approximately $30.

A plot of the layout of the Mosaic C chip appears in Figure 3. (See the lower-right part of Figure 1 for a block diagram.) The dynamic RAM occupies the lower $\frac{3}{4}$ of the core of the chip, and the memory data and address buses extend across the top of the RAM. The other main sections of layout, left to right across the top, are the processor and packet interface, clock driver, ROM, and router.

The chip area is used approximately as follows:

63% The 64KB *dynamic RAM* is organized as 32K 16-bit words. Memory cycles require only a single clock period; the memory bandwidth

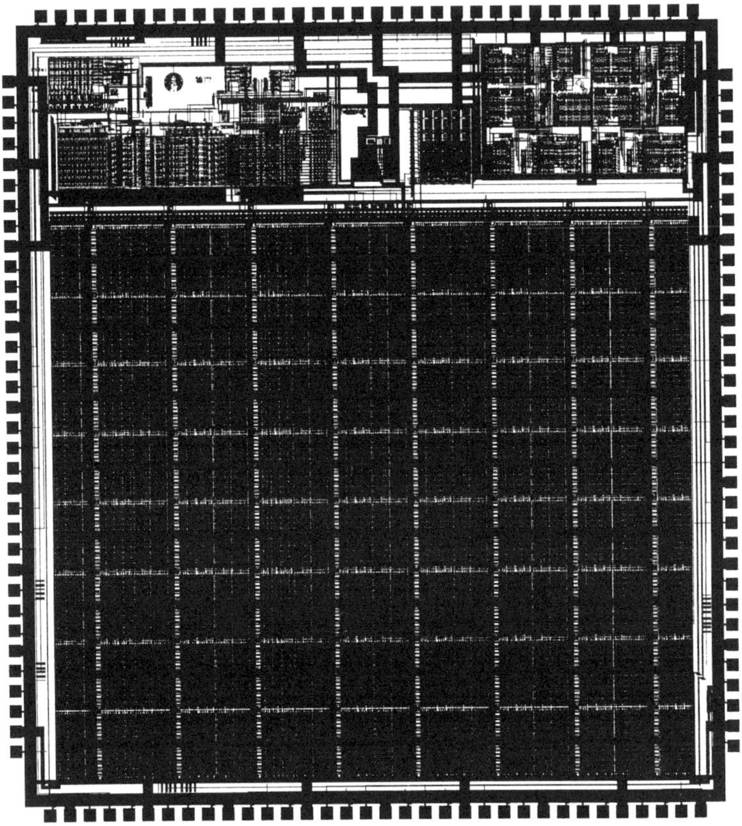

Figure 3: Plot of the layout of the Mosaic C chip.

is, accordingly, 60MB/s. This dynamic memory is refreshed automatically by the processor in otherwise-unused cycles.

1% The *read-only memory* contains initialization, self-test, and bootstrap programs.

3% The 16-bit instruction-interpreting *processor* is programmed in C rather than assembly language, using the Gnu C compiler targeted to the Mosaic instruction set. Standard C benchmarks rate the processor at ≈11MIPS.

3% The 16-bit *packet interface* handles the transfer of packets between the router and memory; performs reliable synchronization between the asynchronous, self-timed router and the synchronous memory; and forms the routing header of packets that are sent from the node. The packet interface is able to send or receive packets at the 60MB/s memory and router-channel bandwidths.

5% The two-dimensional, self-timed, mesh *router* implements the node's part of the message-passing network. The router's 8-bit-wide channels operate asynchronously at a rate somewhat in excess of 60MB/s.

20% The *pad frame* for 136 connections to the package pins, and the wiring between the router and the pad frame. The allocation of pins is 88 for the four bidirectional channels that connect to the north, south, east, and west neighbors; 36 for Vdd and GND, 9 for clock, reset, and refresh timing inputs and two amplified outputs for each to distribute these signals through the mesh; two tachometer outputs; and one LED and test-signature output.

5% *Other wiring* between the parts listed above, internal power distribution, the clock driver, and unused area.

More than one quarter of the silicon area – the packet interface, router, and pad frame – is devoted to communication. Only 3% of the silicon area is used for the processor, and 20 times more area is used for the memory than for the processor. It would be possible to include a much larger processor without greatly increasing the overall chip area.

The Mosaic dRAM is described down to the circuit and layout level in a 1991 paper [21] by its designer, Don Speck. A family of routing and communication chips that are spinoffs from the Mosaic project is described in a companion paper [18] in this volume. The processor and packet interface will be described in the following subsection. The design of the ROM is unremarkable. A small ROM would be required in any case for initialization and bootstrap, but, because the Mosaic chip is a complete computer, we had the opportunity to absorb most of the chip testing into a self-test program in the ROM. The ROM self-test program, described in the second-following subsection, is used in initial wafer testing, in the re-test after inner-lead bonding in the TAB frame, and as part of the board-level testing.

2.2 The Mosaic Processor and Packet Interface

To an experienced chip designer, the design of the Mosaic C processor and packet interface exhibit only a few unusual features. It is how these two parts interact, how they interface to the asynchronous router, and how they utilize the most precious resource on the chip — the memory bandwidth — that makes the Mosaic architecture distinctive and particularly appropriate for a multicomputer node.

Low-latency and low-overhead handling of message packets was imperative for the Mosaic, but, at the same time, message-handling capabilities had to be sufficiently general to allow experiments with different message-handling strategies. The compromise solution used in the Mosaic requires a minimum of three instructions to send a message, two instructions to set up a buffer for message receiving, and from two to five instructions to manage interrupts. For example, a pair of Mosaic nodes, one sending 16B packets and the other receiving these packets, and using the usual context-switching and interrupt-handling routines, communicate at a rate in excess of 1M packets/s.

The Mosaic C processor is a microprogram-controlled rather than a RISC machine, with one-word and two-word instructions, and numerous addressing modes to exploit the single-clock-cycle memory. The number of clocks per instruction (CPI) averages approximately 2.7, but the number of bits required to represent a program is somewhat less than is typical of RISCs. The most unusual feature of the Mosaic C processor is that it is a two-sequence (two-context) machine. Multi-sequence instruction interpretation, in which a processor includes all of the state required for multiple threads of execution, has been used for memory-latency hiding in computers ranging from the Seac to recent synchronous-pipeline multiprocessors, and was first used for the purpose of rapid context switching in the 1950s in the MIT Lincoln Laboratory TX-2 [7]. In the Mosaic, each context has its own program-counter, shadow, and status registers, so that context switching can be performed in zero time. There are also eight general registers for each sequence, and eight shared general registers. Typically, one context is used for running programs, and the other for handling messages under interrupts.

The packet interface manages two DMA channels, one for sending, and one for receiving. Each of those channels includes a small FIFO in the path between the router and the memory. With the 60MB/s asynchronous communication network interfacing to the 60MB/s synchronous memory bus, a 16K-node machine would have a worst case of 10^{12} synchronization events per second, or almost 10^{20} synchronization events per year. Just to be able to reduce the rate of synchronization failure to once per year, the best synchronizers we knew how to build in 1.2μm CMOS technology would require about half of the available clock period. We were clearly very close to a point where small, unexpected, process variations could result in nasty surprises. To deal with this difficult synchronization problem, in which synchronization and data rates are similar, we employed a pipeline synchronizer [19], a technique that can sustain the full bandwidth and achieve arbitrarily low probability of synchronization failure P_f, with the price in both latency and chip area being $\Omega(\log \frac{1}{P_f})$.

The interrupt logic is also part of the packet interface, and it can be enabled to generate interrupts for the following events: a packet has been sent, a packet has been received, or the (program-defined) receive buffer has been exhausted. A single datapath spans both processor and packet interface, and the processor is able to access all of the packet-interface registers: refresh pointer; send pointer, limit, and destination; receive pointer and limit; and interrupt-mask and status registers.

Program execution and the DMA operations are implemented with three nor-nor PLAs that run five concurrent micro-programs: (1) program execution for both contexts, (2a) transferring packets from the memory into the output FIFO, (2b) generating high-priority memory-read requests when the output FIFO is almost empty, (3a) transferring packets from the input FIFO into the memory, and (3b) generating high-priority memory-write requests when the input FIFO is almost full.

The bus arbiter allocates memory cycles on a per-clock-cycle basis according to the following priority scheme: forced dRAM refresh, high-priority send and receive requests, program execution, regular send and receive requests. Any otherwise-unused cycles are used opportunistically for dRAM refresh.

2.3 Wafer Testing

Both the Mosaic chips and the 8×8 boards are tested automatically at their point of manufacture. Despite its complexity, the Mosaic C chip can be thoroughly tested. Because the Mosaic chip is programmable, it is capable of running test programs with no additional hardware support. A basic (self-) test program is included in the ROM; other test programs can be downloaded into the RAM. This testability has been critical to the success of the project.

Testing begins with a finished wafer. The first 4 million test vectors emitted by the Mosaic wafer-test program are used only to supply a clock signal for the self-test program, which computes a sequence of signatures from tests of the ROM, RAM, CPU, and router. Additional vectors are then used to download a short program that emits the signature, which is captured and compared by the tester. Those parts of the chip not covered by the self-test — parts of the router that communicate only with the outside world — are then tested with a sequence of several hundred vectors. Finally, the tester downloads a program that checks for memory bits with marginal refresh-period requirements. The entire test achieves 100% coverage of the dRAM and ROM, and better than 99.99% coverage of the remaining internal circuit nodes.

3 Message-Passing Network

The two-dimensional, self-timed, mesh router implements "wormhole" routing, a blocking form of cut-through routing. The flow-control unit (*flit*) is an 8-bit byte, and flits are conveyed in a queue discipline. Up to five packets may be traversing a single router concurrently, each at 60MB/s. The head of a packet includes $\Delta x, \Delta y$ flits for deadlock-free, dimension-order routing, in which a packet is routed first in x and then in y. The head of a packet forms a path through the network in ≈30ns/router. If the packet encounters a channel that is already occupied, the packet is blocked. The tail flit of the packet breaks the path. See the companion paper [18] in this volume for additional details about the signaling protocols and internal design of the router.

The choice of a two-dimensional mesh for the Mosaic was based on a 1989 engineering analysis. We had planned originally to use a three-dimensional mesh network, because it was thought that a three-dimensional network was more nearly optimal [9, 15] for a 16K-node machine. Why a two-dimensional network?

A Mosaic with 2^{14} nodes could be implemented either as a 128×128 two-dimensional routing mesh or a 32×32×16 three-dimensional routing mesh. Package-pin limitations suggested 8-bit-wide channels that require 11 wires each for the two-dimensional router, or 4-bit-wide channels that require 7 wires each for the three-dimensional router. The three extra wires are required for the request and acknowledge timing signals, and for the tail bit. Thus, the pin counts for the required bidirectional channels are comparable: 88 for the two-dimensional router, and 84 for the three-dimensional router.

A bisection of the 128×128 network is 128 channels in each direction, each operating at 60MB/s, resulting in a bilateral bisection bandwidth of 15.36GB/s. The minimum bisection of the 32×32×16 network is 32×16 = 512 channels in each direction, each operating at 30MB/s, resulting in a bilateral bisection bandwidth of 30.72GB/s. The bisection-bandwidth advantage of the three-dimensional network disappears in smaller configurations, *eg,* a 16×16, 256-node machine has the same minimal bisection bandwidth as a 4×8×8 machine. Although the three-dimensional network would have twice the bisection bandwidth for the 16K-node configuration, it would be more economical to double the bisection of the two-dimensional mesh using bypass channels than it would be to make the entire network three-dimensional for all configurations.

The significance of the bisection-bandwidth figure of merit is that, if message destinations are selected at random, a worst case, then half of the messages must traverse the bisection. However, the localization achieved by the runtime system in process placement reduces the number of messages that cross a bisection to less than $\frac{1}{5}$th of this worst case. Thus, the message volume that the 128×128 network can support is at least 5MB/s per node, *eg,* a 20B message each 4μs (44 instructions), which is consistent with measurements of fine-grain programs. Message latency is not a limitation for either the two- or three-dimensional networks, even with random traffic, and particularly not with traffic that is somewhat localized.

Subsequent, extensive simulations by Michael J. Pertel, who wrote a new, highly efficient network simulator [11] that is available for others to reproduce these results, have demonstrated the optimality of the choice of a two-dimensional network, and of oblivious, dimension-order routing rather than any form of minimal, adaptive routing [12]. Indeed, these simulations reveal that some of the arguments in favor of higher-dimensional meshes are flawed both by using the wrong metric, latency instead of bandwidth, and also by using a functional relationship between latency and dimension that is incorrect for the case of interest, which is large applied loads. One key point is that the bisection utilization under large applied loads increases with radix, and thus favors lower-dimension, higher-radix networks.

Additional, practical considerations are that two-dimensional packaging is considerably simpler, cheaper, and lower risk, and allows easier board replacement than three-dimensional packaging. In detailed packaging designs for both networks, the two-dimensional network required only $\frac{1}{4}$th the number of interboard connections as the three-dimensional network. As illustrated in the closeup of a Mosaic array shown in Figure 4, the mutual fit of the mesh network and the circuit-board medium provides high packaging density, and allows the high-speed signals between routers to be conveyed on short wires. The short wires between neighboring chips take little board area; in addition, they exhibit very low capacitance ($\approx$ 5pF) and inductance ($\approx$ 0.2nH), so that the interchip communication can be accomplished at relatively high speed and low power.

Figure 4: Closeup of a Mosaic array.

The MIT J-machine [10] similarly matches its network and packaging to achieve the same advantages, but in three dimensions, and provides a point of comparison with the Mosaic C. The J-machine is about an order of magnitude larger grain size than the Mosaic C; a J-machine of the same cost as a Mosaic C would, accordingly, have about an order of magnitude fewer nodes, and even less need than the Mosaic to resort to using the third dimension. The Mosaic node requires only $\frac{1}{10}$th the board area (3.6 cm^2 *vs* 36 cm^2) of a J-machine node. The area that would have been required for high-density vertical connections to implement a three-dimensional Mosaic would have seriously degraded the packaging density in the plane of the circuit boards, and the vertical channels would have greatly increased the number of interboard connections.

4 Mosaic Board Assemblies

4.1 The Mosaic C 8×8 Boards

Large Mosaic arrays are constructed using single-board modules tiled in the checkerboard fashion shown in Figure 5, and joined together with stacking connectors. Each board contains an 8×8 array of Mosaic nodes packaged by a surface-mount technique called tape-automated bonding (TAB). TAB packaging achieves substantially lower cost, higher reliability, and higher density than conventional packaging. Instead of bonding chips into packages and packages onto boards, the chips are bonded directly onto boards.

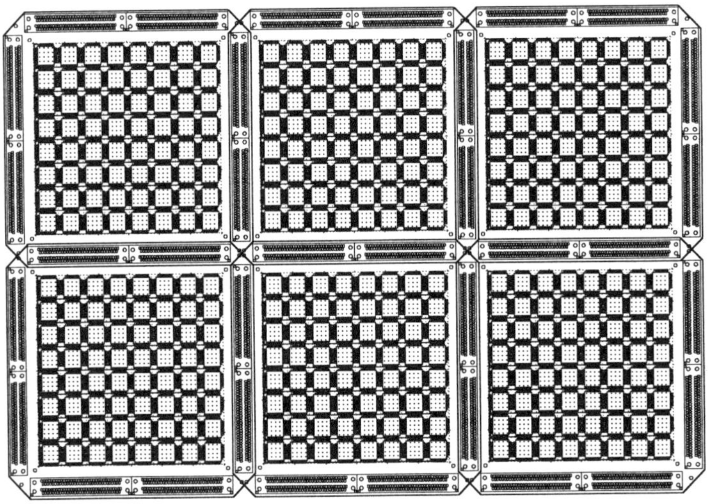

Figure 5: An array of Mosaic 8×8 boards.

After wafer test, functional chips are inner-lead bonded to the TAB frame, and encapsulated to protect them against light and other disturbances. The wafer-test program is then used to retest the packaged chips through test points on the TAB frame. Chips are outer-lead bonded to the board by a robot that removes a chip from the TAB package by trimming and bending the wires into a strain-relief loop to a level below the bottom surface of the chip. The inner-lead bonding is done by Hewlett-Packard in Ft. Collins, Colorado; the testing by Hewlett-Packard in Corvallis, Oregon; and the outer-lead bonding by MCC in Austin, Texas.

The finished board, shown in Figure 6, is about 8 inches on a side, and includes an 8×8 array of Mosaic chips surrounded on all four sides by stacking-connector patterns. The chip pitch in the array is only 19mm. PCB traces

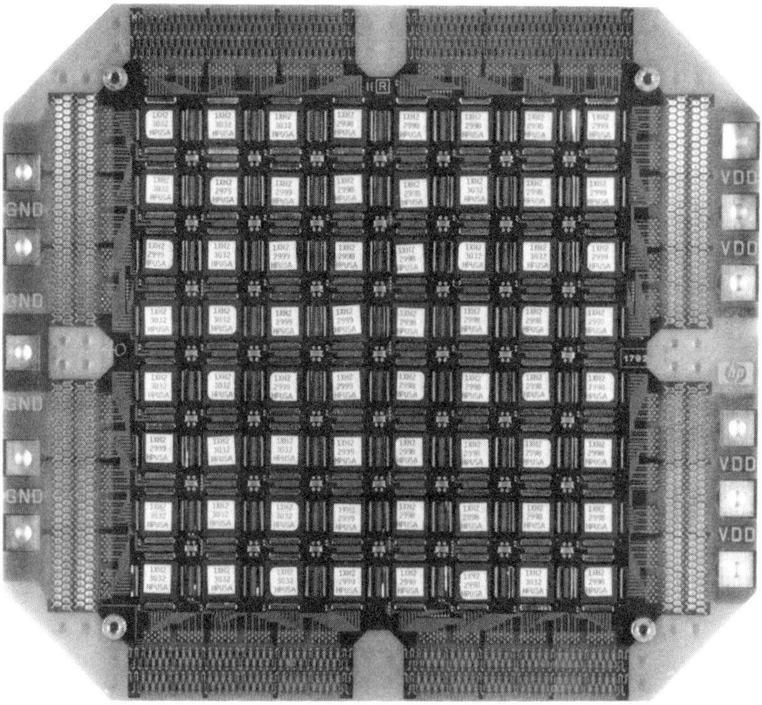

Figure 6: A Mosaic C 8×8 board.

connect channel and control signals between adjacent chips, and between the stacking-connector patterns and chips on the edges of the 8×8 array. The clock, reset, and refresh control signals enter the array on the lower-left corner, and are propagated and amplified from bottom to top along the left-most column, and from left to right along each row. In spite of the long chain of amplifications through CMOS buffers, clock symmetry is maintained.

These 8×8 boards are functional multicomputers. The setup used to test newly completed boards employs the same host interfaces and programs that we use to control operational Mosaic arrays. The test fixture connects the x edge channels of the (0,0) and (7,7) corner nodes through cables to host interfaces in a workstation. All other edge channels are looped back to allow them to be tested. Testing proceeds in two waves starting from the opposite corners, and the workstation displays an animation of the test as it proceeds. The display allows the operator to determine which chips are defective or incorrectly bonded.

A pilot-production run of four 8×8 boards allowed us to verify the complete design with a 256-node Mosaic C system, which is operational and reliable. The Mosaic chips and 8×8 boards are now in modest-scale production under a subcontract with Hewlett-Packard Company. After packaging the

\$30-cost Mosaic C chips onto 8×8 boards, the cost rises to about \$75/node, or \$4800/board. The first 256 boards from the production line will be used to build a 16,384-node, 128×128 machine that will have 1GByte of primary memory, and a peak instruction rate of ≈180GIPS.

4.2 The Mosaic Host-Interface Boards

The Mosaic host-interface boards are versatile, multi-purpose, circuit boards that house a number of Mosaic nodes with Mosaic channels on one side and a microprocessor-bus interface on the other side. The current incarnation of the host-interface board is a two-node Mosaic packaged on the postcard-sized, single-width, SBus board shown in Figure 7.

Figure 7: A Mosaic SBus host-interface board.

The initial purpose of the host-interface boards was to verify the operation of the Mosaic processor, router, and packet interface, fabricated together in memoryless Mosaic chips. The boards provide a bank of 64K×16 SRAM for each memoryless Mosaic, forming a complete Mosaic node. These boards were fabricated through local vendors, and we were able to demonstrate small, functional, Mosaic systems by the Spring of 1990.

The second purpose of the host-interface boards is to provide a software-development platform. The bus interface on the boards presents the Mosaic memory as a bank of ordinary random-access memory. Device-driver and support libraries allow low-level access to the host-interface boards by user programs on the host, which maps the Mosaic memory into its internal address space. This software toolkit allows users to download and monitor programs running on the Mosaic nodes.

The third purpose of the host-interface board is to serve as the interface between hosts and Mosaic arrays, as shown in Figure 8. The ribbon-cable

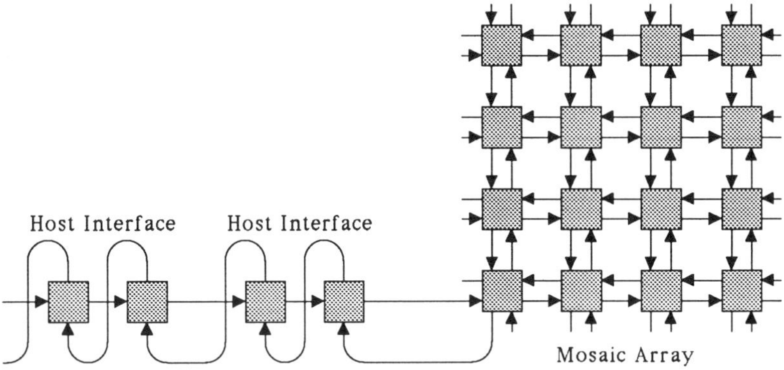

Figure 8: Standard connection of Mosaic host interfaces and arrays.

connectors on the back of the host-interface boards are wired to selected channels on the memoryless Mosaic chips. Programs running on the memoryless Mosaic nodes communicate with the Mosaic array by message passing through the ribbon-cable connectors, and with the host computer by shared memory through the bus interface. Host-interface boards can also be chained together using ribbon cables, forming a linear array of Mosaic nodes, and the basis for using Mosaic components to build local-area networks such as ATOMIC [8].

5 The Mosaic Programming System

5.1 C+-

C+- is the result of an experiment to express reactive-process, concurrent programs [22, 15] in an object-oriented programming notation. Although the C+- system is portable across a wide range of architectures, the Mosaic C has been both the driving force and the reality test behind this effort. Design decisions have consistently been made to avoid compromising the performance of C+- programs on the Mosaic. Higher-level programming systems may be layered on top of C+-, but C+- is intended to serve as the Mosaic's lowest-level, *workhorse* programming system, suitable both for operating-system and application programming.

C+- is an extension of C++, except that its current implementation does not support global variables. The emphasis of C+- is on providing efficient support for the simple abstractions fundamental to the reactive-process computational model: process creation and communication. C+- strives not to impose higher-level policies on synchronization, communication protocols, or process placement.

The emphasis on simplicity and efficiency, together with the well-defined mechanisms to override the default behavior, makes C+- particularly appropriate for expressing operating or runtime systems. The full set of object-oriented features carried over from C++ makes C+- a powerful tool that programmers can use to define and use higher-level constructs.

5.1.1 Extending C++ to C+-

This section assumes that the reader is familiar with the basic concepts of object-oriented languages, and of C++ in particular: classes, inheritance, and access rules.

The C++ *class* concept is carried over intact to C+-: `class` is a user-defined type; an *object* created according to a class definition is a piece of memory with a set of operations defined on it, and a set of access rules. Operations on objects have immediate action, characteristic of sequential programming models.

C+- introduces the *process* concept. Syntactically, the `processdef` keyword replaces the `class` keyword (Program 1). Access rules are associated with data members and/or functions of a process definition, and process definitions can be derived from other process definitions.

```
processdef      P
{
private:
        int     data;
public:
                P()             { data = 0; }         // startup
atomic  void    write(int i)    { data = i; }         // update
atomic  int     read()          { return(data); }     // retrieve
};
```

Program 1: A Process Definition

However, a process created according to a `processdef` is more than a type: it is an independent computing agent, and a unit of potential concurrency. Its `public` interface consists of a set of atomic actions. The invocations of `atomic` functions have delayed action, characteristic of concurrent programming models that tolerate the inevitable latency of communication, rather than attempting to hide the latency from (some) users.

Creating a process is no different from creating an object (Program 2). In most cases, processes are created dynamically (`pp = new P;`), and persist until they are explicitly destroyed (`delete pp;`). One can also create a local instance of a process, just as with any other variable. This local variable is destroyed implicitly when execution leaves its scope.

Invoking an `atomic` function of a process is equivalent to wrapping up the argument list by value, and sending it in a message (`pp->write(i);`).

Message receiving has two forms: *explicit*, when a value is returned from a call to an `atomic` function (`i = pp->read();`); and *implicit*, when an

```
void
f()
{
        int  i;             // declaring an integer
        P*   pp;            // declaring a process pointer

        pp = new P;         // creating a process
        i = pp->read();     // retrieving a value
        pp->write(i);       // updating
        delete pp;          // destroying the process
};
```

Program 2: Programming with Processes

atomic function of a process returns, since that process will remain idle until the next invocation of one of its atomic functions.

The two forms of receive, explicit and implicit, cover the two extremes of the spectrum of possible mechanisms for message discretion: explicit receive accepts only a particular message from a particular process; implicit receive accepts any message from any process. Our experience in programming with processes and messages, particularly the results reported in [5], provide strong arguments that there is a need for a selective-receive mechanism that is more general than the remote procedure call (implicit receive), yet one that imposes minimum runtime-system overhead. The mechanism supported by C+- allows the programmer to enable or disable selectively, at run time, each atomic function of a process.

Pointers to processes form a global name space. They can be exchanged in messages and used in the same way that ordinary pointers can be passed around and dereferenced in C++.

5.1.2 Translating C+- into C++

The reasons for translating from C+- to C++ instead of a full compilation to Mosaic code fall into two categories: First, this was a faster way to build a running system. Second, we had good experience in re-targeting the Gnu g++ compiler to produce excellent code for the Mosaic C processor. The translator's main concerns are global-name-space management, and various optimizations, some of which are listed here:

- Arguments to atomic functions are written and read directly into/from the message buffers, not copied off of the stack.
- Remote Procedure Calls (explicit receives) are implemented in terms of simpler communication primitives.
- Support for communicating arbitrarily complicated, linked data structures.

- Support for communication across heterogeneous machine boundaries. A very simple machine description for each machine that may be involved in direct communication is all that is necessary to handle the alignment, data-size, and data-layout incompatibilities.

5.2 The Mosaic Runtime System

Reactive-process computations expressed in C+- may involve very large numbers of processes, and may be highly dynamic. Except for application programs with regular, predetermined communication topologies, process placement and naming would be a tedious problem for the programmer, and is better performed by the runtime system. Indeed, process placement is the means which the resource allocation performed by a multicomputer runtime system can have a global impact. A multicomputer runtime system also allocates local resources such as node memory, but the distinction between global and local resource allocation blurs when one considers that the memory resources of a single, fine-grain, multicomputer node are so limited that they can easily be exhausted, regardless of the sophistication of the local-memory-allocation algorithm. For example, if a node cannot allocate memory for incoming messages, the computation cannot proceed. However, if resources external to a node can be accessed, there is the potential for improved system reliability, performance, and overall resource utilization.

The Mosaic Distributed Runtime system [6], MADRE, is, accordingly, composed of *components,* one on each node, that communicate and cooperate to manage their collective resources. Each component could be represented as a single process; however, the process abstraction permits the component itself to be implemented as a collection of fine-grain processes. A set of *kernel* processes bound to each node may be aided by a set of *remote* processes created as required on other nodes. For example, MADRE deals with an impending receive-queue overflow by having a kernel process export a portion of the receive queue by creating a remote process. Similar techniques are used for distributed management of memory for code.

Another mechanism for balancing resources is that a component that receives a message instructing it to create a process may forward this message to another component if its node does not have adequate resources. In order to satisfy a requirement of the computational model that references to remote, system processes be generated immediately, MADRE employs different mechanisms for creating system and user processes. The reference to a user process contains the address of its node, whereas the reference to a system process contains the address of the node where placement was first attempted. Subsequent messages to a system process traverse the same path as the message that created this process; there is usually only one such message. The component that accepts placement of a user process returns the reference to the parent, so that subsequent messages to a user process are sent directly.

MADRE's highly streamlined handling of messages is accomplished by using the interrupt context of the two-context Mosaic processor. Other MADRE processes and user processes execute in the other context. Message handling is completely free of copying. The receive channel is set up initially to switch the processor into message handling when the header of an incoming message is received. MADRE determines from the header where to store the payload of the message, and sets up the receive channel accordingly. Send operations are performed directly from memory, and the interrupt on a completed send is used to unlock the memory segment that was sent.

MADRE was written in C+-, and MADRE processes are organized in layers. The structure of messages sent to processes mirrors the process-layering structure. A process operates on a segment of an incoming message, and may then dispatch to a process in another layer. Using C+- process inheritance, processes are provided with access to data and system services from processes in previous layers.

5.3 Preliminary Evaluation

We are witnessing a proliferation of programming systems that attempt to merge concurrent semantics into a particular object-oriented language. Most of these systems are motivated primarily by networks of workstations, and thus favor client-server, Remote Procedure Call (RPC) semantics. Concurrency is usually introduced after the fact, typically [20] by using futures (synchronization variables). Our approach was different in that we started with the reactive-processes computational model, and sought to allow programs formulated in this way to be expressed in an object-oriented notation of choice, a simple derivative of C++.

As in most programming-system experiments, C+- tries to strike a balance between two requirements: *efficiency* and *expressivity*. Our experiences with C+- have so far been positive: Both the Mosaic runtime system and user-application programs exhibit few compromises in efficiency or expressivity.

- *Efficiency:* The C+- environment imposes a low software-overhead penalty on accessing the communication capabilities of the Mosaic. The software overhead for a user process running on top of the runtime system is $\approx$30 assembly instructions per message. Nearly the full communication performance of the hardware is available to the user.

- *Expressivity:* The entire set of features that made C++ the object-oriented notation of choice for a growing number of complex applications is supported: inheritance, protection, overloading, as well as well-defined mechanisms to circumvent default behaviors. The full expressive power necessary for application programs is also available to the system programmer.

- *Portability:* The C+- to C++ translator consists of about 60,000 lines of C++ code, and it is both compile-machine- and target-machine-independent. Various runtime system configurations range from 2,000 to 10,000 lines of C+- code. To support a particular machine or operating system, all that is required is a small library of C+- functions that define access to the communication capabilities.

Retrospective

The Mosaic project has required the efforts of many people over a period of seven years since we started the first, preliminary designs of the principal components of the Mosaic C chip. In addition to the end product, which is being used to explore the application span and programmability of fine-grain multicomputers, this project has been the vehicle and motivation for most of our research group's efforts in VLSI design, routing-chip technology, and reactive-process programming systems. The project could have been done in half the time if the end product were the only goal, but we would have learned less than half as much, and the result would not have been so pleasing.

Acknowledgments

The research described in this paper was sponsored by the Defense Advanced Research Projects Agency, and monitored by the Office of Naval Research. In addition, let us acknowledge the numerous contributions to the Mosaic project made by our operations manager, Arlene DesJardins. Bill Hanna of Hewlett-Packard Ft. Collins developed the TAB tooling, did the detail design of the Mosaic 8×8 board and test fixtures from our conceptual design, and coordinated the packaging work done in the advanced-packaging project at MCC. At HP Corvallis, Kathleen Ryan handled the wafer testing, and John Vietor coordinated the wafer fabrication and the overall HP effort.

References

[1] Tom Alexander. How to Stop Fearing and Start Loving the Parallel Computer. *NSF Mosaic,* pp. 2–11, Spring 1992, National Science Foundation.

[2] William C. Athas. Fine-Grain Concurrent Computation. Caltech Computer Science Technical Report 5242:TR:87 (PhD thesis), 1987.

[3] William C. Athas, Charles L. Seitz. Multicomputers: Message-Passing Concurrent Computers. *IEEE COMPUTER* 21(8): 9–24, August 1988.

[4] Gordon Bell. Ultracomputers: A Teraflop Before Its Time. *CACM,* 35(8): 26–47, August 1992.

[5] Nanette J. Boden. A Study of Fine-Grain Programming Using Cantor. Caltech-CS-TR-88-11, 1988.

[6] Nanette J. Boden. Runtime Systems for Fine-Grain Multicomputers. Caltech-CS-TR-92-10, 1992.

[7] Wesley A. Clark. The Lincoln TX-2 Computer Development. *Proceedings of the WJCC,* pp. 143–145, IRE, New York, 1957. (See also the papers immediately following this introduction.)

[8] Cohen, D., Finn, G., Felderman, R., DeSchon, A. ATOMIC: A High-Speed, Low-Cost, Local Area Network. USC/ISI Technical Report ISI/RR-92-291, Sept-1992.

[9] William J. Dally. *A VLSI Architecture for Concurrent Data Structures.* Kluwer Academic Publishers, Norwell MA, 1987.

[10] Michael Noakes, William J. Dally. System Design of the J-Machine. *Proceedings of the Sixth MIT Conference on Advanced Research in VLSI,* MIT Press, 1990.

[11] Michael J. Pertel. A Simple Network Simulator. Caltech-CS-TR-92-04, 1992.

[12] Michael J. Pertel. A Critique of Adaptive Routing. Caltech-CS-TR-92-06, 1992.

[13] Charles L. Seitz. Ensemble Architectures for VLSI – A Survey and Taxonomy. *Proceedings of the MIT Conference on Advanced Research in VLSI,* pp. 130–135, Artech Books, January 1982.

[14] Charles L. Seitz. The Cosmic Cube. *CACM,* 28(1): 22–33, January 1985.

[15] Charles L. Seitz. Multicomputers. Chapter five in *Developments in Concurrency and Communication,* edited by C. A. R. Hoare, Addison-Wesley, 1990.

[16] Charles L. Seitz. Concurrent Computation and Programming. Chapter one in *VLSI and Parallel Computation,* edited by Roberto Suaya and Graham Birtwistle, Morgan Kaufmann Publishers, 1990.

[17] Charles L. Seitz. The Mosaic C: An Experimental Fine-Grain Multicomputer. in *Future Tendencies in Computer Science, Control, and Applied Mathematics,* Proceedings of the International Conference Celebrating the 25th Anniversary of INRIA, edited by Alain Bensoussan and Jean-Pierre Verjus, LNCS 653, Springler-Verlag, 1992.

[18] Charles L. Seitz and Wen-King Su. A family of routing and communication chips based on the Mosaic. This volume.

[19] Jakov Seizovic. The Architecture and Programming of a Fine-Grain Multicomputer. Caltech Computer Science PhD thesis in preparation; publication expected June 1993.

[20] K. Stuart Smith, Arunodaya Chatterjee. A C++ Environment for Distributed Application Execution. MCC Technical Report ACT-ESP-275-90, 1990.

[21] Don Speck. Fast 512K Scalable CMOS dRAM. *Advanced Research in VLSI 1991: UC Santa Cruz,* MIT Press, 1991.

[22] Wen-King Su. Reactive-Process Programming and Distributed Discrete-Event Simulation. Caltech-CS-TR-89-11, 1989.

RAID-II: Design and Implementation of a Large Scale Disk Array Controller

R. Katz, P. Chen, A. Drapeau, E. Lee, K. Lutz, E. Miller, S. Seshan, D. Patterson

Computer Science Division
Electrical Engineering and Computer Science Department
University of California, Berkeley
Berkeley, CA 94720

Abstract: We describe the implementation of a large scale disk array incorporating over 100 high performance 3.5" disk drives. It is designed to provide 40 MB/s and 40 GB capacity in three 19" racks, attached as a file server on a Gb/s local area network. The controller implements a high bandwidth interconnect between an interleaved memory, an XOR unit, the network interface, and the disk interfaces. The system is now operational, so far obtaining 28 MB/s in loopback tests. We review the design decisions, history, and lessons learned from this three year university implementation effort to construct a truly *large scale system assembly*.

1. Introduction

The University VLSI research community has undergone a dramatic evolution. A VLSI system was originally defined as a "complete system on a single integrated circuit chip." Given the original circuit complexity, few truly interesting systems could be reduced to a single chip. They needed to be surrounded with considerable support circuitry for memory and subsystem interfaces. Initially, this system context was provided by placing the custom VLSI on wire-wrap boards. As the community's design experience grew, the VLSI "systems" spanned multiple chips, placed and interconnected on printed circuit boards. Now the systems we design are significantly more complicated, requiring more complex packaging across modules, printed circuit boards, and racks. We call such systems *large scale system assemblies.*

VLSI systems building at Berkeley illustrates this evolution. Between 1978 and 1981, Patterson, Séquin, and their students developed the Berkeley RISC chips. These were true single chip systems, with just enough of a surrounding context to run some benchmark test programs. SOAR came next, developed by Patterson and his students between 1981 and 1984. They expanded the system from a single chip to a complete microcomputer wire-wrap board and a run-time Smalltalk system.

The SPUR multiprocessor project, from 1984 to 1987, took the system concept one step further. The processing element was a CMOS chip set: a CPU, an FPU co-processor, and an integrated cache controller/memory manager. A large printed circuit board was designed for the cache memory and bus interfaces. Substantial amounts of software were also developed: a distributed multiprocessor operating system, Sprite, a C compiler, and a parallel LISP system.

Our most recent project is RAID-II, a large scale system assembly that implements a high performance disk array and file server. The design and construction of RAID-II began in 1989 under Katz's direction, and is now reaching its completion. Like much of the research community, we have replaced custom VLSI design with commercially available high speed programmable logic devices. We've undertaken significant mechanical design, to support the construction of an assembly able to hold more than 100 small formfactor disk drives and numerous printed circuit boards for disk controllers, network interfaces, and their interconnect. Our design effort focused on building a large, double-sided printed circuit board in surface mount technology that integrates several commercially available subsystems. The distributed file services are being developed within Sprite [Lee 92].

This paper describes the design context, goals, and constraints for RAID-II. The rest of the paper is organized as follows. In Section 2, we describe the design context for RAID-

II, including disk arrays and gigabit networks. Section 3 covers the RAID-II architecture. Section 4 describes the implementation details, including the board design and fabrication of the disk array racks. Section 5 describes the software elements of the controller. The lessons we have learned so far and the project's design history are reviewed in Section 6. Our summary, conclusions, and future plans are given in Section 7.

2. Network-Attached Storage

In this section, we briefly review our design context. We describe redundant arrays of inexpensive disks (RAID) and the major interfaces supported by the controller: SCSI and HIPPI. We also discuss the Ultranet Gb/s local area network and lessons we learned from our first prototype. More details on the underlying technologies can be found in [Katz 92].

2.1. Redundant Arrays of Inexpensive Disks

RAIDs (*Redundant Arrays of Inexpensive Disks*) are disk system organizations that store redundant data to achieve high availability. Alternative RAID organizations tradeoff between availability, I/O performance, and the redundancy overhead [Patterson 88]. The organization best suited for high availability and high I/O operation and data rate is the parity array (RAID-5).

RAID-5: Interleaved Parity

The array is partitioned into recovery groups of N data disks. Parity is computed bit-wise horizontally across the data disks, and is stored on an N+1st parity disk. If a single disk fails, a given bit on the failed drive can be reconstituted from the bits on the surviving disks, by maintaining the sense of the parity. The overhead is only 1/(N+1) bit for each data bit.

For each written data bit, the parity bit must also be updated. The old data bits are read, the difference between them and the new bits is computed, the old parity is read, and parity bits associated with changed data bits are complemented. A logical write becomes four physical I/Os. Thus, the array write rate is reduced to 25% of a conventional disk system. The Log Structured File System developed at Berkeley circumvents this problem by treating disk as an append-only medium to which large segments are written. Parity can be computed in advance for these large, stripe-oriented writes [Rosenblum 91].

Small Computer System Interface

SCSI is the storage interface supported by small formfactor disk drives. It views a disk drive as a linear byte stream. The interface defines a high-level message-based protocol for communications among initiators (masters) and targets (slaves). Initiators manage multiple simultaneous operations. Targets explicitly notify the initiator when they are ready to transmit data or when they need to throttle transfers.

2.2. Gigabit Networks and Diskless Supercomputers

It is now possible to extend the workstation client-server model to higher performance environments. A key goal of RAID-II is to demonstrate the feasibility of *diskless supercomputers*. The network is rapidly emerging as the "backplane" of high performance systems.

High Performance Parallel Interface

The High Performance Parallel Interface, HIPPI, is a high speed unidirectional point-to-point interface. Two-way communications requires two channels, one for commands and write data and one for status and read data. Data is transmitted at 800 Mb/s.

HIPPI implements a simple data transfer protocol. The source asserts a request signal to gain access to the channel. A connection signal grants it. The source begins the transfer when the destination asserts ready, providing simple flow control.

The minimum unit of data transfer is the *burst* of 1 to 256 words. The destination must be able to accept a full burst if it asserts ready. The burst is sent as a continuous stream of words, one per clock period, and continues as long as the channel burst signal is asserted.

When this is unasserted, the sender transmits a CRC checksum for the whole burst.

UltraNetwork

The UltraNetwork is a hub-based Gb/s network. The hubs provide the high speed interconnect for packet routing. They are connected by unidirectional *serial links*, used in pairs. If optical fiber is chosen for the links, data can be transmitted at rates of up to 250 Mb/s and distances to 4 Km. Higher speed is achieved by interleaving transmissions over multiple serial links.

Computers are connected to the network in two different ways: through *host adapters* and *hub-resident adapters*. A host-based adapter resides within a workstation backplane, just like any other peripheral controller. A different approach is needed for RAID-II. It connects to the network through a HIPPI interface, not a standard backplane. The hub-resident adapters place the network interface to the Ultranet within the hub itself.

2.3. RAID-I Prototype and Experiences

RAID-I, the first RAID prototype, consisted of a SUN 4/280 file server with 128 MB of memory, four Interphase dual SCSI channel host bus adapters, and 32 340 MB 5.25" Wren-IV disk drives (4 drives on each of 8 SCSI channels).

[Chervenak 91] evaluated the prototype for performance bottlenecks. The most serious is the server's memory system, which limited application throughput to only 2.3 MB/s. I/O operations caused excessive memory-to-memory copies and cache flushes. The array did better on small random reads, achieving nearly 300 per second before the server becomes CPU-limited. Other bottlenecks include the VME backplane (about 15 MB/s), bandwidth on the disk controller (4 MB/s), and overheads associated with the SCSI protocol. Nevertheless, RAID-I provided an indispensable testbed for experimenting with algorithms for data striping, reconstruction, and hot spare management.

3. RAID-II Architecture

3.1. Design Goals and Constraints

RAID-II is a HIPPI-attached high performance file server that interfaces a SCSI-based disk array to the Ultranet. It packages over 100 IBM 3.5" disk drives (320 MB capacity—1.2 GB are now available) in three 19" racks, providing 40 GB of storage. While the Ultranet provides high bandwidth, its latency is actually worse than slower speed networks. RAID-II also supports Ethernet for small transfers where network latency dominates service time.

We used commercially available components when possible. Thinking Machines (TMC) provided a board set for the HIPPI channel interface. Array Technology (ATC) provided a VME-based multiple SCSI string board. We designed the controller to provide very high bandwidth between the network, disk, and memory interfaces.

The major hardware design is the *XBUS card*, a crossbar that connects the HIPPI boards, multiple VME busses, and an interleaved, multiported semiconductor memory. It provides the high bandwidth datapath between the network and the disks. It is controlled by an external file server through memory-mapped control registers (see Figure 1).

3.2. Controller Architecture

Controller Overview

The XBUS card implements an 8-by-8 32-bit wide crossbar bus. All crossbar transfers involve the on-board memory as either the source or the destination of the transfer. The ports are designed to burst transfer at 50 MB/s, and sustain transfers of 40 MB/s. The crossbar provides an aggregate bandwidth of 400 MB/s.

The controller memory is allocated eight of the crossbar ports. Data is interleaved across the eight banks in 16-word interleave units. Although the crossbar is designed to move large blocks to/from memory and the network and disk interfaces, it is still possible to

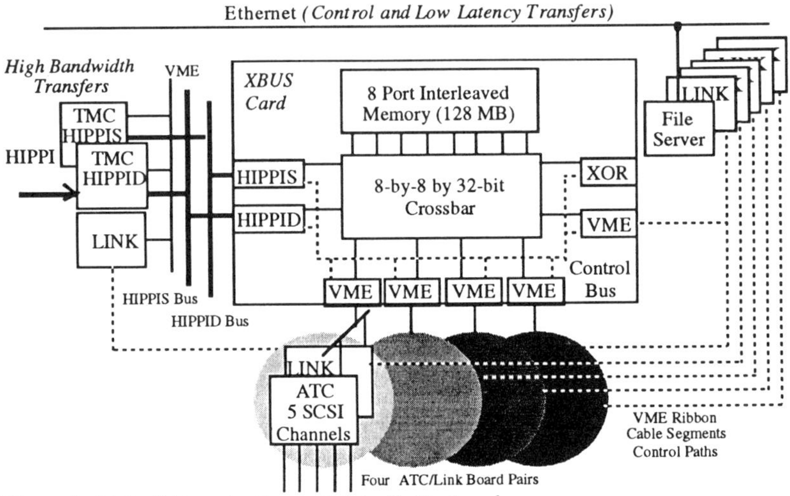

Figure 1: *RAID-II Organization as Originally Designed*
A high bandwidth crossbar ties the network interface (HIPPI) to the disk controllers (ATC) via a multiported memory system. The board incorporates hardware to perform parity calculations. An internal control bus provides access to the crossbar ports, while external point-to-point VME links provide control paths to the surrounding SCSI and HIPPI boards. Due to printed circuit board manufacturing difficulties, the actual prototype was scaled down to a 4-by-8 crossbar with 32 MB of memory.

access a single word when necessary. The external file server accesses the on-board memory through the VME control interface.

Two of the remaining ports interface to the TMC I/O bus (HIPPIS/HIPPID busses). The HIPPI board set also interfaces to this bus. Since these XBUS ports are unidirectional, the controller is limited to a sustained transfer to/from the network of 40 MB/s.

Four additional ports connect the XBUS board to 4 single-board multi-string disk controllers via VME buses. 15 to 30 disks can be attached to each disk controller, in 5 rows of 3 to 6 disks. Thus, 60 to 120 disk drives can be connected to each XBUS card.

Of the remaining two ports, one is dedicated to compute parity and the last links the board to the external file server to provide access to on-board memory and control registers.

XBUS Crossbar

The XBUS is a synchronous multiplexed (address/data) crossbar-based interconnect that uses a centralized strict priority-based arbitration scheme. All paths to memory can be reconfigured on a cycle-by-cycle basis. Each of the eight 32-bit XBUS ports operates at a cycle time of 80 ns.

The XBUS supports reads and write transactions. During each XBUS transaction, 1 to 16 words are transferred over the interconnect. Each transaction consists of an arbitration phase, an address phase, and a data phase. If there is no contention for memory, the arbitration and address phases each take a single cycle; data is then transferred at the rate of one word per cycle. The memory may arbitrarily insert wait cycles during the address and data cycles to compensate for DRAM access latencies and refreshes. The shortest XBUS transaction is a single word write, which takes three cycles (arbitration, address, and data).

Actual memory module contention is infrequent. Most XBUS ports perform large sequential accesses. When they do conflict, the loser of the arbitration deterministically follows the winner around the memory modules, avoiding further conflicts. Each XBUS port buffers data to/from the XBUS (32 KB for HIPPI, 4 KB for VME and XOR).

Before deciding upon the XBUS, we considered using a conceptually simple single wide bus. Because we decided to design in TTL/CMOS, we were limited to a bus cycle of 80 ns. To sustain 40 MB/s over each of the two HIPPI ports, we need a bus bandwidth of 200 MB/s: 80 MB/s for the reads (40 MB/s into and out of memory) and 120 MB/s for writes (as for reads plus 40 MB/s for parity). Since we cannot realistically expect to achieve a bus utilization greater than 70-80 percent, this implied that the bus would have to be 256-bits wide with a peak bandwidth of 400 MB/s. The number of FIFO and transceiver chips to implement such a wide bus port is very large. While we could have used a small number of time-multiplexed 256-bit ports, interfaced to narrower 32-bit and 64-bit busses, the result would have been a more complex system. Eventually, the 256-bit bus was abandoned in favor of a crossbar-based interconnection scheme, the XBUS.

Processing

The configuration of Figure 1 contains no less than seven microprocessors: one in each of the HIPPI interface boards, one in each of the ATC boards, and one in the file server. The HIPPI processors handle some of the network processing normally performed by the server. The disk controller processors manage the details of the SCSI interfaces. The file server CPU does conventional file system processing. Since it is executing file server code, the file server needs access only to the file system meta-data, not user data. The file server cache can be located within the XBUS card, close to the network and disk interfaces.

Performance and Scaling Strategy

Since a single XBUS card is limited to 40 MB/s, our strategy for scaling is to interleave data transfers across multiple XBUS card. These can share a common HIPPI interface through the TMC I/O busses. Two XBUS boards can sustain 80 MB/s, more fully utilizing the available bandwidth of the server's HIPPI interface.

4. RAID-II Implementation

4.1. Disk Array Subsystem

Two disk packaging orientations commonly found in disk arrays are the "wall of disks," in which the drives are mounted in a single vertical plane within a 19" chassis, and the horizontal "shelves of disks." We chose the latter scheme, illustrated in Figure 2.

Our main motivation was to make effective use of the depth of a standard 19" rack. A wall of disks does this only to the extent that the drive housing has depth in its longest dimension. As drive formfactors shrink to 3.5" and below, this dimension also shrinks. Less of the rack depth is used in a wall of disks. The shelf approach give us a high capacity per

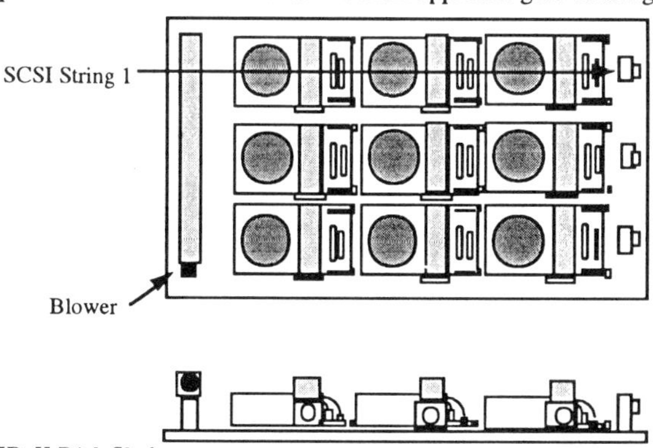

Figure 2: *RAID-II Disk Shelves*
The figure show the disk shelves used in the RAID-II prototype. Drives are placed flat in a 3-by-3 orientation on the shelf. This is in contrast to the "wall of disks," in which drives are placed on one face of the rack. For 3.5" disks, shelves yield a higher storage capacity per cubic foot than the disk wall.

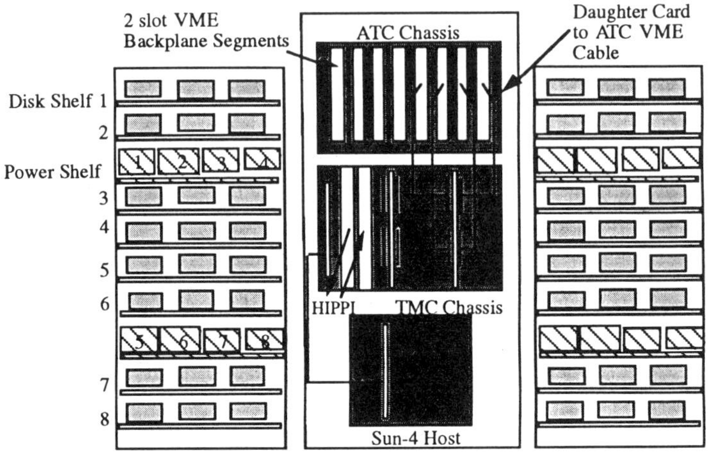

Figure 3: *RAID-II Hardware Racks*
The RAID-II hardware rack contains a SUN-4/280 host at the bottom, a TMC chassis in the middle, and a VME backplane at the top. The TMC chassis holds the HIPPI interface board set, RAID-II XBUS card(s), and a VME-VME link board to allow the host to control the boards in the chassis. The VME backplane is partitioned into independent 2-slot segments for interfacing the disk drives.

square foot. Based on the highest capacity 3.5" drives, our design yields a capacity of 72 1.2 GB drives in a 5 ft. 19" chassis. The total storage is 86.4 GB, an efficiency of 2.34 GB/ft^3.

We accomplished power distribution within the RAID-II chassis with embedded power supply shelves (see Figure 3). Each shelf serves four drive shelves, two above and two below. This symmetric arrangement minimizes the need to cable high current, low voltage power from the supply to the drive. To provide a measure of fault isolation, each drive shelf is powered by an independent AC-DC power converter. Each supply is independently fused, so a failure cannot affect its neighbors via the AC distribution system. Similarly, each SCSI string is independently fused and switched, to minimize fault propagation in the DC distribution system. Finally each drive is individually fused, further isolating power faults, thus protecting the system from the failure of a single drive.

A secondary goal of the subsystem is to support the servicing of the disk array when a drive fails. To allow users to insert and remove drives from the array, without tools or special knowledge, we made the disk drive the basic field replacable unit. We package the drives in a housing that allows simple insertion and removal from the array. The SCSI bus connection, power connection, SCSI address, and a spindle sync signal are connected from the drive to two connectors mounted on our drive housing assembly. This subplate is designed to float to ensure smooth engagement with the mating connectors on the shelf SCSI backplane. The backplane provides geographic addressing for the drive

4.2. XBUS Card

Physical Design

The XBUS card is the single most complex printed circuit board every designed at Berkeley. It makes extensive use of surface mount technology, to achieve high component density and to minimize interconnect lengths, especially in the crossbar. To further increase density, we partitioned the design into a main board and four (identical) daughter cards (see Figure 4). The daughter cards connect the crossbar and the ATC VME interfaces. This efficiently uses the vertical space above and below the main board. In addition, the daughter card physical design is done once and used four times.

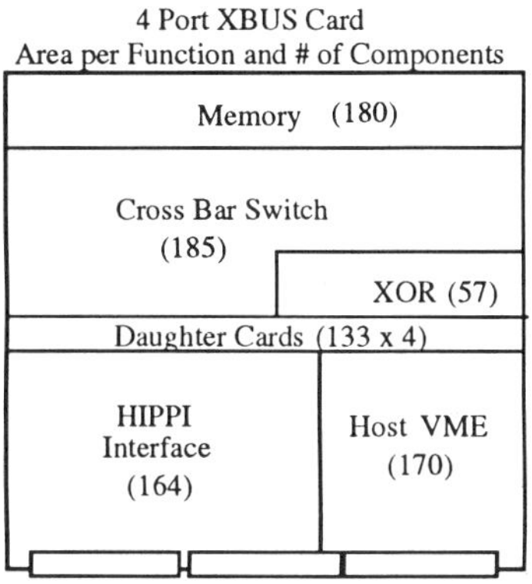

Figure 4: *XBUS Card Layout*
The major subsystems on the board are the interleaved memory system, the crossbar switch, the XOR engine, the daughter card interfaces, the HIPPI source and destination interfaces, and the host VME control interface.

We designed the boards using RACAL's printed circuit board software. While the design of the daughter card proceeded using 8 mil lines/spaces, the XBUS card exceeded RACAL's ability to handle its complexity.

Table 1 summarizes some critical statistics of the two boards we designed, the XBUS and the daughter cards, and compares these with our previous "most complicated board," the SPUR CPU board. In addition, we show the statistics for the original 8-Port version of the XBUS card, which could not be successfully constructed by our board fabricator because of poor yields on blind vias (blind vias connect internal layers without taking space on the top and bottom routing layers). Without them, we were forced to reduce the pins/in^2. Nevertheless, the 4-port board still has dramatically increased pin densities and number of holes on the board.

Table 1: IOC PRINTED CIRCUIT BOARDS

Board	XBUS (4-Port)	XBUS (8-Port)	Daughter Card	SPUR CPU Board
Technology	Double Sided SMT/Thru Hole	Double Sided SMT/Blind Vias	SMT/Thru Hole	All Thru Hole
Size, inches	18.4 x 18.3	18.4 x 18.3	6.5 x 8.2	15.7 x 14.44
Layer Count	18 layers, 14 routing	20 layers, 18 routing	10 layers, 8 routing	10 layers, 8 routing
Line/Space (mils)	6/6	6/6	8/8	8/8
Pin Density pins/sq in	54.7	74.5	47	45
# Nets/Connections	1789/13,203	2147/19,171	162/1211	1472/7041
Total Holes	16,429	23,116	2556	12,702
Via Size (in)	0.016	0.016	0.016	0.025

XBUS Crossbar Implementation

We considered implementing the XBUS with commercial crossbars, Xilinx logic, or custom VLSI. We ruled out the former, designed to support message passing in multiprocessor systems, because of their long configuration latencies. The Xilinx gate arrays are inappropriate for highly interconnected datapaths with guaranteed propagation delays. They are also very expensive. While we can easily fabricate custom VLSI chips, we decided it. This would have increased our risk, because of the long design time and the difficulty of making changes late in the project.

The crossbar-based memory system meets our requirement for high aggregate memory bandwidth while using inexpensive 32-bit ports. However, a single port cannot utilize the full 400 MB/s of memory bandwidth; it is limited to only 40 MB/s. This is not a serious restriction because only the HIPPI ports can sustain more than 40 MB/s.

While the XBUS ports themselves are inexpensive, the crossbar itself uses a large number of highly connected components, complicating the PCB. The datapath uses 196 16-bit transceivers, while the control unit uses eight 3-to-8 decoders, eight 20R8 PAL's and eight 8-bit registers. By using surface mount technology, we implemented the crossbar in 120 square inches or approximately 20% of the RAID controller's total board area.

Memory Subsystem

The memory system of RAID-II must support high bandwidth and large capacity for buffering multiple HIPPI requests. To accomplish this, we chose to use 4 Mb DRAMs supplied by IBM (16 Mb DRAMs have only recently become available) and to interleave across eight memory banks. Each memory bank is independently accessed via the XBUS. Memory is interleaved on a fine-grain basis, 16 words at a time. We chose this approach to smooth bank contention. Ideally, multiple large memory accesses will fall in lock-step, with accesses orderly proceeding from one bank to the next.

Refresh is only performed between XBUS accesses. Refreshing all memory banks at the same time does not raise the maximum power requirement since normal read and write operations keep all memory banks active anyway.

VME Interfaces

The XBUS card implements 5 VME ports: a server control port and 4 ATC ports. The server uses the control port to read and write the memory and control registers. The other ports connect the disk controllers and the XBUS memory. The server port contains additional logic to read and write control registers on the XBUS card: (1) a readable, writable board reset register, (2) a read-only status register, and (3) registers on each VME port for recording the VME address and data during parity errors.

We use three different clock strategies. The interface to the XBUS memory runs at the same clock rate as the other crossbar ports. The VME interface logic runs at twice the XBUS clock rate, to allow for efficient handshaking of data across the VME bus. The VME bus itself is asynchronous. Several VME control signals must be synchronized.

Synchronous FIFOs with independent read and write clocks form the interface between the normal speed and double speed clock regions in the VME ports. Communication from the VME bus interface to the XBUS interface uses two control FIFOs and one address/data FIFO. The opposite direction uses one data FIFO.

Our implementation uses a commercial PAL chip set to provide most of the logic to implement a slave module and generate interrupts. We also used single chip parity transceivers to generate and check parity between the XBUS memory and the VME bus.

HIPPID, HIPPIS Interfaces

All DMA engines are controlled from the file server through control register interfaces, implemented using synchronous FIFOs. The FIFO allows the server to prepare multiple buffers before data actually moves to/from the HIPPI. Unfortunately, these prevent random access to the registers. This would be useful for the HIPPID. If the server sets up a DMA buffer before the HIPPI packet actually arrives (to shorten latency), the server would not know a priori the size of the HIPPI packet. As a result, a DMA buffer would likely be left only partially full. We added a special bit in the reset register to flush the last partial buffer.

We also used FIFOs to interface between the clock worlds of the XBUS and the

HIPPI boards. These also used as speed-matching buffers. Each XBUS port sustains a maximum of 40 MB/s, but a HIPPI board can burst at 100 MB/s. We can generate or receive a 32 KB series of HIPPI packets at full HIPPI speeds by using 32 KB of FIFOs.

Interfacing to the TMC I/O busses was simple. Both buses are without addresses, and transfer a data word every 80 ns. However, control signals are driven at 40 ns intervals. We decided to use very fast (7 ns) PALs clocked on the same edge as our control signal assert clock. With fast PALs and bus drivers, the propagation delay was an acceptable 11 ns.

XOR Engine

The XOR DMA engine is also implemented with FIFOs. Multiple XOR operations can be queued up on the XBUS board. The XOR DMA engine can XOR up to 4095 source buffers and move it into one result buffer. As degenerate cases, the XOR makes it possible to zero memory quickly or to perform quick memory copies.

Hardware Complexity

The complexity of the various state machines within the different portions of the XBUS card are described in Table 2. In general, the state machines are simple, and fit easily within

Table 2: State Machine Summary

Subsystem		# of States	# of PALs	Components Used
Memory Module		9	2	22V10
HIPPD XBUS Port		9	2	22V10
HIPPIS XBUS Port		10	2	22V10
HIPPID I/O Bus		5	1	16R6
HIPPIS I/O Bus		10	1	16R8
XOR XBUS Port	DMA	10	2	22V10
	XOR	12	2	22V10
VME XBUS Port	Parity	4	1	22V10
	VME Control	9	2	22V10
	XBUS Control	15	3	22V10
VME Control Port	Reg. Decode	5	1	22V10

a small number of PALs. However, the VME interfaces proved to be the most complex state machines. The 6 PAL implementation is replicated for each of the 5 VME interfaces.

Table 3 summarizes the hardware "level of effort." We have broken down the subsystems into the memory, HIPPI interfaces, XOR engine, VME interface, XBUS crossbar, system design, and mechanical/RAID chassis design. System design consists of time spent on high level design decisions, like partitioning and clocking of the major subsystems, and thus has no schematic, simulation, and debug elements. The mechanical and chassis design effort includes all activities related to the design and construction of the physical design of the RAID system. This involves the electrical design of the disk shelves, including a SCSI backplane and power isolation logic, and the mechanical design of the disk shelves. In addition, we have lumped the implementation of the daughter card into this category.

Of the individual hardware subsystems, the most complex by far was the VME interfaces. Our implementation supports full VME functionality while optimizing single word and block transfers. The XBUS memory interface is particularly complex. We cannot know

Table 3: Hardware System Complexity (person-hours)

Section	Design	Schematic Capture	Simulation and Debug	Hardware Debug	Total
Architectural Design	960				960
Mechanical/Chassis Design	1120	907	160		2187
Memory	100	60	80	160	400
HIPPI Destination	40	50	50	60	200
HIPPI Source	40	50	70	60	220
XOR	30	20	20	10	80
VME Interface	400	150	300	280	1130
XBUS	80	20	20	10	130
Total	1810	1257	700	580	5307

in advance how much data will be accessed in a VME block mode operation. Because the XBUS protocol requires that the bus request line must be deasserted once cycle before the end of a transaction, we are forced to read ahead in order to predict the end of a transfer.

The table does not include approximately 1000 hours of design time invested in the wide bus interconnection strategy that we later abandoned, 1800 CPU hours of logic simulation, and 2000 CPU hours of printed circuit board routing time. The latter two figures were run in a distributed manner over several workstations, and do not represent elapsed time. It is interesting to note that the routing time of our scaled-down 4-port version of the XBUS card was only 60 hours of CPU time in comparison.

5. RAID Controller Software

5.1. Software Overview

Because of the high communications latency between the server CPU and the network interfaces, the software for RAID-II is partitioned between the server and the processors embedded in the HIPPI interfaces. The critical performance issue is interrupt handling, especially since low latency for HIPPI response is essential. Moving software functionality to the HIPPI boards allows the server to field fewer interrupts per packet.

We partitioned the driver software according to the hardware being controlled or functionality it provided, as shown in Figure 5. The RAID driver maps blocks within a multi-disk RAID into logical blocks on each individual disk, using the XOR engine (via the XBUS driver) to do parity calculations and reconstruction. The XBUS driver manages the XOR hardware directly, hiding the hardware interface details from the RAID driver. Similarly, the ATC driver provides the RAID driver with a device-independent view of the disks. The VME link driver performs two functions. First, it initializes the VME link boards, and provides access to individual memory locations on the VME for debugging purposes. Second, it runs the DMA engine on the link boards to give higher-level software a relatively fast (8 MB/s) data path between the XBUS board and the server. However, the link driver does not orchestrate single-word transfers or interrupts over the VME links; these are done transparently in hardware. The driver works with the HIPPI board code to provide high-level network software (Ultranet virtual circuits over the HIPPI interface).

This has several advantages. First, the device drivers could be debugged as the hardware became available, allowing us to bring up the software in parallel with the hardware.

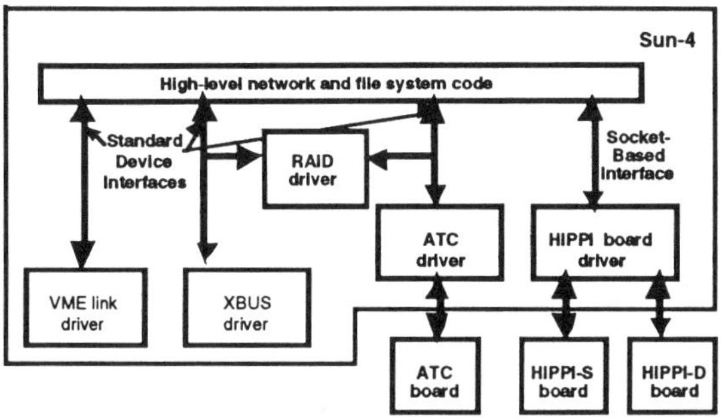

Figure 5: *Interconnections Between RAID-II Software Modules*
This diagram shows the communications paths between the various software modules. Several of these do not run on the file server host; their control and data goes over the VME backplane and the VME link boards.

Second, the hardware interface was isolated from the high-level code, which managed network protocols and the file system. Thus, the high-level code was tested even before the hardware was ready, using lower-performance driver stubs to test functionality.

5.2. The RAID Device Driver

The RAID device driver implements the logical abstraction of RAID-5 disk arrays through a Sprite block device driver. The RAID driver is a "meta" device driver that maps logical requests to physical I/O requests processed by lower level device drivers.

The RAID driver stripes I/O requests over multiple disks. This improves performance by increasing data transfer rates and providing automatic load balancing. It also maintains a parity checksum for the data stored on the disks so that in the event of a disk failure, data is not lost. In the event of a disk failure, it reconstructs a failed disk to a spare.

Each parity block in a RAID-5 is associated with several logical blocks. A system crash during the update of a given block can make the parity inconsistent for other blocks sharing the same parity block. The consistency of parity blocks must always be maintained.

After a system crash, it is difficult to know the state of the array. Either all parity must be regenerated or a logging mechanism must be used to identify just those parity blocks that may be inconsistent at the time of the crash. It is dangerous to operate a RAID-5 with a failed disk, since a system crash can make the parity inconsistent and make it impossible to reconstruct the data on the failed disk.

Most implementations of RAID-5s perform some status logging to stable storage. In our case, we use an extra disk as this log. Careful tuning has made the logging overhead negligible.

5.3. Software Control Flow

All network information, whether read or written, goes over an Ultranet virtual circuit. The first step in any file operation is to establish such a circuit (if one does not already exist for that file). Most of this setup is handled by Ultranet hardware and the HIPPI boards; higher-level software receives a circuit identifier when the setup completes. Next, the client sends the file name to the server over the virtual circuit. The connection is now ready to process file reads and writes. Data can move in two directions—disk to network (client read) or network to disk (client write), as shown in Figure 6 and Figure 7 respectively. To be brief, we have included the detailed operational steps in the figure captions.

5.4. Software Complexity

Table 4 summarizes the number of lines of code and estimated level of effort associated with each major software component of the design. Most of the driver software was rather

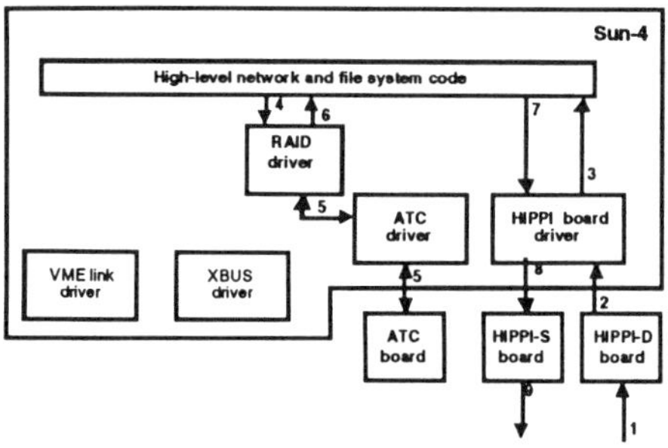

Figure 6: *Read Processing*
1. Read request arrives at the HIPPID board.
2. HIPPID board copies the message into the XBUS board and interrupts the Sun-4
3. HIPPI driver calls back the high-level code to service the request.
4. High-level code calls the RAID driver to read blocks from disk into XBUS memory.
5. RAID driver maps calls ATC driver (possibly multiple times) to read the desired blocks from disk into XBUS board.
6. RAID driver calls back high-level code when done.
7-8. Reply header is passed down to the HIPPIS board.
9. HIPPIS board sends out header followed by data from XBUS board.

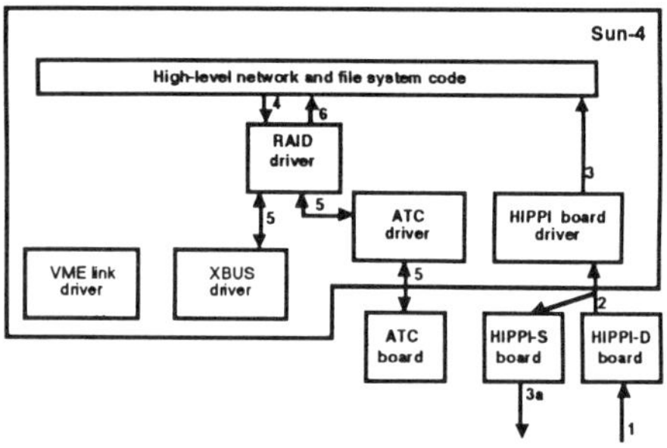

Figure 7: *Write Processing*
For the write diagram, here are the steps:
1. Write request arrives at the HIPPID board. Data and headers are put into preallocated buffers in XBUS board memory.
2. HIPPI-D board notifies both the HIPPI driver and the HIPPIS board.
3a. HIPPI-S board acknowledges receipt of the write.
3. HIPPI driver calls back high-level code.
4. High-level code tells RAID driver to write certain buffers in the XBUS memory.
5. RAID driver coordinates ATC driver and XBUS driver to write data and parity. Parity is calculated on the XBUS board using the XOR engine.
6. RAID driver notifies high-level software when the write is complete so the buffers may be recycled.

Table 4: System Software Complexity

System Component	Subsystem	Lines of Code	Estimated Person Hours
VME Link Driver		1916	200
XBUS Board Driver		1560	100
HIPPI Drivers		6634	400
RAID Driver		7086	500
ATC Driver		2576	200
Embedded HIPPIS and HIPPID	29000 Assembly Code	700	500
	High Level Code	8500	
Totals		28,972	1900

straightforward. However, two elements were very complex: the RAID driver and the software associated with controlling the HIPPI-based network interfaces. The HIPPI driver communicates with the TMC HIPPI boards over a VME-link. Unlike the other drivers, its interface to higher level software is based on sockets. The embedded code on the HIPPIS and HIPPID manages connections over the Ultranet. A remarkably small amount of this code is written in assembly language.

6. Lessons Learned and Design History

6.1. Lessons Learned

We still have much to learn over time about the effectiveness of our design and implementation. Nevertheless, we have already learned several important lessons.

First, logic simulation tools are very mature; they work and they work well. We were able to write reasonably complete VHDL models for complex OEM VLSI chips, such as our FIFOs and the VME PAL set, and to use these effectively within our simulation environment. We had very few logic bugs in the actual implementation.

Second, we continue to underestimate the time it takes to complete logic simulation. A good rule of thumb is that we spend as much time in simulation for verification and debugging as we do for logic design and capture. In the case of this project, it was about six months of logic design and six months of simulation.

Third, physical design tools, even industry standard tools, are inadequate for complex, two-sided, surface mount boards. The tools we used provided no coupling between the logic design and physical board placement. For example, Viewplace performs an initial placement that is then lost when exporting the netlist to RACAL.

Part of our CAD problem was that commercial autorouters, originally designed for thru hole boards, do not perform well on very dense surface mount designs. More emphasis must be place on getting every routed signal from surface mount pads to a vias from which they can be routed. RACAL did not make the correct tradeoff between routing a signal, thereby creating blockages for later nets not yet routed, and getting the pads connected to vias before routing. We created many hand routes from pads to vias, often using diagonals.

Fourth, many aspects of system building can be performed by subcontractors. This is critical for the success of university system building projects, because few groups can do it all themselves. We used different local subcontractors to design our sheet metal, to fabricate our shelves and disk holders, and assemble our surface mount printed circuit boards.

Fifth, we were effective at leveraging our industrial contacts to obtain major subsystems: the TMC HIPPI boards and the ATC SCSI string boards. However, such interactions have drawbacks. Much of our complexity comes from interfacing to boards for which we have no schematics or access to internal firmware.

Sixth, we used a locally written operating system, Sprite, as the basis of our server software architecture. This decision has its pluses and minuses. On the positive side, we had considerable local expertise, rapid turnaround on support (most of the time), and ready access to the source code. The negatives were that we could not leverage existing device drivers, and that the long-term future of our underlying software base remains unclear.

6.2. Design History

We began our research on high performance I/O subsystems in early 1987, with a variety of discussions on the tradeoffs between availability and performance in disk systems. The paper originally defining the RAID taxonomy was written in Fall 1987. NSF sponsored the implementation of the first "proof of concept" RAID-I between 1987 and 1990. That design effort focused on choosing the appropriate commercially available technology for disk drives, disk interfaces, and I/O controllers, and the development of the fundamental algorithms for data striping and reconstruction after disk failures. We began putting the prototype together in late 1988. It became operational by mid-1989 and was stable by year end.

A DARPA-sponsored project to scale up the design for HIPPI interconnects and substantially larger numbers of disk drives, RAID-II, started in mid-1989. During much of 1990, we gained operational experience with RAID-I while designing RAID-II. RAID-I became a main file server for our research group.

We had planned to design our own HIPPI and SCSI interfaces. As we learned more about their complexity, we sought industrial partners to help us implement them. In early 1990, TMC agreed to give us with their two-board HIPPI interface, while ATC provided us with their VME five-string SCSI boards.

We knew that RAID-II would require much greater bandwidth between the HIPPI interfaces and the disk drives than a conventional file server's backplane. Our original plan was to use a single wide, high bandwidth bus. This is similar to the organization of the strategy HIPPI RAID-3 product [Katz 92]. We progressed on this design, but abandoned it in mid-1990 when the OS designers could not force network headers to be aligned on 8 byte boundaries. Ed Lee proposed the crossbar interconnection network, and sold it to the rest of the group in several stormy design reviews. This became the basis of the final design.

The logic design began in Fall 1990 and was essentially complete by September 1991. We used Viewlogic for schematic capture and simulation. While a year may appear long, it was to be expected. The hardware design team consisted of only four students. This was their first major logic design. The VME state machines where quite complex, requiring frequent revision as pieces of the design came together. In retrospect, the design effort could have used more personnel.

Ken Lutz began the physical design in the Fall of 1991, and it quickly became clear that the logic would not fit on a single board. The logic was partitioned into an XBUS card and four daughter cards containing the VME interfaces for the ATC cards. The latter was fabricated by January 1992, and provided an early learning experience with surface mount technology.

Even after partitioning, the XBUS card remained too complex to be routed by RACAL. While waiting for new beta software, we took the opportunity to perform more extensive system simulations of the board. This was worthwhile, because several obscure timing problems were uncovered and fixed.

We spent the first three months of 1992 fighting RACAL to complete the XBUS physical design. Iterative routing passes typically took hundreds of hours, and still required considerable hand completion. To minimize the risk, we decided to overlap the development of the original 8-by-8 port board and a simpler 8-by-4 port board. The latter greatly reducing the interconnect complexity of the crossbar. In fact, the original board, though finally routed, could not be fabricated by the PCB manufacturer, in part due to poor yields on blind vias. Our operational prototype is based on the second design.

The boards were submitted for fabrication in April 1992. The 4-port board was fabricated, assembled, and ready for testing by the end of May. It came up very rapidly. Only a small number of logic problems were encountered, attesting to the completeness of our simulation strategy. Most of the time was spent improving its electrical performance and on repairing the usual interfacing problems.

Figure 8 summarizes the bugs found between 8/4 and 11/30/92. Figure 9 shows their sources. Many bugs are found initially, but then trail off. The major errors were due to the manufacturing and assembly processes: bent pins or the mounting of the wrong component. The logic and software errors have been modest compared with our previous designs.

At the end of 1992, the HIPPI boards can transfer data at 28 MB/s in loopback tests. We have transferred data to SUN-3 clients at 3.3 MB/s, the limit of the client memory system. This compares to 700 KB/s for a conventional Ethernet-attached server. A good deal of software tuning remains to be done to achieve the original performance goals.

7. Summary, Conclusions, Future Plans

The RAID-II disk array storage server represents a natural evolution in system building projects within the universities. We have moved away from designing custom LSI chips towards constructing large scale assemblies of mechanical and electrical components. In

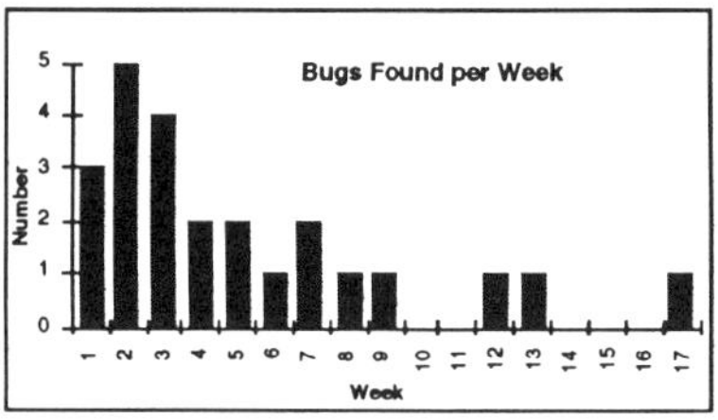

Figure 8: *Bugs Found Per Week*
The number of bugs found are shown as a function of the week of debugging. The shape of this curve is very familiar to system designers: a rapidly rising number of bugs with a long tail of falling numbers of bugs per week.

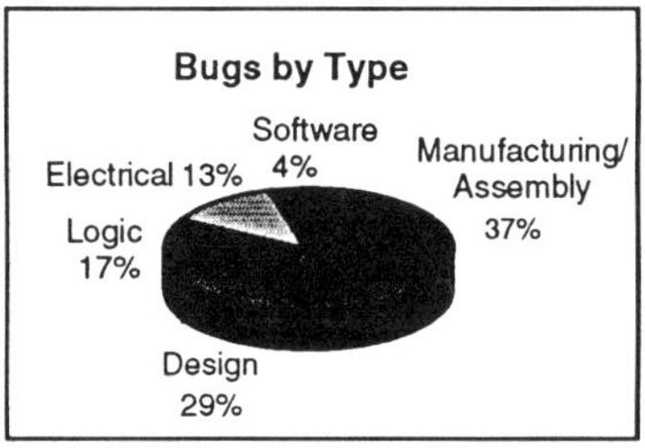

Figure 9: *Sources of Bugs*
The largest single source of bugs is from the manufacturing and assembly process, such as a bent pin or a misplaced component. Design flaws are mainly interface protocol errors while logic flaws are mistakes in the schematics. Electrical flaws are noise problems that require us to run at slower clock rates. Software problems are self-explanatory.

these kinds of projects, we must worry as much about sheet metal and mechanical design as we used to worry about circuit design. RAID-II is network-attached hardware, adding another level of software complexity beyond the operating system.

In 1987, we developed the original RAID taxonomy. Yet even today, there are few commercially available RAID-5 systems, and none that would give us the open environment for the kinds of experimentation with network-attached storage we plan to pursue next. Universities can still operate at the forefront of technology, given generous help from industry. We offer our special thanks to the companies that supported us: IBM for disk drives and memories, ATC for disk controllers, TMC for their HIPPI interfaces, IDT for their high speed FIFOs, and UltraNetwork Technologies for their networking equipment.

We have several major projects underway that plan to make use of RAID-II as a storage server on a high speed network. We are developing application software to use RAID-II as a high capacity video server. The prototype will be used as the destination for the real-time data capture of high resolution video microscope images over a HIPPI-based network between campus and the Lawrence Berkeley Laboratory, In addition, we plan to use RAID-II as the disk frontend for a distributed mass storage system spanning multiple robotic storage systems, including optical jukeboxes and 8 mm and VHS tape robots.

This research has been supported by DARPA/NASA Contract NAG2-591, and the California MICRO Program in conjunction with Array Technology Corp., DEC, HP, IBM, Intel, Seagate, StorageTek, Sun, and Thinking Machines.

8. References

[Chervenak 91] A. L. Chervenak, R. H. Katz, "Performance Measurements of the First RAID Prototype," Proceedings ACM SIGMETRICS Conference, San Diego, CA, (May 1991).

[Katz 92] R. H. Katz, "High Performance Network and Channel-Based Storage," *Proceedings of the IEEE*, V. 80, N. 8, (August 1992).

[Lee 92] E. K. Lee, P. M. Chen, J. H. Hartman, A. L. Drapeau, E. L. Miller, R. H. Katz, G. A. Gibson, D. A. Patterson, "RAID-II: A Scalable Storage Architecture for High-Bandwidth Network File Service," Technical Report UCB/CSD 92/672, (February 1992).

[Patterson 88] D. A. Patterson, G. A. Gibson, R. H. Katz, "The Case for RAID: Redundant Arrays of Inexpensive Disks," Proceedings ACM SIGMOD Conference, Chicago, IL, (May 1988), pp. 106–113.

[Rosenblum 91] M. Rosenblum, J. Ousterhout, "The Design and Implementation of a Log-structured File System," Proc. ACM Symp. on Operating Systems Principles, (October 1991).

Formal Specification of Abstract Memory Models

David L. Dill Seungjoon Park Andreas G. Nowatzyk
Stanford University Sun Microsystems

Abstract

The Murφ description language and verification system for finite-state concurrent system is applied to the problem of specifying a family of multi-processor memory models described in the Sparc Version 8 architecture manual.

The description language allows a straightforward operational description of the memory model which can be used as a specification for programmers and machine architects. More importantly, the automatic verifier can be used to check the correctness of assembly language synchronization routines, when executed in a particular memory model.

This paper describes the memory models and their encoding in the Murφ description language, and describes how synchronization code can be verified. We also present some interesting findings from doing the verification.

1 Introduction

At this time, there is a great deal of effort going into the design and implementation of shared-memory multiprocessors. It is widely (though not universally) believed that shared-memory multiprocessors are easier to program than message-passing multiprocessors because they present programmers with a familiar model: a common global memory. The actual implementation of a multiprocessor will be much more complicated than a collection of processors with a huge multi-port memory (which would be impossible to implement efficiently), in order to achieve reasonable performance; however, parallel programs executed on such machines should always have behavior consistent with the behavior of a collection of processors attached directly to a single ideal shared memory.

If made precise, the idea that a multiprocessor should simulate atomic reads and writes to a common global memory would be *sequential consistency* [Lam79]. However, in practice, much greater performance can be achieved if some weaker model than sequential consistency is enforced [DSB86, Goo89, GLL+90]. The result should be a compromise which yields a machine that is still reasonably programmable, but can be implemented more efficiently. There is reason to hope that this can be successful, because in any well-written large-scale concurrent program, the low-level synchronization code (which may even be generated by a compiler) is the only part that depends on the details of the memory model; this code can be carefully crafted to work in a particular memory model, then used elsewhere by programmers who need not be deeply familiar with its internals.

The purpose of an *abstract memory model* is to define possible program behaviors on a multiprocessor. The actual implementation of the memory system of a multiprocessor could be completely different from the abstract model, as long as the multiprocessor presents the *illusion* of executing a parallel program in the abstract memory model (but with better performance than a direct implementation would give).

An abstract memory model insulates programmers from needing to understand the detailed implementation of a multiprocessor. More importantly, programs written with the abstract model in mind can be ported without modification to other machines that use different approaches to implement the same model.

However, dealing with concurrency is never easy, and even sequential consistency can be counter-intuitive at times. Hence, it is crucial to be able to specify these weakened memory consistency models precisely, and to be able to check that the synchronization primitives constructed on top of those models functions correctly.

Our approach to these problems is to describe the memory model operationally, using a simple general-purpose description language for concurrent systems called Murφ [DDHY92]. Such a description provides a precise specification of the machine architecture, both for implementers and programmers.

Although Murφ as a description language is well-suited for describing memory models, it is not unique in this regard. The major advantage of using Murφ is that it is also an *automatic formal verification system.* There is a tool that support exhaustive checking of all of the reachable states of the system for deadlocks or violations of user-specified properties; in conjunction with a description of the memory model, this tool can be used to check whether small parallel programs (especially synchronization routines) have the desired properties in that memory model.

To illustrate this method, we examine memory models described in the SPARC Architecture Manual Version 8 [Int92]. The SPARC is an open architecture; there are many vendors who are planning to implement it. Hence, it is vital to have a precise specification of the memory model, in order to guarantee object-code compatibility among these different implementations. Currently, there are two specifications of the memory model: an English language description, with diagrams, and an axiomatic description that constrains the possible orderings of reads and writes that each processor can observe.

The English description is subject to the same risk of ambiguity and misinterpretation as all informal specifications.

The axiomatic description is precise, and it has been used to prove the correctness of synchronization routines, including those in the SPARC Architecture Manual [SFC91]. However, it is written in a form that is unfamiliar to programmers and machine architects, and there are no automatic methods for reasoning with the axioms (for example, to prove that a synchronization routine is correct). Our specification method is intended to be precise, yet understandable to programmers and architects, while supporting automatic verification.

The approach here is different from that used by Collier [Col92], who infers the behavior of programs from a set of ordering relations, which are not necessarily easy to convert into an operational form (our descriptions are executable). Our method more closely resembles that of Gibbons, et al. [GMG91], who give *I/O automata* specifications of memory models. The primary differences here are the description languages, and more importantly, our emphasis on support for automatic verification (verification with I/O automata is generally by hand).

There have been other efforts to apply automatic verification methods to multiprocessor cache consistency problems [BG85, MS91, DDHY92]. These methods are attacking a different problem from this paper: in each case, a *protocol* implementing the memory model was verified by checking certain properties, such as "if a cache has a dirty copy of a memory location, no other cache has a copy." The results reported here are at a higher level of abstraction: we are specifying the "user view" of memory behavior and verifying that programs behavior properly in that model. Conceivably, methods which have been applied to protocol-level implementations of cache consistency could be applied to the memory model problem, as well.

2 Memory Model

The SPARC memory models describes the behavior of *loads* and *stores* to the global memory, along with some special atomic synchronization instructions such as *load-store* and *swap*. The SPARC memory models share several principles. One is *global ordering*: all processors "see" (through the results of their loads) non-local stores in the same order. Another principle is that a single-processor system should behave as though it were using the conventional memory model.

All SPARC implementations must support the standard memory model, called *Total Store Ordering* (TSO). In TSO, store operations from a processor P_i must update the global memory in the same order they were issued by the processor. However, a load issued by P_i may return a value recently stored by P_i before the store updates the global memory. Hence, a processor P_j may load an old value *after* P_i has loaded a new value. This property would not be allowed by a shared multi-port memory, so TSO is weaker than sequential consistency. Equivalently, any synchronization code that is guaranteed to work properly in TSO will also work properly in a sequentially-consistent model, but not the reverse.

There is an even weaker model called Partial Store Ordering (PSO), which is similar to TSO except that it allows stores from processor P_i to update the global memory without regard to the order in which they were issued. Because PSO is weaker than TSO, PSO code works properly in TSO, but TSO code may fail in PSO.

We describe each of these models more precisely below.

2.1 Total Store Ordering (TSO)

A block diagram of the abstract memory model appears in Fig. 1. In the informal descriptions of the SPARC memory models, each processor has a *store buffer,* which lies between the processor and global memory. The store buffer allows the processor to issue a *store* and then proceed without waiting for the store to complete. There is also a switch that connects the global memory to one of the store buffers at all times. The switch can change at random, although it should connect to each store buffer infinitely often (so that any pending stores can eventually update the memory).

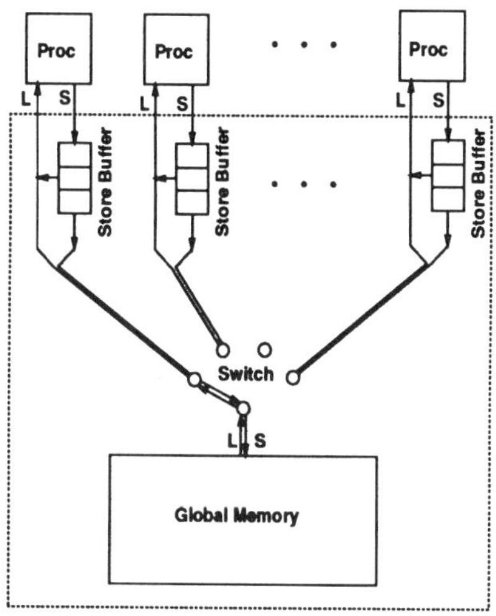

Figure 1: Diagram for the informal description of TSO and PSO memory models.

In the TSO model, the store buffer is a FIFO queue: stores from processor P_i update the memory in the order they were issued by P_i (of course, stores from different processors may interleave arbitrarily). Load operations from P_i first check the store buffer to see if there has been a store to the memory location being loaded. If so, the load returns value associated with the most recent store to that location. Otherwise, the load returns the value associated with the location in main memory.

TSO is weaker than sequential consistency because a load from P_i does not check the store buffers of the *other* processors in the system. In order to do any synchronization at all, it is necessary to have some special instructions. There are two generic instructions: *load-store* and *swap*. Load-store has the effect of atomically loading the value of an address into a register while storing -1 in the address. A load-store is entered in the store buffer exactly like a store, but all further loads and stores by the processor are blocked until the load-store has completed (i.e the global memory and the register have been updated). *Swap* is similar to load-store, except that it exchanges the contents of a designated register and memory, instead of writing -1 in the memory.

2.2 Partial Store Ordering (PSO)

Partial Store Ordering (PSO) is a weaker model than TSO that is intended to allow more concurrency in an implementation. PSO is similar to TSO except that the store buffer is not FIFO. Operations in the store buffer can be executed in any order by the global memory, providing that stores to the same address are executed in the same order as they were issued (this requirement ensures that a single processor adheres to the conventional memory model). For example, suppose that store S1 precedes store S2 in program order. In a directory-based scheme, it is possible that performing S1 will require invalidating entries in the caches of other processors. In TSO, S2 cannot be performed until S1 has been performed, but in PSO, it is legal to perform S2 without waiting.

To give programmers explicit control over the ordering of memory operations in PSO, there is a special instruction called *stbar* ("store barrier"). This instruction behaves as though it were inserted in the store buffer when issued by the processor. The behavior of the store buffer is constrained so that none of the operations inserted after an *stbar* can be performed by the global memory until all of the operations issued before the *stbar* have been performed. The load instruction and atomic load-store instructions operate in the same manner as in TSO.

3 Murφ description language and verifier

The description language we use for the memory models is Murφ, a simple high-level programming language to describe finite-state asynchronous concurrent systems. It serves as common front end for different verifiers and simulators. The language is intended to be a testbed for rapidly trying out new ideas for verification methods, so we have placed a premium on simplicity, flexibility, and ease-of-implementation. This has resulted in some restrictions on the expressiveness and convenience of the language that may seem unnecessary. Some of these restrictions may be lifted in the future.

Murφ allows the declaration of familiar data types, including subranges of integers, arrays, records, and user-defined enumerations. Additionally, procedures and global variables can be declared. The operational part of the language is based on *iterated guarded commands* (Murφ was inspired by Misra and Chandy's Unity language [CM88]). A *state* is an assignment of values to the global variables. A Murφ program consists of a collection of *rules*, each of which has a *condition*, which is Boolean expression referring to the global variables, and an *action*, which is a statement that modifies the values of the variables, yielding a new state.

A Murφ program is executed as follows: an initial state is specified by the user. Then the following loop is executed forever: some rule whose condition is satisfied by the current state is chosen and its action evaluated, yielding a new current state. If there are no rules whose conditions are true, the execution halts. Although the action may be a compound statement consisting of a sequences of smaller statements, conditionals, and loops, it is executed *atomically*: no other rule can be executed before the action completes.

There may be many possible executions of a Murφ program, each resulting from a different choice when several rules have true conditions in the current state. The Murφ simulator chooses at random from among these rules, and the verifier tries them exhaustively by depth-first or breadth-first search (in either case, states are stored in a table when they are first generated, so that a state will not be processed more than once).

The operation just described is an *interleaving* model of *asynchronous* concurrency, with communication and synchronization through *shared variables*. One of the advantages of this model is the simplicity of modeling the parallel composition of two processes: the combined process is the union of the rules of the individual processes. In the resulting description, each process can take any number of steps between the steps of the other.

One essential construct in Murφ is the *ruleset*, which is used to describe a collection of rules that vary over a parameter. Rulesets are very important for *scalability*: the number of processes in the system can be changed by embedding the rules for a generic process inside a ruleset that varies over a process identifier parameter. Hence, the number of processes can be changed by modifying a single global constant declaration.

There are three types of errors can be detected in a Murφ description. There is an *error* statement that can appear in an action. Error statements are particularly useful in conditionals for cases that are not supposed to occur. Also, the user can provide *invariants* separately. An invariant is a Boolean expression which is intended to be true in every reachable state of the system, so a state that violates the invariant is in error. Finally, the system can detect *deadlock states*, which are states that have no other states as successors. If a problem of any type is detected, the verifier prints out a *diagnostic trace*, which is a sequence of states that leads to a state exhibiting the problem. Breadth-first search is preferred because it yields the shortest such trace.

Compared with other languages for specifying correctness conditions, such as temporal logic, error statements, invariants, and deadlocks are unsophisticated. However, they are generally sufficient for verifying our synchronization examples. The one major deficiency is in the specification of *fairness properties*: for example, if two processes are competing for a lock, we would like to be able to show that each eventually obtains the lock. Checking this property is currently beyond the capabilities of the Murφ verifier, unless the problem results from a deadlock. We are now implementing a facility for checking fairness properties.

4 Memory model

The Murφ models of TSO and PSO are direct implementations of the informal description, since the axiomatic definition does not lend itself to an easy operational description. The memory model describes everything except the processors themselves (the components inside the dashed box in Fig. 1). To verify assembly-language routines, rules must be added to model the processors executing the appropriate code.

The memory and switch are modeled as a single process that nondeterministically chooses a non-empty storebuffer and executes an instruction from it. In TSO, this instruction is always the oldest entry; in PSO, it is chosen nondeterministically according to the constraints of PSO.

Here we provide excerpts from the TSO description. The PSO description is almost the same except for the choice of instruction from the store buffer. The syntax of Murφ is derived from various standard programming languages, especially Pascal and C.

For scalability, the number of processors, size of the store buffer, size of memory, number of bits of memory, number of registers, and so on, are under the control of declared constants whose value can be changed in a single place. For verification, these constants should be kept very small in order to bound the size of the state space that must be explored.

```
Const
  ProcessorNum  : 3;      -- number of processors.
  AddressNum    : 8;      -- size of address space.
  RegisterNum   : 8;      -- number of registers.
  ValueNum      : 8;      -- number of distinct values.
  StoreBufferSize : 4;    -- capacity of StoreBuffers.
```

With a few exceptions, small values for the scaling parameters do not change the validity of the model. For example, there is no purpose in modeling 32 registers and megabytes of memory for synchronization procedures that only change one register per processor and only mention one global memory location. However, the descriptions of the memory models do not bound the size of the store buffer, so fixing it at any finite value may, in theory, risk overlooking some problems that could be demonstrated with a larger store buffer. We believe, but have not proved, that this is not a problem for the synchronization algorithms.

There are data structures for the states of the processors, store buffers, and global memory. Following are the type declarations we actually use:

```
Type
  Processor   : 0 .. ProcessorNum - 1;
  Address     : 0 .. AddressNum - 1;
  Register    : 0 .. RegisterNum -1;
  Value       : 0 .. ValueNum - 1;
  Instruction : Enum{ Inone, Istore, Ildstore, Iswap, Istb };
  Label       : 0 .. MaxLabel ;
  IssueIndex : 0 .. StoreBufferSize - 1 ;
  IssueRec : Record               -- StoreBuffer entry
      Instr : Instruction;        -- instruction code
      Addr  : Address;            -- address to load or store
      Reg   : Register;           -- register to load or store
      Val   : Value;              -- value to be stored
    End;                          -- StoreBuffer keeps instructions
                                  -- in program order.
  StoreBuffer_t : Record              -- StoreBuffer can be partially full
    Count : 0 .. StoreBufferSize;  -- number of entries in StoreBuffer.
    Ar : Array [IssueIndex] of IssueRec;
  End;
```

Next, we give some variable declarations. The state of the memory system consists of the store buffers (one for each processor), which queue up stores, atomic load-stores, swaps, and other transactions that alter the state of the memory, and the memory itself, which stores a value for each address.

```
Var
  -- memory system variables
  StoreBuffer : Array[Processor] of StoreBuffer_t;
  Memory     : Array[Address] of Value;
  -- processor state
  PC         : Array[Processor] of Label;
  Registers : Array[Processor] of Array [Register] of Value;
  RegDep     : Array[Processor] of Array[Register] of Boolean;
  CCR        : Array[Processor] of Boolean;    -- Condition Code Register
  -- other
  CritProc   : Processor;
```

The processor state is also stored in the variables, including a program counter, registers, and condition code registers. The RegDep flag is set when an atomic load-store transaction is put into a store buffer, to prevent further instructions which depend on that register from executing before the load-store is performed (this is an implicit requirement of the TSO and PSO memory models).

`CritProc` is an auxiliary variable that we added in order to verify a locking routine. Whenever a processor enters its critical section, `CritProc` is set to the processor ID. This abstracts the use of an actual resource controlled by the lock, and and is used in the specification to catch problems that result when the lock is cleared prematurely.

There are procedures which model basic operations that processors can perform. These procedures are called in rules modelling the processor behaviors. There is not sufficient space to show all of these, so we show one: the procedure that loads data from a memory location into a register.

Newer transactions appear at lower indices in the store buffer. The procedure searches linearly from lower indices to higher indices for a store instruction to the address being loaded. If such an instruction is found, the stored value is loaded into the register; if no store is found in the store buffer, the value is loaded from main memory. Registers and PCs are indexed by processor numbers, and `nPC` is to model delayed branching.

```
Procedure Load(Const p:Processor; Const a:Address; Const r:Register);
Var found : Boolean;
Begin
  found := false;
  For i := 0 To StoreBuffer[p].Count Do
    Alias j: StoreBuffer[p].Count - i Do
      If ( (!found) & StoreBuffer[p].Ar[j].Addr = a &
     StoreBuffer[p].Ar[j].Instr = Istore )
      Then
         Registers[p][r] := StoreBuffer[p].Ar[j].Val;
         found := true;
      End;
    End;
  End;

  If (!found) Then  Registers[p][r] := Memory[a];  End;
  PC[p] := nPC[p]; nPC[p] := nPC[p] + 1;
End;
```

For both TSO and PSO, there is a core set of rules that are the same in the verification of all the synchronization routines. These rules nondeterministically choose one of the store buffers and then choose an entry in it to process, according to the ordering constraints of the memory model. In TSO, the entry that is chosen must be the oldest (`Ar[0]`); in PSO, the constraints are more complicated, as explained above.

Here we show the rules for TSO. The `Ruleset` at the beginning can be thought of as choosing a processor arbitrarily (in reality, it defines a collection of rules, one for each processor, and one of the rules is chosen nondeterministically). Note that, unfortunately, this does *not* require the memory/switch to execute instructions from every store buffer infinitely often, reflecting our inability to express fairness constraints.

```
Ruleset p:Processor Do
  Alias R  : StoreBuffer[p],
        ar : R.Ar[0] Do
    Rule "Execute one of the oldest instructions of each buffer''
      ( 0 <  R.Count )
    ==>
    Begin
        If ar.Instr=Istore Then Memory[ar.Addr] := ar.Val;
        Elsif ( ar.Instr=Ildstore | ar.Instr=Iswap ) Then
                Registers[p][ar.Reg] := Memory[ar.Addr];
                Memory[ar.Addr] := ar.Val;
                Clear RegDep[p][ar.Reg];
        End;
        ArrangeBuffer(p,0);    -- dequeue the instruction
    End;
  End;
End; -- Processor
```

For PSO rules, we use one more nested ruleset which nondeterministically index into the selected store buffer. The conditional on the rule checks if the chosen entry can legally be executed before the others, that is, it ensures that there are no store barriers issued before in that buffer and no previous store instructions to the same address. If the condition holds, it can process the store; otherwise, the rule does nothing.

5 Verifying Synchronization Routines

5.1 Modeling processor behavior

The SPARC architecture manual gives several assembly language routines for standard synchronization paradigms, including spin locks (two versions: one using load-store one using swap), produce-consumer with a bounded buffer, and Dekker's algorithm. There is a TSO version and a PSO version for each algorithm. Figure 2 shows the assembly language code for a spin lock. This is taken verbatim from the SPARC Architecture Manual, except that two instructions have been inserted to store the processor number (p) in the global memory location cr_m. These instructions were added to improve error detection, as explained below.

Note that in figure 2, a conditional branch consists of three instructions: a tst instruction which sets condition code register (CCR) if the tested register is zero, followed by the branch itself (be or bne), followed by another instruction (in this case nop) that is executed even if the branch is taken (the SPARC has "delayed branching"). These non-memory operations are

```
-----------------------------------------------
retry:  ldstub [lock], %l0    -- load-store
        tst    %l0            -- branch if 0
        be     out
        nop
loop:   ldub   [lock], %l0    -- load
        tst    %l0            -- branch if non-0
        bne    loop
        nop
        ba,a   retry          -- jump
out:    nop { CritProc := ID (p) for invariant }
-----------------------------------------------
crit:   mov    p,  %l1        -- l1 := processor
        st     %l1, [cr_m]    -- store
-----------------------------------------------
unlock: stub   %g0, [lock]    -- store value zero
-----------------------------------------------
```

Figure 2: Assembly language program for spin lock in TSO mode.

written as procedures and called from rules, just as the memory operations are.

For each processor, there is a collection of rules modelling the behavior of the processor. The program counter is represented by a variable with a small range of values: just enough to have a distinct address for each instruction. For readability, we have given each PC value a symbolic name. A rule modelling an assembly language instruction has a condition that checks whether the PC value corresponds to the address of the instruction, that there is an empty entry in store buffer if the instruction is a store, that there are no pending atomic instructions if it is a load, and that `RegDep` is not set if it is `tst` (the failure of any of these requirements prevents the instruction from executing). The action of the rule performs the operation associated with the instruction and increments the PC (sometimes this is done in a procedure and sometimes explicitly in the rule).

Here is a typical rule, modeling the instruction at label `loop:` of figure 2.

5.2 Correctness specification

The invariant below is used to specify the correct behavior of the spin lock. When a processor is about to enter a critical section, the invariant assures that the value written to the location `[cr_m]` is the identifier of the last processor to hold the lock. This ensures that a lock is not released too early, before all writes to lock-protected memory have completed and that mutual exclusion is satisfied.

```
Alias OK : StoreBuffer[p].Count!=StoreBufferSize &
             Forall i:IssueIndex Do
               i<StoreBuffer[p].Count -> (
               StoreBuffer[p].Ar[i].Instr != Ildstore |
               StoreBuffer[p].Ar[i].Instr != Iswap )
             End Do
Rule
  PC[p] = loop & OK
  ==>
  Begin
    Load(p,lock,10);        -- read the value of the memory
                            -- location [lock] into register 10.
    PC[p] := PC[p] + 1;
  End;


Invariant  "Mutual Exclusion of Memory Access in Critical location"
  Forall p:Processor Do
     ( PC[p] = out ) -> CritProc = Memory[cr_m]
  End;
```

5.3 Results

The spinlock example described was computational the most difficult, although all examples required the same order-of-magnitude time and space.

On a SparcStation 2 on the spinlock example modeled with three processors and a storebuffer size of three, the verifier explored 81,185 states in less than 200 seconds using, while the same example in TSO visited 74,701 states in 160 seconds. The time is not proportional to the number of states because a state may be visited several times (depending on the number of incoming edges in the state graph), and because the amount of time for each rule varies with the complexity of the rules.

Ultimately, no errors were found in the synchronization routines when combined with the appropriate models. Also, as expected, the TSO routines all failed when combined with the PSO memory model.

However, there were several points along the way when it seemed that problems had been discovered. In each case, the problem turned out to be a misunderstanding. For example, at first it appeared that Dekker's algorithm could fail under PSO. However, the problem turned out to be an error in the specification of the memory model: we omitted the PSO requirement that stores *to the same address* update the memory in the same order they were issued.

Also, as is usually the case with formal specification, preparing the specification leads to extremely careful reading of the written specification. In the process, several opportunities for clarification in the description were identified.

The most important issue was the handling of register dependencies. The SPARC Architecture Manual says that a load-store or swap blocks all future *loads and stores* until it has been performed. At first, we interpreted this literally to mean that register-only instructions, such as the `test %l0` in figure 2 are *not* blocked. The spin lock code fails under this assumption.

As was later clarified, it is an undocumented assumption that accesses to registers that depend on pending atomic instructions are also blocked until the atomic instruction is performed [Sin91]. In fact, there is a strong argument that an such an assumption does not belong in a *memory* model, given that it is an assumption about *processor* behavior. However, the memory model is not very useful for predicting the behavior of programs without considering register dependencies, so we now believe that these assumptions must be stated as part of the model.

More recently, we have been specifying the new proposed *Relaxed Memory Order* (RMO) for the SPARC V9 architecture soon to be specified by SPARC International. RMO is weaker than PSO in that load operations do not block store operations; in other words, a store can be performed before a load even though the store follows the load in the program instruction order. Explicitly stating assumptions about processor behavior, such as the effects of register dependencies, seems to be crucial for precisely specifying RMO. Our experiments with TSO and PSO in Murφ first alerted us to this fact.

6 Conclusion

We believe that this type of operational description strikes an appropriate balance between formality and understandability by programmers and machine architects. Moreover, the availability of formal verification tools allows users to experiment with the effects of the memory model on small assembly-language routines. Also, as we have learned in this experiment, developing an operational description and running the verifier can be very effective at clarifying the subtle details of the models and synchronization routines.

Abstract memory models other than TSO and PSO have been proposed. We conjecture that the same approach can be applied to these models. However, some of these models are defined very abstractly in terms of orderings between loads and stores. We cannot be sure that a simple operational description exists for these models. Moreover, the problem of formally comparing these partial order models with operational models remains to be solved.

There are several limitations to our approach. The verifier requires finite-state descriptions, so fixed values are required for parameters such as the number of processors and the sizes of store buffers. Also, the verifier used here cannot cope with fairness properties, such as the requirement that every store issued by every processor eventually be performed by the memory. The *state explosion problem,* which is usually regarded as a central problem in finite-state verification, has not been an issue in this effort, because of the small size of the assembly language routines. However, it would become a problem for verification of larger programs.

Once a formal description of a memory model has been defined, it would be very helpful to be able to compare that description with a high-level description of a multiprocessor architecture with a particular method for maintaining cache consistency. We have implemented a facility in Murφ for checking *simulation preorder* between two Murφ descriptions, which could be used for this purpose [DHW91]. However, the state spaces that need to be explored in simulation preorder checking are large because *pairs* of states (one from each description) must be enumerated instead of individual states. We are currently investigating ways of reducing the state explosion in simulation preorder checking which may make feasible automatic comparison of two memory models or of an implementation and a memory model.

Acknowledgements

We would like to thank Pradeep Sindhu for help with understanding details of the memory models.

This research was supported by the National Science Foundation (grant number MIP-8858807), the Defense Advanced Research Projects Agency (contract number N00014-87-K-0828) and by the Powell Foundation. Most of this work was done using equipment generously donated by Sun Microsystems.

References

[BG85] J-L. Baer and C. Girault. A petri net model for a solution to the cache coherence problem. In *Proceedings of the First Conference on Supercomputing Systems.* IEEE Computer Society, 1985.

[CM88] K. Mani Chandy and Jayadev Misra. *Parallel Program Design — a foundation.* Addison-Wesley, 1988.

[Col92] William W. Collier. *Reasoning about Parallel Architectures.* Prentice-Hall, 1992.

[DDHY92] David L. Dill, Andreas J. Drexler, Alan J. Hu, and C. Han Yang. Protocol verification as a hardware design aid. International Conference on Computer Design: VLSI in Computers, 1992.

[DHW91] David L. Dill, Alan J. Hu, and Howard Wong-Toi. Checking for language inclusion using simulation preorders. 1991.

[DSB86] Michel Dubois, Christoph Scheurich, and Faye Briggs. Memory access buffering in multiprocessors. In *Proceedings for the*

13th Annual International Symposium on Computer Architecture, pages 434–442, June 1986.

[GLL+90] Kourosh Gharachorloo, Daniel Lenoski, James Laudon, Philip Gibbons, Anoop Gupta, and John Hennessy. Memory consistency and event ordering in scalable shared-memory multiprocessors. In *Proceedings for the 17th Annual International Symposium on Computer Architecture*, pages 15–26, May 1990.

[GMG91] P.B. Gibbons, M. Merritt, and K. Gharachorloo. Proving sequential consistency of high-performance shared memories. In *Proc. 3rd ACM Symposium on Parallel Algorithms and Architectures*, pages 292–303, 1991.

[Goo89] James R. Goodman. Cache consistency and sequential consistency. Technical Report 61, SCI Commitee, March 1989.

[Int92] SPARC International. *The SPARC Architecture Manual Version 8*. Prentice Hall, 1992.

[Lam79] Leslie Lamport. How to make a multiprocessor computer that correctly executes multiprocessor programs. *IEEE transactions on computers*, C-28(9):690–691, September 1979.

[MS91] K. L. McMillan and J. Schwalbe. Formal verification of the gigamax cache-consistency protocol. In *Proceedings of the International Symposium on Shared Memory Multiprocessing*, pages 242–251. Information Processing Society of Japan, 1991.

[SFC91] Pradeep S. Sindhu, Jean-Marc Frailong, and Michel Cekleov. Formal specification of memory models. Technical Report CSL-91-11, Xerox Palo Alto Research Center, December 1991.

[Sin91] Pradeep Sindhu, 1991. Private communication.

An Area-universal VLSI Circuit

Paul Bay
Thinking Machines Corporation
245 First Street
Cambridge, MA 02142

Gianfranco Bilardi
Dipartimento di Elettronica ed Informatica
Università di Padova
Via Gradenigo 6/A, 35131
Padova, Italy.

Abstract

An *area-universal* VLSI circuit is proposed which has area $O(A)$ and can simulate any synchronous digital VLSI circuit of area A with only $O(\log A)$ slowdown. The universal circuit is a hybrid of a mesh and a concentrator fat-tree and accomplishes its simulations by a combination of routing techniques. It could be viewed as a Field Programmable Gate Array (FPGA). The layout of the simulated circuit is "compiled" into a "program" that is loaded into control registers of the universal circuit. The compilation takes $O(A)$ sequential time.

1 Introduction

For over a decade, Very Large Scale Integration (VLSI) has provided an attractive medium for the realization of computational structures. Dedicated VLSI structures have been developed for a variety of fundamental operations in arithmetic, matrix algebra, signal processing, sorting, etc. (For a sample, see [13].) These structures have been analyzed in terms of their area-time performance in a VLSI model of computation first described in [12]. Efficient designs for different operations are based on quite different architectures, such as arrays, trees, hypercubes and their derivatives, and a number of hybrid combinations of these.

It is natural to wonder how crucially area and time performance depend upon a close match between the architecture and the computation to be performed. In Section 3, we give the following nearly tight characterization of this dependence:

Theorem 1 *There exists a VLSI circuit C_u of area $O(A)$ that, given any VLSI circuit C of area A, can be programmed to simulate C with slowdown $O(\log A)$.*

The terms in the above statement will be defined in Section 2. One intuitive interpretation of the theorem is that a "special-purpose" architecture can be better than the "general-purpose" circuit C_u by at most a constant factor in area and a logarithmic factor in time.

We refer to C_u as a *universal* circuit. Universal computing devices, which can simulate any member in a class of computing devices characterized by a bound on some resource, play a central role in computer science. Researchers have investigated universality in various models of parallel computation, e.g., [14, 4, 6, 5]. In particular, using the VLSI model, several papers explore the design of area-universal networks for message routing, e.g., [9, 7, 8, 1]. As we shall see below, there is an intimate connection between universal VLSI routers and universal VLSI circuits.

Versatile circuits that can be programmed to perform specific functions are of considerable interest and are commercially available. In this category fall programmable logic arrays (PLAs) and field programmable gate arrays (FPGAs) [3]. Our universal circuit C_u can be viewed as an FPGA. Theorem 1 provides an analytic characterization of a class of computations that can be programmed efficiently on C_u.

FPGAs are frequently used as a fast and inexpensive way to implement a chip design C; the chip will be fabricated only after adequate testing. To support this and similar types of application, it would be useful to have a tool capable of compiling a suitable description of C into a program that makes C_u simulate C. In Section 4, we describe an efficient algorithm for such a compiler, proving the following result:

Theorem 2 *There is an algorithm that, given a description of a VLSI circuit C of area A, can compute in time $O(A)$ all control bits needed for C_u to simulate C.*

2 Problem Formulation

A VLSI circuit is a layout of a boolean machine, a directed graph with nodes representing boolean gates and arcs representing wires. We assume the machine is synchronized by a global clock. For simplicity and without loss of generality, we assume that, in the circuit to be simulated, each gate consists of a combinational stage, which computes a boolean function of the inputs, followed by a one-bit storage device, which stores the output of the combinational stage. In-degree ("fan-in") and out-degree ("fan-out") are at most two for each node.

The *layout* is an embedding of the boolean machine in a two-dimensional square grid, with nodes mapped to grid vertices and arcs mapped to edge-disjoint grid paths (see [12] for the details of the model). The *area* of a VLSI circuit is defined to be the area of the smallest rectangle enclosing its layout. We shall consider only circuits with square layouts, since any rectangular layout can be transformed into a square one with at most a constant factor increase of the area. For ease of presentation, we assume that the input and output of the circuit are encoded by its initial and final state, respectively. However, our universal circuit could easily be adapted to simulate circuits with I/O ports active throughout the computation. We assume that a signal takes one time step to propagate from the output buffer of one gate to the output buffer of the following gate, regardless of the length of the wire joining the gates and regardless of the function computed by the second.

Circuit C_u is said to *simulate* circuit C with *slowdown* t if (1) there is a one-to-one correspondence between the storage devices of C and a subset of the storage devices of C_u, (2) the corresponding storage devices have the same initial state, (3) the remaining storage devices in C_u are initialized to a fixed state that depends only on the structure of C, and (4) for each tick of C's clock, the storage devices of C_u that correspond to storage devices in C are updated with the new values from C in t ticks of C_u's clock. The initial values of the surplus storage devices of C_u can be thought of as the "program" of the universal circuit that permits it to simulate C. The program is computed externally from a suitable description of C.

3 The Area-Universal Circuit

This section establishes Theorem 1. The general idea behind our universal circuit C_u is as follows. A gate of the target circuit C is simulated by a *universal gate* of C_u. A universal gate is one that can be programmed to implement any boolean function of two inputs by setting four control bits, and is equipped with a one-bit storage cell on each input and output line. A wire of C is simulated by routing a one-bit message from the output of the gate where the wire starts to the input of the gate where the wire terminates. The setting of the control bits of the universal gates and of the switches of the router constitute the "program" that, when loaded into C_u, makes it simulate C.

A straightforward implementation of the outlined approach could be based on any of the universal VLSI routing networks mentioned in the introduction. Since we are interested in deterministic simulations, and off-line routing is sufficient for our purposes, the concentrator fat-tree (CFT) of [9] would yield the best result. After working out some details, one would get a circuit of area $O(A \log^2 A)$ capable of simulating any circuit of area A with slowdown $O(\log A)$.

Below, we develop a routing algorithm for a network consisting of a modified CFT with $A/\log^2 A$ leaves, with each leaf connected to a $\log A \times \log A$ mesh of finite automata, as shown in Figure 1. (For simplicity, we assume A and $A/\log^2 A$ are powers of two.) As we will see below, this choice will reduce the area of the universal circuit to $O(A)$ while maintaining an $O(\log A)$ slowdown.

The area-A circuit C to be simulated is partitioned into $A/\log^2 A$ square

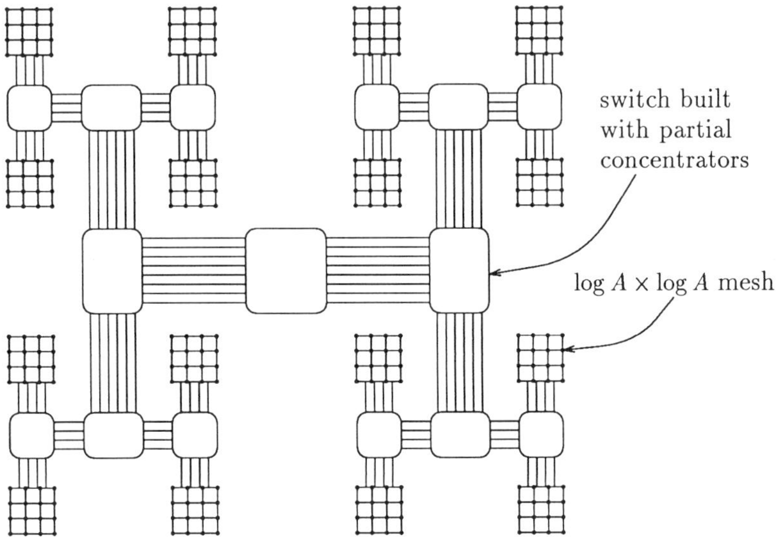

Figure 1: The area-universal circuit C_u is a hybrid of a fat-tree and mesh.

cells, each corresponding in a natural way to a mesh of C_u, as shown in Figure 2. A given mesh of C_u routes messages associated with all wires completely contained in the corresponding cell of C. Moreover, it routes a message to or from the boundary for each wire that has its source and/or destination in the cell, but is not completely within it. The portion of wire from the boundary of the source cell to the boundary of the destination cell is simulated by the CFT. In Subsections 3.1 and 3.2 we shall give a detailed description of the mesh and the CFT routing respectively.

3.1 Mesh Routing

The goal is to design a $\sqrt{N} \times \sqrt{N}$ mesh of finite automata (FAs) that can be programmed to route a set of single-bit messages that correspond to the wires in a cell of the circuit being simulated. The routing must be carried out in $O(\sqrt{N})$ time and $O(N)$ area. (In our case, $N = \log^2 A$.) At first, this goal might seem achievable by well-known mesh routing techniques such as Batcher's sorting or Benes's routing. However, such algorithms typically require $O(\log N)$ bits to be stored at each node and/or $O(\log N)$ bits to be transmitted across an edge at each step. Indeed, any algorithm that can route an arbitrary permutation off-line must store $\Omega(N \log N)$ bits just to identify the permutation. In our case, however, the source-destination pairing of the messages to be routed is not arbitrary: the layout of the circuit to be simulated is an $O(N)$-bit description of the permutation. We

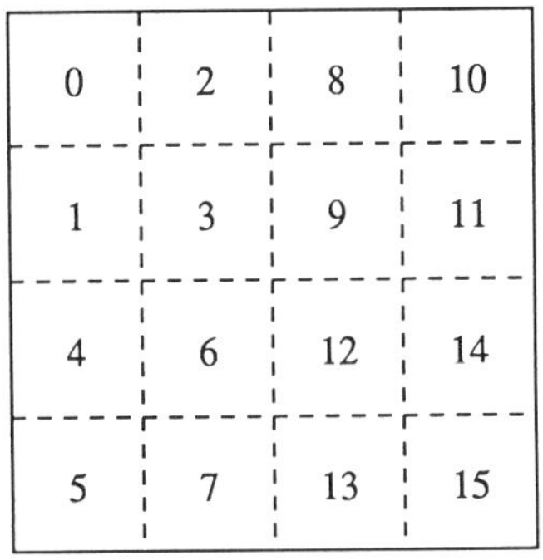

Simulated circuit C

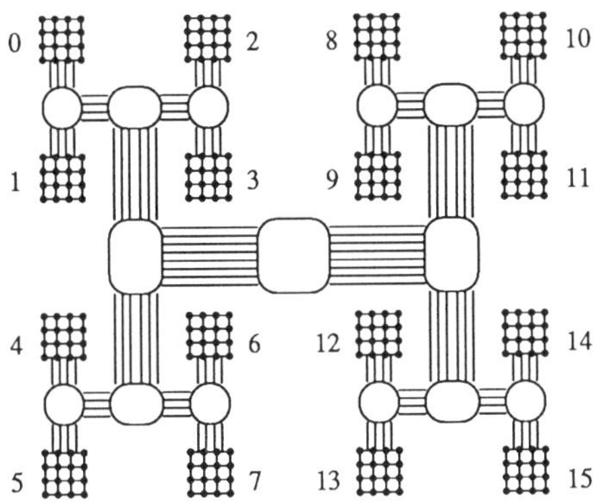

Area-universal circuit C_u

Figure 2: Square cells of the simulated circuit are mapped to the meshes at the leaves of the area-universal circuit C_u.

exploit this constraint directly by routing messages along paths in the mesh that correspond to the wire layout. This strategy is adequate for all wires shorter than, say, $2\sqrt{N}$, but would be too slow for the at most $\sqrt{N}/2$ longer wires. For them, and for routing messages to/from the boundary for those wires not completely contained in the cell being simulated, we resort to a constant number of applications of the following lemma.

Lemma 1 *There is a $\sqrt{N} \times \sqrt{N}$ mesh of finite automata which can be programmed to route, in $O(\sqrt{N})$ time, any set of $\sqrt{N}/2$ single-bit messages where at most a constant number of messages have the same source or destination.*

Proof: The routing is accomplished in five steps, each taking $O(\sqrt{N})$ time. For steps 1 and 5 it is convenient to assume that the top and the bottom of each mesh column are connected to form a cycle.

Step 1: Within each column, compress together the message bits from their (potentially dispersed) sources using a circular shifting movement. The movement is best visualized as a "train" of message bits, initially empty, that moves up the column. Source FAs add their bits to the end of the train as it goes by. The train starts and ends at the row whose index is equal to the number of message sources in columns to its left; this position is computed externally prior to the routing and stored as a flag in the relevant FA. The train does not pick up any message bits until it wraps around from top to bottom the first time, and makes exactly two revolutions before returning to its starting position. All columns execute this process simultaneously, and the net effect is to distribute exactly one message to each of the first $\sqrt{N}/2$ rows.

Step 2: Within each row, move the message bit to the diagonal. This is easily accomplished if each FA stores a bit specifying whether the diagonal is to the left or right.

Step 3: Using the mesh as a crossbar switch, route the messages on edge-disjoint paths so that they appear in the second half of the diagonal, ordered by nondecreasing destination column and, within each column group, by decreasing destination row. This permutation is calculated off-line, and control bits are stored in each FA to specify between which two of its four connections to the mesh it is expected to transmit a message bit.

Step 4: Within each row, use a process symmetric with Step 2 to move the message bit to its destination column.

Step 5: Within each column, move the message bits to their final destinations. The process is symmetric with Step 1 with the exception that the destination finite automata remove their bits from the front, rather than the tail of the train as it goes by. □

3.2 CFT routing

We now focus on the part of C_u devoted to simulating wires of the target circuit with source and destination in different meshes.

Network structure. We begin by defining a useful building block. An (r, s, α) *partial concentrator* (PC) is a directed graph with r input vertices and $s \leq r$ output vertices such that any $k \leq \alpha s$ inputs can be connected to ("concentrated" onto) some k outputs by k vertex-disjoint paths. Figure 3 illustrates a $(15, 6, 1/2)$-PC.

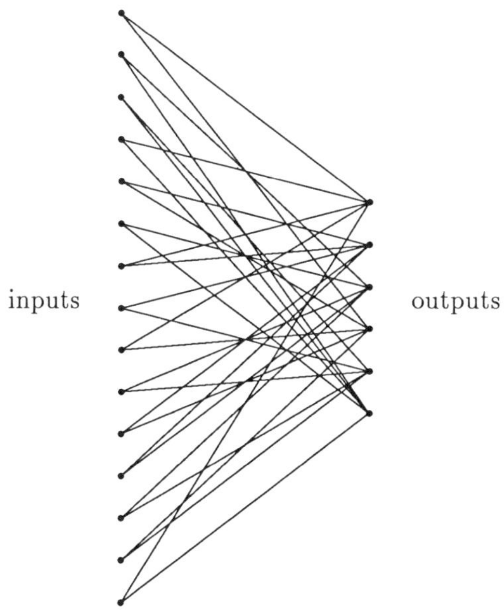

Figure 3: A $(15, 6, 1/2)$ partial concentrator: any 3 inputs can be connected to some 3 outputs. The graph is bipartite, every input has outdegree 2, and every output has indegree 5.

A *concentrator fat-tree (CFT)*, introduced in [9], is a complete binary tree whose nodes are stages of switches organized as concentrators. Here, we define a specific CFT with parameters chosen to optimize the performance of C_u. The number of leaves is $L = A/\log^2 A$. For a given node of the CFT, let n denote the number of leaves of the subtree rooted at that node. A leaf ($n = 1$) consists of a set of $w(1) = \lceil 4 \log L \rceil$ 1-bit registers that hold messages being input to the fat-tree from the mesh, and by as many output registers. Except for the root, each node is connected to its parent by a group of $w(n) = \lceil \max(4 \log L, \sqrt{n}) \rceil$ wires in each direction, referred to as a *channel* of *capacity* $w(n)$. The structure of a non-root internal node

ν $(1 < n < L)$ is illustrated in Figure 4. The two input channels from ν's children are connected to the output channel leading to ν's parent via a $(2w(n/2), w(n), \alpha)$-PC. Similarly, the input channels from the parent and from one child are connected to the output channel to the other child via a $(w(n)+w(n/2), w(n/2), \alpha)$-PC. The root of the CFT consists of a straightforward connection between the two children with a channel of capacity $w(L/2)$ in each direction.

In a partial concentrator, an input vertex is essentially a fan-out element; an output vertex is implemented as a constant-size combinational switch followed by a 1-bit buffer. Under the control of a constant number of bits, the switch establishes a connection between one of the input arcs and the output buffer. To each switch is attached a shift register of $O(\log A)$ cells used to store control information computed off-line, as specified below.

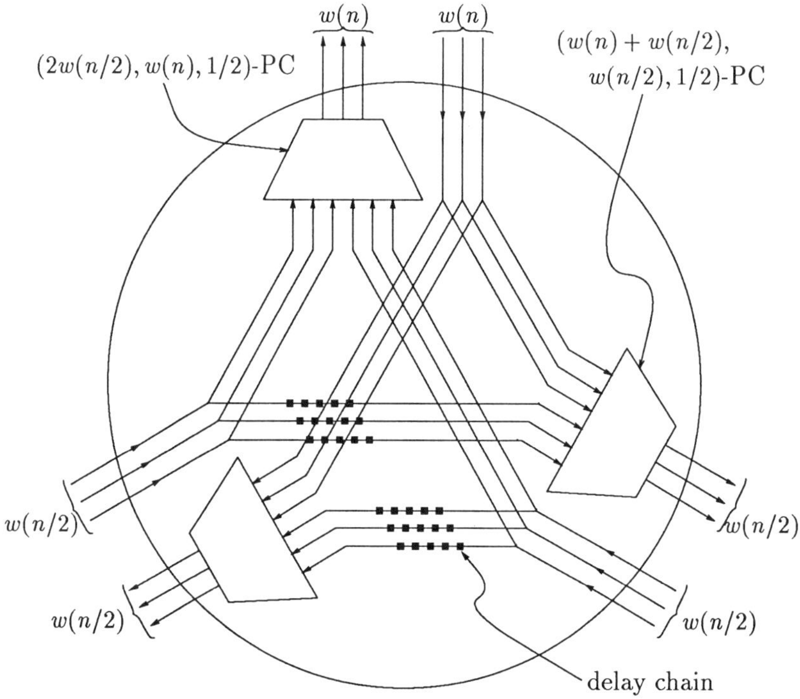

Figure 4: A node of the concentrator fat-tree

The connections from one child of ν to the PC going to the other child are realized through chains of $2\log(L/n) - 1$ delay elements. We have added these chains to the CFT of [9] to equalize all leaf-to-leaf paths and make the network *pipelinable.* (The technique was used in previous constructions, e.g.,

[2].) With the chains, the CFT can be viewed as the cascade of $2\log L - 1$ stages of buffers interleaved with $2\log L - 2$ stages of combinational logic. For $1 \le k \le \log L - 1$, the k-th combinational stage includes the PCs with output toward the parent of those nodes with $n = 2^k$. For $\log L \le k \le 2\log L - 2$, the k-th combinational stage includes the PCs with output toward the children of those nodes with $n = L/2^{k-\log L+1}$.

It is easy to verify that all partial concentrators needed in the CFT are $(\beta w, w, \alpha)$-PCs with $\beta \in \{\sqrt{2}, 2, \sqrt{2}+1\}$. By modifying a proof in [11] we have shown that, for all positive even w, there exist bipartite $(5w/2, w, 1/2)$-PCs where every input has outdegree at most 6 and every output has indegree at most 15. PCs with $\beta < 5/2$ for use in C_u are obtained by simply deleting the appropriate number of input vertices and their adjacent edges.

Network area. A node of the CFT with parent channel capacity $w(n) = w$ consists of three $O(w)$-vertex PCs, $O(w)$ delay chains of length $O(\log L) = O(w)$, and $O(w)$ controlling shift registers of length $O(\log A) = O(w)$. Thus such a node can be laid out in a $O(w) \times O(w)$ region. To evaluate the area of the entire circuit, let $S(n)$ be the length of one side of the square layout of a subtree with n leaves. The recurrence relation for $S(n)$ is

$$S(n) = \begin{cases} O(\log A) & \text{if } n = 1 \\ 2S(n/4) + O(\log A) & \text{if } 1 < n \le \log^2 A \\ 2S(n/4) + O(\sqrt{n}) & \text{if } \log^2 A < n \end{cases}$$

By solving this recurrence, we get $S(A) = O(\sqrt{A})$; hence the total area of C_u is $O(A)$.

Routing time. The CFT routes sets of messages among its leaves. Messages always follow the unique shortest path in the underlying binary tree, but may be routed on different wires within a channel depending on the network's congestion. A useful measure capturing the bandwidth requirements posed by a message set M is the *load factor* $\lambda(M)$. Let the *load* $\ell(M, c)$ placed on a channel c by a message set M be defined as the number of messages that must cross c. Let $|c|$ denote the capacity of c. Then $\lambda(M)$ is the maximum over all channels of the quantity $\ell(M, c)/|c|$.

Simple adaptations of arguments from [9] lead to the following two lemmas:

Lemma 2 *A message set M with $\lambda(M) \le 1/2$ can be routed by the concentrator fat-tree of C_u on vertex-disjoint paths.*

Lemma 3 *A message set M for the concentrator fat-tree of C_u can be partitioned into $O(\lambda(M))$ subsets, each with load factor at most* $1/2$.

With these lemmas in hand, we are ready to argue that the CFT part of C_u can route a set M of single-bit messages corresponding to the inter-mesh wires in $O(\log A)$ time. Observe that the number of wires that enter or leave any $\sqrt{m} \times \sqrt{m}$ region of C is at most $4\sqrt{m}$. The root channel of each subtree of C_u to which such a square region has been mapped has capacity at least $\sqrt{m}/\log A$, so $\lambda(M) \leq 4 \log A$.

Using an off-line algorithm described in Section 4, M is partitioned according to Lemma 3 into $q = O(\log A)$ message batches $B_0, ..., B_{q-1}$, each with load factor at most 1/2. Messages from a given batch are routed on the vertex-disjoint paths guaranteed to exist by Lemma 2; these paths are computed off-line and stored in order of batch number in the controlling shift registers of the partial concentrators. The q batches are pipelined through the $2\log L - 2$ stages of the CFT in time $2\log L - 2 + q - 1 = O(\log A)$.

4 Compiling for the Area-universal Circuit

This section describes a compiler that translates a suitable description of a circuit C of area A into settings for the control bits of C_u. We also show that such a compiler works in time $O(A)$, thus establishing Theorem 2.

Compiler input. We assume the circuit C to be simulated is described by a $\sqrt{A} \times \sqrt{A}$ array of constant-size records. Array element (i, j) indicates whether there is a gate at the intersection of row i and column j of C, and if so, its type and orientation. The record also indicates the presence or absence of wires in the adjacent unit-length segments of the ith horizontal and j^{th} vertical layout tracks. Such a description might be the output of a layout editor or automatic layout program.

Compiler output. The compiler computes the control bits for the universal gates, for the FAs in the mesh nodes, and for the switches in the fat-tree concentrators.

Compiler structure. Our compiler has the following main stages, each outlined and analyzed below.

1. Wire preprocessing, to determine which wires are simulated by meshes and which by the CFT.

2. Mesh processing, to determine control bits for the universal gates and the FAs.

3. CFT processing, to determine the settings for the switches in the partial concentrator graphs.

Stage 1: Wire preprocessing. This stage converts wires of C with source and destination in different meshes into a set of messages M stored as an array of $L = A/\log^2 A$ linked lists. Each list contains the messages with source or destination at a particular mesh. A similar structure is formed for the wires with source and destination in the same mesh. To produce these data structures, the compiler scans the description of C in, say, row major order. Each time the source of a wire is encountered, the wire is traced to its destination, and its length is computed in the process. If the wire goes between meshes, it is added to the linked lists for its source and destination in the first array, otherwise it is added to the second array. Both the row-major scan and the wire-tracing detours take time $O(A)$.

Stage 2: Mesh processing. Recall that each $\sqrt{N} \times \sqrt{N}$ mesh of finite automata is required to simulate the gates and wires in a local region of C ($N = \log^2 A$). The wires pertaining to a given mesh, identified in Stage 1, are easily partitioned into those that are shorter than $2\sqrt{N}$ and those that are longer. The control bits for the gates and for the wires shorter than $2\sqrt{N}$ are available, almost directly, from the given two-dimensional array description of C. For emulating long wires using the method described in the proof of Lemma 1 the compiler proceeds as follows.

For Steps 1 and 5, a simple column-major pass through the relevant region of C is sufficient to determine which FA in each column initiates or terminates the movement of a "train". To set up the path taken by the messages in Steps 2, 3, and 4, we first initialize two $\sqrt{N}/2$ element arrays S and D. For a given message m, $S[m]$ will be the index where the routing path intersects the top half of the diagonal ($0 < S[m] \leq \sqrt{N}/2$), and $D[m]$ will be the index where the routing path intersects the bottom half of the diagonal ($\sqrt{N}/2 < D[m] \leq \sqrt{N}$). Assuming the source and destination of each message have already been labeled with the same message id, make a column-major pass through C, scanning each column from bottom to top. If m is the id of the ith source encountered in the scan, let $S[m] = i$. Similarly, if m is the id of the ith destination encountered, let $D[m] = \sqrt{N}/2 + i$. The column-major pass can clearly be completed in $O(N)$ time.

Once S and D are initialized, it is a simple matter to determine control bits so the FAs transmit the messages along the (edge-disjoint) routing paths required for Steps 2, 3, and 4. Message m moves horizontally to the diagonal in row $S[m]$ in Step 2, then horizontally in row $S[m]$ and vertically in column $D[m]$ in Step 3, and finally horizontally in row $D[m]$ from the diagonal to its destination column in Step 4. Since the paths are of length $O(\sqrt{N})$, computing the control bits takes $O(\sqrt{N})$ time per message, or $O(N)$ overall. So Stage 2 takes $O(\log^2 A)$ time per mesh and $O(A)$ time for the entire circuit.

Stage 3: CFT processing. This stage is the most complex and is split into two substages, the first to determine the partition of the messages into batches, and the second to compute the control bits for the concentrator switches relative to each batch.

Stage 3.1: Partition into batches. This stage partitions the message set M produced by Stage 1 into messages sets $B_0, ..., B_{q-1}$, with $q = O(\lambda(M))$ and $\lambda(B_i) \leq 1/2$ for each i. The partitioning algorithm was described in [9] without a time analysis; we tersely review it here to show that it runs in time $O(A)$ for our application. Figure 5 summarizes the algorithm.

Number of batches. The number of batches is chosen to be $q = 2^{\lceil \log(4\lambda(M)) \rceil}$. To obtain q, the load factor $\lambda(M)$ is first computed. The loads on the channels are obtained in $O(|M| \log L)$ time by tracing the path of each message through the tree. Then the load on each channel is divided by its capacity and the maximum over all channels is found in $O(L)$ time.

Partitioning by peak height. Next M is partitioned into $M_1, ..., M_{\log L}$, where M_h represents (in the same format as M) the subset of messages with source and destination having their lowest common ancestor (*peak*) at height h. The partition is computed in $O((|M| + L)\log L)$ time: $O(L)$ time to initialize the representation of each of $\log L$ subsets and, for each message, $O(\log L)$ time to compute h and to insert the message in M_h.

A procedure for a load-balanced split. The heart of the partitioning algorithm is a procedure *Bipartition* (Q, Q_a, Q_b). It takes as input a set Q of messages with the same peak height h, and outputs a partition of Q into subsets Q_a and Q_b such that, for every channel c, the loads $\ell(Q_a, c)$ and $\ell(Q_b, c)$ differ by at most one. *Bipartition* consists of a pairing phase and a splitting phase, to be described next.

The *pairing phase* forms pairs of message sources and pairs of message destinations, leaving at most one unpaired source and one unpaired destination in each subtree of height h. Specifically, for $k = 0, 1, ..., h$, successively, each subtree of 2^k leaves is examined. Within the subtree, the currently-unpaired sources are matched into pairs, with one left over in case of an odd number of sources. A similar process is applied to destinations.

The result of the pairing phase can be conveniently viewed as an undirected graph where to each message of Q there corresponds an edge, with a source vertex and destination vertex. Source (destination) vertices of two edges coincide exactly when the sources (destinations) of the corresponding messages are paired. By construction, the connected components of such a graph are either cycles of even length, or simple paths of odd length.

In the *splitting phase*, the graph is traversed and messages corresponding to edges are alternately placed in Q_a and Q_b. The reader may verify that the overall running time of procedure *Bipartition* is $O(L + |Q|)$.

Batch formation. For each peak height h, the algorithm calls *Bipartition* to divide M_h into first two subsets, then four, then eight, and so on, ending with a partition into $q = 2^p = 2^{\lceil \log(4\lambda(M)) \rceil} = O(\lambda(M))$ message subsets $Q^h_{p,0}, ..., Q^h_{p,q-1}$. Then, corresponding subsets from different peak heights are recombined to form the final batches $B_0, ..., B_{q-1}$. The load balancing property of *Bipartition* ensures that $\ell(Q^h_{p,i}, c) < \ell(M_h, c)/2^p + 1$ on any channel c. Hence, $\ell(B_i, c) < \ell(M, c)/2^p + \log L$. Dividing by $|c|$, taking the maximum over all channels c, recalling that $|c| \geq 4 \log L$, and observing that $2^p \geq 4\lambda(M)$, we obtain $\lambda(B_i) < 1/2$. Thus, by Lemma 2, each batch can be routed by our CFT on disjoint paths.

Time of Stage 3.1. Initializing a message set (line 5) and computing the union of two message sets (line 12) can be done in time $O(L)$. Thus the loops of lines 4 and 11 each require $O(\lambda(M)L)$ time. As noted above, a single call to *Bipartition* (line 10) takes $O(|Q_{j,i}| + L)$ time. Since the 2^j calls to *Bipartition* in the innermost **for** loop (line 9) operate on disjoint subsets of M_h, they take total time $O(|M_h| + 2^j L)$. Summing over $j = 0, 1, ..., p-1$ and considering that $p = O(\log \lambda(M))$, we see that the **for** loop of line 8 takes

1. compute $\lambda(M)$
2. partition M by peak height into $M_1, ..., M_{\log L}$
3. $p := \lceil \log(4\lambda(M)) \rceil$
4. **for** $i :=$ 0 **to** $2^p - 1$ **do**
5. $B_i := \emptyset$
6. **for** $h :=$ 1 **to** $\log L$ **do**
7. $Q^h_{0,0} := M_h$
8.   **for** $j :=$ 0 **to** $p - 1$ **do**
9.   **for** $n :=$ 0 **to** $2^j - 1$ **do**
10.   *Bipartition* $(Q^h_{j,i}, Q^h_{j+1,2i}, Q^h_{j+1,2i+1})$
11.   **for** $i :=$ 0 **to** $2^p - 1$ **do**
12. $B_i := B_i \cup Q^h_{p,i}$

Figure 5: Leiserson's algorithm for partitioning a message set

time $O(|M_h| \log \lambda(M) + \lambda(M)L)$. Finally, since the different M_h's are disjoint subsets of M, the **for** loop of line 6 executes in time $O(|M| \log \lambda(M) + \lambda(M)L \log L)$. Thus, assuming $\log \lambda(M)$ is $O(\log L)$, the cost of the entire partitioning algorithm can be summarized as $O((|M| + \lambda(M)L) \log L)$. Since $|M| = O(A/\log A), L = A/\log^2 A$, and $\lambda(M) = O(\log A)$, the running time is $O(A)$.

Stage 3.2: Control bits of concentrator switches. We compute the paths (and the corresponding control bits) for a batch of messages by starting at the input terminals and augmenting them stage by stage. When augmenting the paths through a particular (r, s, α)-PC, say G, the set U of the $k \leq \alpha s$ input vertices of G that belong to the paths is already known. In time $O(k)$, we construct the bipartite subgraph H of G induced by the edges incident on U. Then, the paths are simply extended along the edges of a maximum matching of H. Such matching is guaranteed to have size k and, since G has bounded degree, can be found in $O(k^{3/2})$ time [10].

Time of Stage 3.2. If the output of G is connected to channel c, the control bits pertaining to the $\ell(B_i, c)$ messages of batch B_i can be computed in time $O(\ell(B_i, c)^{3/2})$. Summing over all batches and using the relation $\ell(M, c) = \sum_i \ell(B_i, c)$ yields time $O(\ell(M, c)^{3/2})$. Since there are $A2^{-h}/\log^2 A$ channels at height h, each with load $O(2^{h/2} \log A)$, the total time for determining all concentrator switch settings is

$$\sum_{h=1}^{\log(A/\log^2 A)} O(A2^{-h}/\log^2 A)(2^{h/2} \log A)^{3/2} = O(A/\sqrt{\log A}).$$

In conclusion, the entire compiler runs in time $O(A)$, as claimed.

5 Conclusions

The main objective of this work has been to investigate how important special-purpose architectures are to the performance of VLSI circuits. The main conclusion is that, at least in asymptotic terms, there is relatively little room for designing special-purpose circuits with higher performance than the general-purpose circuit described here. It would be interesting to reduce the simulation time penalty below $O(\log A)$ and/or prove a lower bound for it.

We have mentioned that our universal circuit has the functionality of an FPGA. Further investigation is required to determine the extent to which the ideas behind our universal circuit could be exploited in actual FPGAs. The basic ideas should prove useful; however, focusing on circuits of a specific size, rather than on asymptotic behavior, would probably suggest different design choices. For example, in the mesh portion of the simulation, it may turn out to be more efficient to rely more on combinational than on sequential logic. For concentrators, designs with logarithmic depth may be preferable to those with constant depth adopted in this paper, as they are more regular and easier to obtain. Also, the tradeoff between mesh size and CFT height could be adjusted according to a more detailed estimate of the times spent routing in each.

Acknowledgements

This work was carried out while the first author was a student in the Computer Science Department at Cornell University, Ithaca, NY. The second author's work was supported in part by the Italian National Research Council and the Italian Ministry of University and Research.

References

[1] P. Bay and G. Bilardi. Deterministic on-line routing on area-universal networks. In *Proceedings of the* 31st *Annual Symposium on Foundations of Computer Science*, pages 297–306, October 1990.

[2] R. P. Brent and H. T. Kung. A regular layout for parallel adders. *IEEE Transactions on Computers*, C-31(3):260–264, March 1982.

[3] S. D. Brown, R. J. Francis, J. Rose, and Z. G. Vranesic. *Field-Programmable Gate Arrays.* Kluwer Academic Publishers, Boston, MA, 1992.

[4] S. A. Cook and H. J. Hoover. A depth-universal circuit. *SIAM Journal on Computing*, 14(4):833–839, November 1985.

[5] Z. Galil and W. J. Paul. An efficient general-purpose parallel computer. *Journal of the ACM*, 30(2):360–387, April 1983.

[6] L. Goldschlager. A universal interconnection pattern for parallel computers. *Journal of the ACM*, 29(3):1073–1086, October 1982.

[7] R. I. Greenberg and C. E. Leiserson. Randomized routing on fat-trees. In S. Micali, editor, *Randomness and Computation*, pages 345–374. JAI Press, Inc., 1989.

[8] T. Leighton, B. Maggs, and S. Rao. Universal packet routing algorithms. In *Proceedings of the* 29th *Annual Symposium on the Foundations of Computer Science,* White Plains, New York, October 1988.

[9] C. E. Leiserson. Fat-trees: Universal networks for hardware-efficient supercomputing. *IEEE Transactions on Computers*, C-34(10):892–900, October 1985.

[10] C. H. Papadimitriou and K. Steiglitz. *Combinatorial Optimization: Algorithms and Complexity.* Prentice-Hall, Inc., Englewood Cliffs, NJ 07632, 1982.

[11] N. J. Pippenger. Superconcentrators. *SIAM Journal on Computing*, 6(2):298–304, 1977.

[12] C. D. Thompson. *A Complexity Theory for VLSI.* Ph.D. dissertation, Carnegie-Mellon University, August 1980.

[13] J. D. Ullman. *Computational Aspects of VLSI.* Computer Science Press, Rockville, MD, 1984.

[14] L. G. Valiant. Universal circuits. In *Proceedings of the* 8th *Annual ACM Symposium on the Theory of Computing*, pages 196–293, 1976.

Delay Fault Testing: Trading Fault Coverage, Test Set Size, and Performance

William K. Lam
Alexander Saldanha
Robert K. Brayton
Alberto L. Sangiovanni-Vincentelli

University of California - Berkeley CA

Abstract

We prove that 100% delay fault testability is not necessary to guarantee the speed of a combinational circuit. There exist path delay faults which can never impact the circuit delay (computed using any correct timing analysis method) unless some other path delay faults also affect it. These are termed robust dependent delay faults and need not be considered in delay fault testing. Necessary and sufficient conditions under which a set of path delay faults is robust dependent are proved; this yields more accurate and larger delay fault coverage estimates than previously used. Next, assuming only the existence of robust delay fault tests for a very small set of paths, we show how the circuit speed can be selected such that 100% robust delay fault coverage is achieved. This leads to a quantitative tradeoff between the test set size for a circuit and the verifiability of its performance. Finally, under a bounded delay model, we show that the test set size can be reduced while maintaining the delay fault coverage for the specified circuit speed.

1 Introduction

The need for ensuring the temporal (dynamic) correctness of circuits has led to the development of delay fault testing. Two fault models have been proposed for delay fault testing, namely gate delay faults and path delay faults. Of these the latter is more attractive since it models a larger set of defects and can give guarantees on circuit performance [4]. However, because there can be an exponential number of paths in a circuit, path delay fault testing often requires very large test sets; a large number of paths also renders current analysis and synthesis techniques for path delay fault testing infeasible on large circuits. This paper addresses the issue of guaranteeing circuit performance using small (possibly linear size) test sets. For circuits where a 100% guarantee is not possible, we also address the question of providing a large fault coverage guarantee with a small test set.

A **path delay fault** occurs along a path P if the delay along P falls outside its specified limits (in general, either upper or lower bounds). Each path has two possible long path delay faults associated with it, a rising delay fault and a falling delay fault. [1] The goal of delay fault testing is to certify the absence of delay faults that cause the manufactured circuit to fall outside its specified delay bounds.

In this paper we focus on upper bounds on circuit performance. We show that the path delay faults in a circuit can be partitioned into two sets. For the first set, the occurrence of one or more delay faults can cause an increase in circuit delay. This set must be tested for delay faults to certify correct operation. It includes all robust testable path delay faults (RPDF). The delay faults in the second set are termed **robust dependent (RD) delay faults**. The occurrence of these delay faults cannot increase the circuit delay unless some non-RD delay faults also occur. Thus, RD delay faults need not be tested to ensure that a manufactured circuit operates at the desired speed.

We define the **delay fault coverage** of a test set as the percentage of all non-RD delay faults guaranteed not to occur if the circuit passes the test set. This is different from the percentage of all delay faults that are robust testable. The latter is commonly called the **delay fault testability** of the circuit. There may be delay faults in a circuit for which no robust test exists, yet the absence of the delay fault is guaranteed by the test set (since the circuit passed the delay tests). Thus, delay fault coverage is a function not only of the test set size but also of the number of faults implied absent by the test set. The fault coverage of a test set can also depend on the delay bounds used in the testing procedure.

This paper contributes to the question of how much delay fault testing is necessary in three ways:

1. Necessary and sufficient conditions for a set of delay faults to be RD are derived. We show that the delay faults on any set of long false paths that can exist for any particular delay assignment to connections in the circuit forms an RD set. This does not affect the test set size since every RPDF still must be tested. However, it does give a more realistic and better estimate of the delay fault coverage.

2. We show how to trade-off delay fault coverage and circuit speed. By relaxing the operating speed for the circuit (this is the clock period in synchronous circuits), a larger delay fault coverage can be achieved. We call the minimum clock rate for which the coverage is 100% the **robust delay** of the circuit. Using this concept, we present path delay fault testing methodologies whose test set sizes are **linear** in the size of the circuit (as opposed to 100% RPDF testing which can be exponential in size).

3. Using a bounded delay model (the delay of a tested path is guaranteed to lie between both upper and lower bounds) we formulate a simple linear program to reduce the delay fault test set size without decreasing the fault coverage.

Algorithms implementing these ideas are described and a small set of examples is used to illustrate these effects on delay fault testing.

[1] Each path delay fault is associated with a path. We interchangeably refer to either the delay fault or the associated path.

2 Verifiable delay analysis

A **connection** is just a wire or a pin-to-pin path through a gate. A **path** is denoted as a sequence of connections or points. The delay of a connection $e = (a, b)$ between points a and b is denoted either $\delta(e)$ or $\delta(a, b)$, and the delay of path P is denoted $\delta(P)$. Delays may depend on whether a signal is rising or falling in which case a delay is denoted $\delta^r(e)$ or $\delta^f(a, b)$. For a path P, $\delta^r(P)$ means the delay of P when its output is rising; similarly for falling.

A circuit design C is a net list of gates. In analyzing a design, typically one assumes some delay model, for example a connection e has rising delay $\delta^r(e)$ such that $d^r_{\text{min}} \leq \delta^r(e) \leq d^r_{\text{max}}$ (called a **bounded delay model**). Usually an analysis of the performance of a circuit design derives properties about the design **assuming** that all connections meet their delay bounds. For example, in timing verification [3], an upper bound τ on the delay the circuit is computed, where τ is some complex function of the assumed delay bound parameters. Note that this analysis is valid only for **acceptable** designs and manufactured circuits whose connections meet the delay bounds.

To guarantee that a **manufactured circuit** C_m operates at some specified delay we must make different assumptions since we know nothing a priori for the circuit under test. The following information is known if C_m passes the test. Let $\mathcal{Q}$ be the set of paths tested such that for each $P_i \in \mathcal{Q}$,

$$\tau^i_{\text{min}} \leq \sum_{e \in P_i} \delta(e) \leq \tau^i_{\text{max}},$$

where τ^i_{min} and τ^i_{max} depend on the type of testing done. Note that there are no assumptions about delays on connections; only verifiable delays of paths are used. Only implicitly is there an assumption about connection delays, *i.e.* there exists some fixed but unknown delay $\delta(e)$ on each connection e. Thus the circuit under test (the manufactured circuit) is $C_m = (C, \Delta)$ where Δ is a vector of delay values, one value for each connection. We say Δ is a **delay assignment**. We assume $\Delta^e \geq 0$ for connection e but no upper bounds are assumed; just that certain sums are bounded:

$$\tau^i_{\text{min}} \leq \sum_{e \in P_i} \Delta^e \leq \tau^i_{\text{max}}.$$

The fact that we have no delay model assumptions on connections does not preclude delay analysis on C. We shall show that the following kind of analysis is possible. Let $D \subseteq \mathcal{P}$, where $\mathcal{P}$ is the set of all complete paths of C, *i.e.* paths from primary inputs to primary outputs (such as those tested). It is possible that the following can be determined through analysis: for any τ, if $\delta(P_i) \leq \tau$, $P_i \in \mathcal{P} - \mathcal{D}$, then $\delta(P_j) \leq f(\tau)$, $P_j \in \mathcal{D}$. This implies that even before a manufactured circuit C_m is tested, we know that paths in $\mathcal{D}$ need not be tested for guaranteeing C_m as good or bad. Here $f(\tau)$ is some function determined by analysis from the net list C.

We show two types of this kind of analysis in the paper. The first is where $f(\tau) = \tau$; the set of paths $\mathcal{D}$ for which such a property holds is called **robust**

dependent. The second type of analysis is where $f(\tau)$ is a complex function evaluated by solving a linear program.

Delay analysis has been the subject of intense research in the past few years. Although our delay model for $C_m = (C, \Delta)$ is a fixed but unknown delay, the model for establishing $\delta(P_j) \leq f(\tau)$ can be obtained by any analysis that is a relaxed version of this. Thus even though $\delta(e) = \Delta^e$ for connection e of C_m, analysis methods assuming $\Delta^e_{\text{min}} \leq \delta(e) \leq \Delta^e_{\text{max}}$ also give legitimate bounds. Of course all the analysis methods used in this paper start only with the assumption that $\tau^i_{\text{min}} \leq \sum_{e \in P_i} \Delta^e \leq \tau^i_{\text{max}}$ for those paths P_i in the test set. Thus if the circuit passes verification of lower and upper bounds on tested path delays, we are guaranteed that the circuit will operate correctly anywhere between the tested ranges. This accounts for both static (due to process variation, operating temperature, etc.) and dynamic (due to slope factors, crosstalk, etc.) delay variations that normally occur during the life of the circuit. Correct operation is certified if the delay variations do not cause any circuit delays to fall outside the tested delay bounds. Of course, nothing can be guaranteed if sometime during its life C_m operates outside the tested bounds. It is also instructive to realize that delay analysis based on verifiable path delays is less likely to be impacted by dynamic delay variations than delay analysis based on gate delays. This is because it is likely that a small delay variation on a connection will not get reflected on the path delay.

In summary, the delay analysis of this paper differs from the usual in that we perform delay analysis using only verifiable path delays instead of non-verifiable connection delay assumptions.

3 Definitions

3.1 Robust delay-fault testing

A **controlling value** for a gate f is a value at its input that determines the value at the output independent of the other inputs, and is denoted $A(f)$. For example, 0 is a controlling value for an AND gate. A **non-controlling value** for a gate f is a value at its input which is not a controlling value for the gate, and is denoted $I(f)$. For example, 1 is a non-controlling value for an AND gate. A **simple gate** is any one of AND, OR, NAND, NOR, and NOT. All the results described here apply only to simple gates. Only these gates have controlling vs. non-controlling values for each input.

Let $P = \{f_0, f_1, ..., f_m\}$ be a path where each $f_i, i > 0$ is the output of a gate. The inputs of f_i other than f_{i-1} are called **side-inputs** of f_i along P and denoted $S(f_i, P)$. A path that starts at a primary input and ends at a side-input of P is a **side-path** of P. A **path delay-fault** occurs along path P if the delay along P falls outside its specified limits (in general, either upper or lower bounds).

Definition 3.1 *A test for a delay-fault is* **robust** *if and only if the test is valid independent of delays on all the connections in the circuit. A circuit has* **100% robust delay-fault testability** *if and only if every path delay-fault has a robust test.*

Definition 3.2 *Let* $P = \{f_0, f_1, ..., f_m\}$. *A delay-fault for the rising (falling) transition at* f_m *is said to be* **robust testable** *by the vector pair* $< v_1, v_2 >$ *if and only if at each node* f_i, $< v_1, v_2 >$ *yields the desired transition being tested, and for each* $g_j \in S(P, f_i)$:

1. $g_j(v_2) = I(f_i)$, *and*

2. *if* $f_{i-1}(v_1) = I(f_i)$, *then there is no transition on* g_j.

The vector v_2 *applied after* v_1, *delayed by an amount greater than the longest path in the circuit.*

3.2 Delay analysis

Definition 3.3 *A* **leaf-dag** *is a circuit composed of AND and OR gates with multiple fanout and inverters only permitted at the primary inputs, and with each inverter allowed only single fanout.*

Every circuit (composed of simple gates) can be converted to a leaf-dag, albeit with possibly an exponential number of gate duplications. There is a one-to-one correspondence between each path in a circuit and its leaf-dag. The delay fault and timing properties of each path in a circuit are thus retained by the transformation to a leaf-dag.

Definition 3.4 *The* **I-edge** *of a path in a circuit refers to either the connection from the primary input if no inverter is there, or else the connection immediately after the inverter.*

Let η be the leaf-dag of C. Any delay assignment in C also exists in η; simply use the same delay on each connection in η as the corresponding connection in C. However, the converse is not true.

The following definition is adapted from [1]. Given a delay assignment and assuming an initial unknown value on each connection of a circuit C, a path $P = \{f_0, f_1, ..., f_m\}$ in C is **true** under a single vector v if and only if the following conditions hold at each gate $f_i, i > 0$:

1. If all inputs of f_i are non-controlling, f_{i-1} must be the last side-input to present the non-controlling value.

2. If at least one input of f_i is controlling, f_{i-1} must be the first side-input to present the controlling value.

This definition extends naturally to sensitization under a multiple vector input denoted $v_0, v_1, ..., v_n$ with v_n applied last. The single vector **delay of a circuit** is determined by the longest true path under any input vector for the given delay assignment; similarly for multiple vector delay. The **delay of a manufactured circuit** refers to the actual operating delay of the circuit and is determined by the longest true path for the unknown delay assignment that exists in the manufactured circuit.

The following theorem relates the testability of a multiple stuck-0 (stuck-1) fault on the I-edges of each path of length at least L to the existence of a true path under a single vector of the primary inputs whose rising delay is of length at least L. Given a set M of I-edges, the multiple fault M stuck-0 (stuck-1) corresponds to a stuck-0 (stuck-1) fault on each I-edge in M.

Theorem 3.1 (adapted from [2]) *Let C be a circuit with a given delay assignment and η its leaf-dag. Let $\mathcal{P} = \{P_1, P_2, ..., P_n\}$ be the set of all paths in C of rising (falling) length at least L, and let P_η be the corresponding paths in leaf-dag η. At least one $P_i \in \mathcal{P}$ is true under the vector v for the rising (falling) transition at an output in C if and only if v is a test in η for the multiple stuck-0 (stuck-1) fault on the I-edge of each path in $\mathcal{P}_\eta$.*

Theorem 3.1 describes the exact condition for computing the delay of a circuit under a single input vector. A tighter delay may be achieved by using a multiple vector criterion. However, the single vector criterion is an upper bound on the delay achieved by any correct multiple vector criterion. Thus, the existence of a test in η for the multiple stuck-fault on the I-edges of each path in $\mathcal{P}_\eta$ (in Theorem 3.1) is a necessary condition for a path in $\mathcal{P}$ in C to be true under multiple vector criterion. However, it is unknown if this is sufficient under a multiple vector criterion.

Corollary 3.1 *Let C be a circuit with a given delay assignment and η its leaf-dag. Let $\mathcal{P} = \{P_1, P_2, ..., P_n\}$ be the set of all paths in C of length at least L, and let $\mathcal{P}_\eta$ be the corresponding paths in leaf-dag η. At least one $P_i \in \mathcal{P}$ is true under the vectors $v_0, v_1, ..., v_n$ for the rising (falling) transition at an output in C only if v_n is a test in η for the multiple stuck-0 (stuck-1) fault on the I-edge of each path in $\mathcal{P}_\eta$.*

The condition described by Theorem 3.1 and Corollary 3.1 is **robust**; the delay estimate remains valid for any decrease in connection delays [3].

4 Robust dependent delay faults

4.1 Functional interactions between paths

We demonstrate with some examples the impact that the functional interaction between paths can have on the delay fault coverage.

Example: Consider the simple circuit of Figure 1. There are six delay faults (a rising and falling transition delay fault for each path). The paths $\{b, y, f\}$ and $\{b, x, y, f\}$ are robust testable for falling transition delay faults, but only path $\{b, y, f\}$ is robust testable for the rising transition delay fault. Path $\{a, x, y, f\}$ is robust untestable for both transitions and $\{b, x, y, f\}$ is untestable for the rising transition.

First consider the case of a falling transition at the output f. We argue that whenever a falling transition propagates along $\{a, x, y, f\}$ a longer falling delay

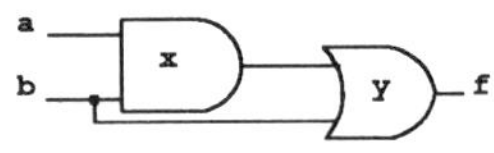

Figure 1: Redundant circuit with 100% robust delay fault coverage

exists along $\{b, x, y, f\}$. Assume that a falling delay propagates along $\{a, x, y, f\}$. This implies the final value of b is 0. The output of x changes (after some delay) to 0 when any input switches to 0. b must be at 1 if the falling edge on a propagates through x. Thus, b too has a falling transition from 1 to 0 in order for this to happen under any delay assignment. b must be 1 at the moment a reaches x, since $\{a, x, y, f\}$ is the longest falling path. Since b eventually changes to 0, the path $\{b, x, y, f\}$ is longer than $\{a, x, y, f\}$ for the falling transition. Since a robust test exists for $\{b, x, y, f\}$, the occurrence of this longer delay fault can be detected. Hence $\{a, x, y, f\}$ need not be tested for the falling delay fault if $\{b, x, y, f\}$ is tested for a falling delay fault.

Now consider the rising transition at the output of f. Assume that path $\{b, y, f\}$ is tested for the rising delay fault. To propagate the last rising edge to f along $\{a, x, y, f\}$, b must have a rising transition. Thus, if $\{a, x, y, f\}$ is the last rising edge, then the rising transition from b through the output of y is later than that from x to the output of y. Hence $\{b, y, f\}$ is longer than $\{a, x, y, f\}$. Similarly, it can be shown that $\{b, x, y, f\}$ cannot propagate the last rising edge at f unless $\{b, y, f\}$ has a longer rising delay. Thus neither $\{a, x, y, f\}$ nor $\{b, x, y, f\}$ can affect the rising delay at f without implying that the rising delay of $\{b, y, f\}$ is longer. ■

There are several conclusions to be drawn from the example:

1. The requirement of robust delay fault testing is related to the sensitization of the path. However, unlike path sensitization which is determined under a specific delay assignment to the connections of the circuit, the requirement for testing a delay fault along a path is a delay-insensitive condition and must be valid for all possible delay assignments.

2. 100% robust delay fault testability is not necessary for 100% robust delay fault coverage.

3. Even 100% single stuck-fault testability is not necessary for 100% robust delay fault coverage. In Figure 1 the stuck-0 fault on the AND gate is redundant. This is a particularly significant since it is widely believed that a circuits without complete stuck-fault coverage are not amenable to testing for delay faults.

Of course, 100% RPDF is a sufficient condition for 100% delay fault coverage. However most synthesis techniques for 100% RPDF typically involve a large area penalty. Often the performance of the circuit is also degraded. We will show sufficient conditions under which a path need not be made robust testable. While we cannot claim that the condition is necessary, we show topological conditions under which it is necessary.

4.2 Conditions for RD paths

Definition 4.1 *Let $\mathcal{D}$ be the set of all path delay faults in a circuit C and $\mathcal{R}$ a subset of $\mathcal{D}$. If for all τ, the absence of delay faults ($> \tau$) in $\mathcal{D} - \mathcal{R}$ implies the delay of C is $\leq \tau$, R is said to be* **robust dependent** *(**RD**).*

A RD set is insensitive to the delay assignment, *i.e.* $\mathcal{R}$ can be eliminated from consideration in delay fault testing under every delay assignment. This is in contrast to some delay faults whose absence is implied under only some and not all delay assignments; this is considered in Sections 5 and 6. Another significant point is that the implied delay of the circuit ("delay of C" in Definition 4.1) may be computed by any correct timing analysis method such as the single or multiple vector criteria referred to earlier. For example, an RD set using a single-vector criterion means that the implied circuit delay is computed according to Theorem 3.1.

With every rising (falling) path delay fault is an associated path for the rising (falling) transition. A set of paths is called a **RD path-set** if the delay faults on the paths is an RD set.

We now state a sufficient condition for RD paths. An internally non-inverting circuit is one that has inverters only at the primary input leads of the circuit. Any circuit can be converted to an internally non-inverting circuit with at most a single duplication of each gate of the circuit.

Lemma 4.1 *Let η be an internally non-inverting circuit and let M be a set of I-edges. If M stuck-0 (stuck-1) is redundant, then every subset of M is also stuck-0 (stuck-1) redundant.*

Proof: Express each output as a function of the literals corresponding to I-edges of η. Now each output may be viewed as a monotonic increasing Boolean function of these Boolean literals. A stuck-0 (stuck-1) fault on a subset of I-edges cannot increase (decrease) a monotonic increasing function. Since M is a redundant multiple stuck-0 (stuck-1) fault, it does not decrease (increase) any output function; if it did M would be testable. Thus, neither can any subset of M, which proves each stuck-0 (stuck-1) subset of M is redundant. ∎

Since every leaf-dag is an internally non-inverting circuit, the above result holds for all leaf-dag's.

Theorem 4.1 *Let C be a given circuit and η its leaf-dag. Let $\mathcal{R}_\eta$ be the paths in η corresponding to a set of paths $\mathcal{R}$ in C. Let M_η be the I-edges of $\mathcal{R}_\eta$. If M_η stuck-0 (stuck-1) is redundant in η, then $\mathcal{R}$ is a rising (falling) RD path-set in C.*

Proof: Let M_η be stuck-0 redundant. Suppose path $P \in \mathcal{R}$ of length L is true for the rising transition for some delay assignment. By Corollary 3.1, the multiple fault corresponding to a stuck-0 fault on the set F of I-edges of all paths of length at least L (for the given delay assignment) is testable in η. F includes at least one edge $e \notin M_\eta$ since M_η and all its subsets are stuck-0 redundant (Lemma 4.1). Thus some path outside $\mathcal{R}$ is a true path of length at least L. By the definition of robust dependent delay faults, this implies that $\mathcal{R}$ is an RD path-set in C.

The proof for the falling transition is similar. ■

Theorem 4.1 yields a sufficient condition under which a set of paths cannot affect the circuit delay under any delay assignment. Even though a path $P \in \mathcal{R}$ may become sensitizable under some delay assignment, it does not determine the circuit delay, since there exists another path $Q \notin \mathcal{R}$ at least as long as P. A delay fault along P causes a delay at the circuit outputs if and only if a delay fault exists along Q. Thus, delay faults for paths in $\mathcal{R}$ need not be tested.

The condition of Theorem 4.1 is sufficient. However it is necessary under the single vector criterion for delay analysis and the assumption that C has the following property: given any ordering of path lengths there exists a delay assignment to the connections of C that realizes the ordering. In general, all circuits do not meet this requirement (leaf-dag's do).

Theorem 4.2 *Assume that each ordering of path lengths in a given circuit C can be realized by some delay assignment. Let η be the leaf-dag corresponding to C, and $\mathcal{R}_\eta$ the paths in η corresponding to a set of paths $\mathcal{R}$ in C. Let M_η be the 1-edges of $\mathcal{R}_\eta$. $\mathcal{R}$ is a rising (falling) RD path-set using the single vector criterion in C if and only if M_η stuck-0 (stuck-1) is redundant in η.*

Proof:

If part: Given by Theorem 4.1.

Only if part: Assume the multiple fault M_η stuck-0 is testable. Consider any delay assignment that makes the paths outside $\mathcal{R}_\eta$ shorter than all the paths in $\mathcal{R}_\eta$. By assumption, this partial order between the path lengths can also be satisfied in C by some delay assignment. By Theorem 3.1, the delay of C is at least the length of the shortest path in $\mathcal{R}$, since M_η is testable in η. Consequently, $\mathcal{R}$ is not an RD path-set for the single vector criterion in C. ■

It is possible to develop a tighter necessary condition which takes into consideration only those path orderings which exist for some delay assignments in a given circuit; however we do not have any application for it as yet. We do not know a necessary condition for RD delay faults under multiple vector criteria for delay computation since the sufficient condition for a path to be true under any of these criteria is unknown.

Example: We use the circuit of Figure 2 to explain the difference between path sensitization and robust dependent delay faults. Upper bounds on delays are shown within each gate; lower bounds are assumed 0. Since the output is constant at 0, the last transition under any input change is a falling edge. Thus the rising delay of the circuit is 0; hence we are not concerned with testing for rising transition delay faults. None of the falling transition delay faults along the three paths in the circuit (denoted P, Q, and R in the figure) is RPDF. The sets $\{P\}$ and $\{Q\}$ are both RD but the set $\{P, Q\}$ is not RD; neither is $\{R\}$ an RD set. This can be easily verified by checking the testability of the appropriate stuck-fault given by Theorem 4.1. However, under different delay assignments each path is responsible for the last falling transition at the output f. In the circuit shown on the left R is the longest sensitizable path; P and Q are the longest sensitizable paths in the circuit on the

Figure 2: Sensitizable paths and RD delay faults

right. Thus, although a path may be sensitizable under some delay assignment it may still be part of an RD set. ■

We note that the statement of Theorem 4.1 is different from previous work in timing analysis [2]. The distinction is that the previous work provides a necessary delay insensitive condition under which a path is sensitizable (called *static co-sensitization*). In contrast, Theorem 4.1 provides a sufficient delay insensitive condition (Theorem 4.2 shows when this condition is also necessary) under which a set of paths (even possibly sensitizable) cannot independently affect circuit delay.

4.3 False paths and RD delay faults

It is doubtful that identification of a maximum RD set will be computationally tractable on most circuits. In practice, therefore, we are interested in an estimate of the RD set which can be used to compute a lower bound for fault coverage. A technique for identifying a large RD set is related to the identification of long false paths in timing analysis [3].

Theorem 4.3 *Let C be a circuit whose longest true path on all single vectors is less than L for some delay assignment. Then the set of delay faults on all paths of length $\geq L$ is an RD set.*

Proof: Let η be the leaf-dag corresponding to C. Without loss in generality consider the rising transition delay of η. By Theorem 3.1 the multiple fault M corresponding to a stuck-0 fault on the I-edge of each path of length $\geq L$ is redundant. By Theorem 4.1, the delay faults on these paths is an RD set in η. As stated earlier, since there is a one-to-one correspondence between delay faults in η and C, the result follows. ■

There exist well-known circuits with many long false paths and the above result eliminates consideration of these paths in robust delay fault testing. The theorem extends the RD property to every set of long false paths under any delay assignment to the wires of the circuit. Since a set of delay faults is RD if the corresponding paths are long false paths, for any given delay assignment, we have the flexibility of choosing a delay assignment to try to maximize the number of long false paths. An algorithms based on this is explored in the next section.

4.4 Finding an RD set

Theorem 4.1 states the sufficient condition for an RD set in a given circuit C in terms of the testability of a multiple stuck-fault in the leaf-dag η corresponding to C. In practice, the identification of redundant multiple stuck-faults and the transformation to a leaf-dag are difficult to perform - the former requires large compute times, the latter usually requires enormous space.

We state a result that allows highly developed single stuck-fault redundancy identification methods to be exploited in determining a maximal RD set.

Theorem 4.4 *Let C be a non-inverting circuit and M a redundant multiple stuck-0 (stuck-1) fault in C (faults considered on or after I-edges). Let C_M be the circuit obtained by replacing each connection in M by 0 (1). If P is rising (falling) RPDF in C, then P is rising (falling) RPDF in C_M.*

Proof: Assume P is rising RPDF in C but is not rising RPDF in C_M. P cannot be in any rising RD set of C. There are two cases to consider since P is not rising RPDF in C_M. If some connection in P is set to 0 in obtaining C_M from C, then the rising RD set (determined by M) includes P, thus contradicting. Thus P is non-RPDF in C_M and no connection of P is set to 0. Let $< v_1, v_2 >$ be the robust rising transition test for P in C. Consider any OR gate f along P. By Definition 3.2 each side-input to f is at 0 under both v_1 and v_2 in C and there is no transition on these side-inputs. Since C is non-inverting, on asserting a multiple stuck-0 fault, these side-inputs to f remain at 0 under v_1 and v_2 with no transition in C_M. Hence propagation of the rising transition at any of the OR gates along P in C_M is not affected. Consider any AND gate along P. By Definition 3.2 each side-input to an AND gate is at 1 under v_2 in C. Since P is non-RPDF in C_M, at least one side-input to one or more AND gates in P is 0 under v_2 in C_M. The output of P is 1 in C under v_2 while the output of P is 0 in C_M under v_2. Hence v_2 is a test for M, thus contradicting that M is redundant.

The proof for the falling transition is similar. ■

The converse is not true; if P is RPDF in C_M, it may not be RPDF in C. The theorem states an important result about robust delay fault testability; a path P that is rising RPDF remains rising RPDF under removal of stuck-0 redundancies in a non-inverting circuit. This directly gives us a strategy for identifying a maximal RD set. This technique identifies redundant multiple stuck-faults by iteratively identifying redundant single stuck-faults. It eliminates the need to unfold a given circuit into a possibly exponential size leaf-dag.

The algorithm operates on a non-inverting circuit C' derived from a given circuit C by duplicating each gate at most once. The main loop of the algorithm iteratively performs two steps. First, redundant stuck-0 connections in C' are identified and replaced by 0. C' is irredundant at the termination of this step. In the second step selective duplication of gates is performed in C'. This possibly creates new redundant connections. These two steps are iterated until the resulting C' is fanout-free and irredundant. The paths in C' not passing through any constant connection form the non-RD set; the testability of these paths is determined in C to compute the delay fault coverage.

Simple heuristics are used to select the order of redundancy removal and duplication. Redundant connections are determined in decreasing order of the number of paths passing through them. This aims at eliminating the largest set of paths possible at each step. Duplication is performed on a level by level basis from primary

```
Convert C to an internally non-inverting circuit C'
R = {}
While (C' has any gate with fanout > 1) {
   While (C' has a redundant stuck-0 connection e) {
      Replace e by constant 0
      R = R ∪ {e}
   }
   Find gate g with fanout > 1
   Duplicate g and move fanout connections
      so each duplicate has less fanout than g
}
All paths with some connection in R form rising RD set.
```

Figure 3: Algorithm to find a maximal rising RD set

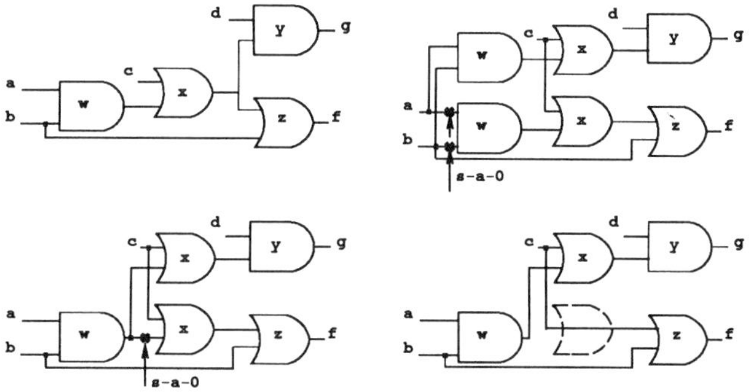

Figure 4: Algorithm on example circuit

outputs to primary inputs. This allows redundant connections to be created close to the primary outputs, thus potentially allowing the elimination of large numbers of paths. While the resulting RD set is maximal, no claims are made on whether this strategy yields a set that is close to the maximum RD set.

We illustrate the working of the algorithm on the circuit of Figure 4. The given circuit (top-left in Figure) is non-inverting. Since there are no redundant single stuck-0 faults, we reduce the fanout of gate x by duplicating it (bottom-left in Figure). There is now one redundant stuck-0 fault in the circuit. On removing it, the resulting circuit has no internal fanout greater than one (bottom-right in Figure) and the algorithm completes. The rising RD set for this circuit consists of the paths $\{a, w, x, z, f\}$ and $\{b, w, x, z, f\}$ which pass through the redundant connection. Note that if we had directly operated on the leaf-dag of the given circuit (top-right in Figure) the multiple stuck-0 fault is identified on the I-edges of these same two paths.

5 Robust delay

In Section 4 it is shown that for 100% delay-fault coverage there should only be either RPDF or RD delay-faults. But what about circuits that do not meet this criterion? For these circuits, one would like to get as much information about the circuit performance as possible from the limited RPDF set. Testing only the RPDF set in such a circuit is not sufficient to guarantee the circuit operates at the specified speed; some untestable non-RPDF and non-RD delay-faults could cause the circuit delay to exceed specification. This raises the question of whether there is a minimum τ such that the passing of the robust tests for a RPDF subset will guarantee the circuit delay will never exceed τ under all possible delay faults. If such a τ exists, then 100% delay-fault coverage (relative to τ) is achieved by testing a fraction of the robust testable delay-faults. Obviously, the existence of τ depends on the topological distribution of the subset of RPDF tested. Here, we give a mild condition that guarantees the existence of finite τ. A set of paths is called a **RPDF path-set** if every delay-fault on each path in the set is RPDF.

Definition 5.1 *The* **robust delay** τ_R *of a circuit is the minimum circuit delay* τ *that can be guaranteed by testing all the RPDF delay-faults.*

Theorem 5.1 *Let* $\mathcal{D}$ *denote the set of all paths in a circuit* C*,* $\mathcal{P} \subseteq \mathcal{D}$ *an RPDF path-set and* $\mathcal{R} \subseteq \mathcal{D}$ *a RD path-set. If every connection* $\mathcal{D} - \mathcal{R}$ *appears along some path in* $\mathcal{P}$*, then* C *has finite robust delay.*

Proof: Let d be the maximum delay of any path in $\mathcal{P}$ and let n denote the maximum number of edges in any path in C. Each path $P_i \in \mathcal{P}$ is tested to ensure that its delay does not exceed d. Let $\tau = n * d$. If the delay of some path Q in $\mathcal{D} - \mathcal{R}$ exceeds τ, at least one connection e in Q has delay exceeding d. But e is in some path $P_i \in \mathcal{P}$ (since e appears in some path in $\mathcal{P}$), and the delay of P_i would exceed d contradicting our assumption. Hence no path in C exceeds τ. Since the robust delay is no larger than τ, the result follows. ■

Theorem 5.1 relates the distribution of robust testable paths in a circuit and verifiability of the circuit's performance. If a circuit has a RPDF set that "covers" all the connections along non-RD paths, the robust delay of the circuit is finite. Otherwise, it may be infinite, since delay faults on some paths can cause the circuit's delay to be any value without being detected on RPDF paths. Hence, the robust delay is a measure of the verifiability of a circuit's performance under its available testability.

Although the robust delay may be too large to be useful, upper bounds on robust delay can be easily computed. How close the upper bounds are to the best circuit performance will depend on the testing strategy used, *e.g.* fixed delay or variable testing using upper and lower bounds. This is explained below.

5.1 Computing robust delay

To obtain an upper bound on robust delay, one can select any RPDF path-set $\mathcal{P}$ and test the delays in $\mathcal{P}$. Assume the (fixed delay) testing scheme latches the circuit outputs at some time d. Let $e \in P$ refer to a connection in path P and $\mathcal{R}$ represent a RD set. An upper bound on robust delay is:

$$\max\left(\left(\sum_{e \in Q} \delta(e), \forall\, Q \notin (\mathcal{P} \cup \mathcal{R})\right), d\right)$$

subject to:

$$\sum_{e \in P} \delta(e) \leq d, \; \forall P \in \mathcal{P}.$$

This is because a path can not have delay longer than the solution of the above linear program without violating a constraining inequality, which implies a delay-fault on a tested path. Note that the RD delay-faults are ignored.

In the **bounded delay model** the delay of each connection c is specified as $[d^c_{\text{min}}, d^c_{\text{max}}]$. The delay of a path P in this model is specified as $[d^P_{\text{min}}, d^P_{\text{max}}]$; these bounds could be determined by summing up the minimum and maximum delays of connections in the path respectively. Another issue of delay-fault testing is whether a variable or fixed clock speed is used during application of the tests. A fixed clock period means that each path delay is verified to be less than the (fixed) clock period of the circuit. However, a variable clock allows greater flexibility to the testing process. In particular, paths delays can be verified against their lower and upper bounds. We consider both situations; in particular, the added flexibility of a variable clock rate helps in reducing the test set size after the identification of a dependent set.

Definition 5.2 *Under the bounded model a* **short-path delay-fault** *occurs along P if and only if the actual delay of $P < d^P_{\text{min}}$. A* **long-path delay-fault** *occurs if and only if the actual delay of $P > d^P_{\text{max}}$.*

Although a long-path or short-path delay-fault causes a delay to fall outside specified bounds we are only interested in detecting those delay-faults that slow down the circuit. Long-path and short-path delay-faults will be exploited only in reducing the size of the test set and implying bounds on the delays of other paths, thus increasing the delay-fault coverage. [2]

The upper bound on robust delay may be improved by using bounded delays for the paths in $\mathcal{P}$ and testing to check that the upper and lower bounds are met (variable delay testing).

Theorem 5.2 *An upper bound on the robust delay τ_R of a circuit is:*

$$\tau_R \leq \max\left(\left(\sum_{e \in Q} \delta(e), \; \forall\, Q \notin (\mathcal{P} \cup \mathcal{R})\right), d\right)$$

where

$$d = \max(d^P_{\text{max}}) \; \forall P \in \mathcal{P}$$

[2] A robust test for short-path delay-faults must be generated. It is known that a vector pair $< v_1, v_2 >$ tests for a long-path delay-fault for a rising or falling transition if and only if $< v_2, v_1 >$ tests for a short-path delay-fault for the opposite transition along the same path [5].

and

$$d^P_{\text{min}} \leq \sum_{e \in P} \delta(e) \leq d^P_{\text{max}}, \; \forall P \in \mathcal{P}$$

where $\mathcal{P}$ = set of all RPDF paths and $\mathcal{R}$ = a maximal RD set[3].

Proof: Let τ be the optimal value of the above linear program. Then the only paths longer than τ form a RD set. By definition, delay-faults in $\mathcal{R}$ can only be responsible for the delay if the delay on some non-RD paths is at least as much. Since the latter are accounted for in the linear program, the delay of the circuit is less than τ. Further, τ is the minimum delay that can be guaranteed by testing the RPDF paths given the selected RD set $\mathcal{R}$; so τ_R is no more than τ. ■

It can be seen that the closer d^P_{min} and d^P_{max} are, the tighter the bound will be, because the feasible space is more restricted. Another way to improve the bound is to use a larger RPDF set, so that more constraining inequalities are introduced to further reduce the feasible space, hence, tightening the upper bound. This will be illustrated in the examples in the next section.

An interesting case is when the circuit is 100% robust delay-fault testable and the entire RPDF set is tested. Then, the constraining inequalities are lower and upper bounds on all the paths. Obviously, the linear program yields the maximum path delay. Hence, in this case robust delay reduces to the usual notion of delay.

It is also interesting to note that the notion of the delay of a circuit is now tied to its testability. If 100% delay-fault coverage is achievable for a specified delay, it is "safe" to use, since it can be guaranteed. Thus, regardless of the true delay computed by accurate timing analysis, the delay value that can be verified is the "usable" delay; verifiability of this delay depends on the amount of available testability. With 100% robust delay-fault testability, the true delay can be verified and thus is usable. For circuits with less than 100% delay-fault testability, the verifiable delay is the robust delay, possibly greater than the true delay. Thus, the notion of robust delay captures the relationship between the amount of testability of a circuit and its "usable" delay. This usable delay varies with the amount of the circuit's delay-fault testability.

5.2 Robust delay examples

Example: The circuit of Figure 5 is 100% single stuck-fault testable and but only 16 out of the 26 (rising and falling) path delay-faults are RPDF (about 62% testability). Although some paths are non-RPDF and non-RD, fortunately the circuit has finite robust delay. Assume each gate contributes bounded delay $[0.9, 1.0]$ which is assigned to the output connection of the gate. The delay of each path is bounded by the sum of the lower and upper bounds of the connections in the path. Assume the RPDF faults pass the tests for short-path and long-path delay-faults.

---

[3]In general, we need to replace each $\delta(a_i, a_j)$ in the inequalities by $\delta^r(a_i, a_j)$ or $\delta^f(a_i, a_j)$, as appropriate to each rising or falling path, if $\delta^r \neq \delta^f$.

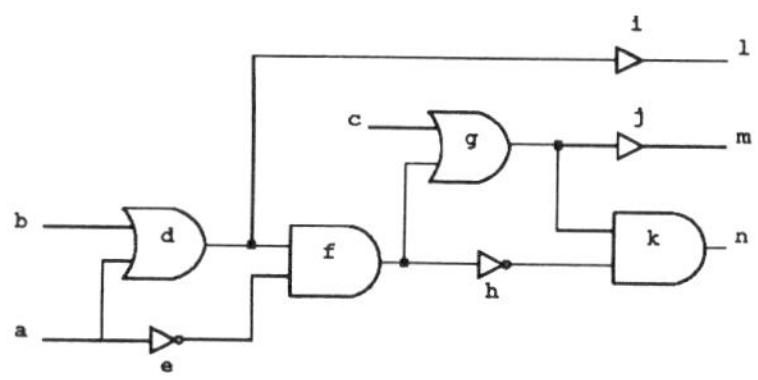

Figure 5: Computing robust delay

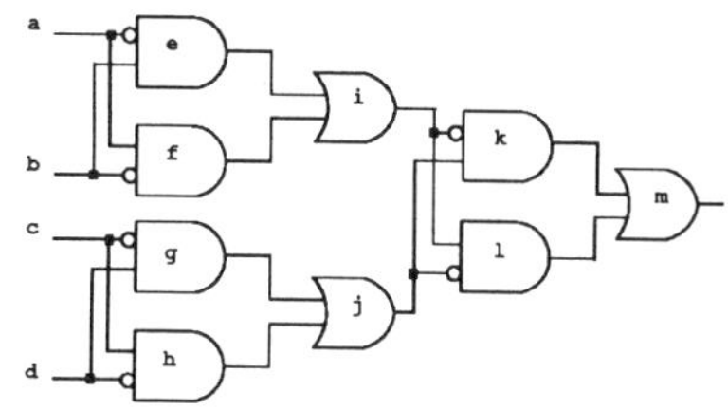

Figure 6: 4-bit parity checker

The RPDF paths for both rising and falling transitions, denoted by $\mathcal{P}$ are: $\{a,e,f,h,k,n\}$, $\{a,e,f,g,j,m\}$, $\{b,d,f,h,k,n\}$, $\{b,d,f,g,j,m\}$, $\{c,g,k,n\}$, $\{c,g,j,m\}$, $\{b,d,i,l\}$, $\{a,d,i,l\}$. The robust delay is obtained as:

$$\max\ (\sum_{x \in Q} \delta(x), 4.0),\ \forall\ Q \notin \mathcal{P}$$

such that

$$\begin{array}{lll}
3.6 & \leq \delta(a,e)+\delta(e,f)+\delta(f,h)+\delta(h,k)+\delta(k,n) & \leq 4 \\
3.6 & \leq \delta(a,e)+\delta(e,f)+\delta(f,g)+\delta(g,j)+\delta(j,m) & \leq 4 \\
3.6 & \leq \delta(b,d)+\delta(d,f)+\delta(f,h)+\delta(h,k)+\delta(k,n) & \leq 4 \\
3.6 & \leq \delta(b,d)+\delta(d,f)+\delta(f,g)+\delta(g,j)+\delta(j,m) & \leq 4 \\
1.8 & \leq \delta(c,g)+\delta(g,k)+\delta(k,n) & \leq 2 \\
1.8 & \leq \delta(c,g)+\delta(g,j)+\delta(j,m) & \leq 2 \\
1.8 & \leq \delta(b,d)+\delta(d,i)+\delta(i,l) & \leq 2 \\
1.8 & \leq \delta(a,d)+\delta(d,i)+\delta(i,l) & \leq 2
\end{array}$$

The linear program result is 4.40. Therefore, with only 62% testability, 100% fault coverage can be achieved with the robust delay $\tau_R = 4.40$, only 10% more than the computed delay of 4.00. Note that this is exactly the robust delay of the circuit (not a bound) if the upper and lower bounds on each path were precise.

Under a floating delay model where only the upper bound is available, this scheme produces a higher robust delay:

$$\max\ (\sum_{x \in Q} \delta(x), 4.0),\ \forall\ Q \notin \mathcal{P}$$

such that

$$\begin{array}{ll}
\delta(a,e)+\delta(e,f)+\delta(f,h)+\delta(h,k)+\delta(k,n) & \leq 4 \\
\delta(a,e)+\delta(e,f)+\delta(f,g)+\delta(g,j)+\delta(j,m) & \leq 4 \\
\delta(b,d)+\delta(d,f)+\delta(f,h)+\delta(h,k)+\delta(k,n) & \leq 4 \\
\delta(b,d)+\delta(d,f)+\delta(f,g)+\delta(g,j)+\delta(j,m) & \leq 4 \\
\delta(c,g)+\delta(g,k)+\delta(k,n) & \leq 2 \\
\delta(c,g)+\delta(g,j)+\delta(j,m) & \leq 2 \\
\delta(b,d)+\delta(d,i)+\delta(i,l) & \leq 2 \\
\delta(a,d)+\delta(d,i)+\delta(i,l) & \leq 2
\end{array}$$

The result is 8.00, twice as much as the computed delay. ■

Example: Figure 6 shows a more realistic circuit, a 4-bit parity checker in which all 32 delay-faults are RPDF. We assume each connection has delay $[0.9, 1]$; these are only used to derive bounds on path delays. If we test all delay-faults, we get the computed delay as the robust delay. By testing only a fraction of the delay-faults we can observe how the magnitude of robust delay varies with the percentage of the tested faults. In the table below the first column is the number of delay-faults (out of a total 32) that are tested; the second is the percentage tested; the third is the robust delay.

Test set size	% testability	Robust Delay
32	100	4.00
30	94	4.40
28	88	4.40
26	81	4.40
24	75	8.00
22	69	8.00
20	63	8.00
18	56	8.00
16	50	8.00
≤ 15	≤ 46	∞

From the table, it can be seen that the computed delay of 4.00 can be verified by testing all RPDF faults; call this 100% testing. With 81% testing, 90% of the computed delay can be verified, i.e. 4.40 vs. 4.00; with as little as 50% testing, twice the computed delay can be verified. Below 44% testing, some delay faults may lengthen the circuit delay yet remain undetected in the testing, thus yielding infinite robust delay.

The test set of size 16 is linear in circuit size since each RPDF path includes some connection not covered by any other RPDF path. ■

The above examples illustrate that the amount of testing can be tailored to fit the required performance. Higher performance requires more testing. Fairly tight upper bounds on robust delay can be obtained with low testing effort. In particular, a test set linear in circuit size allows a remarkably high verifiable performance.

6 Test set size under bounded delays

In this section we illustrate the use of a bounded delay model in reducing the test set size for robust delay-fault testing.

Not all non-RD delay-faults need be tested under the bounded delay model. The fact that tested paths have delays within specified lower and upper bounds can be used to reduce the number of delay-faults tested. There are two aspects that deserve attention. First, the clock period that is applied while performing the delay-fault testing must be variable. Second, only bounded delays on complete paths are exploited in reducing the test set. This is because bounded delays on connections cannot be verified by testing; typically, bounds on connections are only used in determining bounds on path delays.

The goal is to perform two (one long-path and one short-path) tests on a few selected paths rather than a single long-path delay-fault test on a larger set of paths.

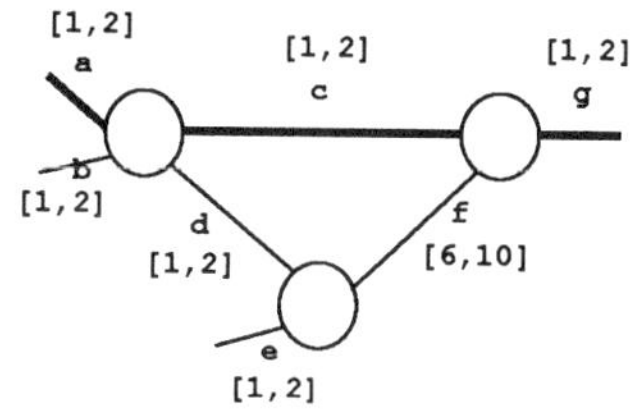

Figure 7: All paths need not be delay-fault tested under bounded delay model

This is possible if the occurrence of each of the untested delay-faults is prohibited by the absence of delay-faults on the tested paths. Under these conditions, this test strategy yields 100% delay-fault coverage.

Example: Consider the example shown in Figure 7. Each edge has bounded delay specified as $[1,2]$ except the edge f which has delay $[6,10]$. Let the specified clock period be 16. Assume that all but the bold path have been tested and each falls within its lower and upper bounds (taken as the sum of the lower and upper bounds, respectively, of each edge in the path). We argue that the bold path does not have to be delay-fault tested. This is done by proving that any assignment that slows down the chosen path (causing its delay to be greater than 16) causes some other tested path to also speed up or slow down outside its specified bounds. The proof is expressed as a linear program expressing the condition that the chosen path be the only one to exceed the clock period.

$$\begin{array}{lll} 9 & \leq \delta(a) + \delta(d) + \delta(f) + \delta(g) & \leq 16 \\ 9 & \leq \delta(b)b + \delta(d) + \delta(f) + \delta(g) & \leq 16 \\ 3 & \leq \delta(b) + \delta(c) + \delta(g) & \leq 6 \\ 8 & \leq \delta(e) + \delta(f) + \delta(g) & \leq 14 \\ & \delta(a) + \delta(c) + \delta(g) & > 16 \end{array}$$

This set of constraints is not feasible (under the additional constraints that the delay of each edge be non-negative); this demonstrates that the path $\{a, c, g\}$ cannot have the only delay-fault that slows down the circuit. Note that the bounds on the connections are not used, e.g. $1 \leq \delta(a) \leq 2$ is not used since it is not verified. ■

Based on the above example, given a set of non-RD delay-faults, implications using the bounded delays may be used to reduce the test set size by formulating a series of linear programs. While an exact version of such an algorithm appears intractable, a heuristic algorithm that is intuitively promising is under development.

7 Results

We show results on MCNC benchmark circuits in Table 1. Each circuit is first optimized using the standard script (called *script.rugged*) in the SIS system. The lower bound on delay fault coverage, determined using the algorithm described in Section 4.4, is compared against the robust delay-fault testability in the optimized circuits. The first column is the circuit name; the second is the total number of delay-faults (rising and falling transitions); the third gives the size of the RD set

Name	# PDF	# RD	# RPDF	Testability	FC-1	FC-2
misex2	758	317	373	0.49	0.85	0.99
b9	1044	224	791	0.76	0.97	1.00
vg2	1576	0	1576	1.00	1.00	1.00
sao2	2706	1548	626	0.23	0.54	0.83
apex6	9898	3815	4743	0.48	0.78	0.96
apex5	20118	11162	8343	0.42	0.93	0.99
seq	48380	29451	16517	0.34	0.87	0.99
bw	49864	46873	627	0.01	0.21	0.66
cbp.32.4	22278702	22260784	7474	0.0003	0.42	0.83

PDF = All path delay-faults in optimized circuit; RD = Chosen RD set ; Non-RD = Path delay-faults which are not in chosen RD set ; RPDF = Robust testable delay-faults ; FC-1 = (# RPDF) / (# Non-RD) ; FC-2 = (# Tested faults) / (# Non-RD) ; Tested faults = Non-RD faults whose absence is guaranteed if circuit passes RPDF tests at finite speed

Table 1: Lower bound on delay-fault coverage

found; the fourth lists the size of the RPDF set; the fifth gives the robust delay-fault testability. The column FC-1 gives the fault coverage computed using only the RD set. Note the substantial increase in this compared to the testability. For example, 93% of the total delay-faults in *bw* are identified as an RD set resulting in an increase from 0.01 to 0.20. The column FC-2 gives the fault coverage achieved by using a variable testing speed but only using upper bounds on the path delay (lower bounds assumed 0 for all paths). A unit delay is assumed on each connection in the circuit. In this case, any non-RD path whose delay is implied to be a finite number if the circuit passes the RPDF tests is considered tested. Most of the circuits do not have finite robust delay since 100% fault coverage cannot be achieved for any finite delay. *b9* is the only example shown that does not have 100% RPDF testability yet has finite robust delay. Despite this, all the circuits show very high fault coverage. For example, the coverage in *bw* is now raised from 0.21 to 0.66. Considering that the testability is 0.01, this is a significant increase. Note however that the coverage in column FC-2 ignores the speed of the circuit. To get the coverage reported, the circuit may have to be slowed down considerably.

8 Conclusions

We have shown that the delay-fault coverage achievable by a test set is relative to the speed at which operation is desired. This observation helps address the issue of how much delay-fault testing is required to guarantee that a circuit operates correctly at its specified speed in three ways.

For a specified circuit speed, there exist path delay-faults which never need to be tested, since they cannot affect the circuit delay without some other delay-faults (which are included in the set to be tested) also impacting the circuit delay. Necessary and sufficient conditions for the existence of these delay-faults, called

RD sets, are proved. We show that delay-faults on any set of long false paths that occurs, given any delay assignment to the connections of the circuit, forms a RD set.

The next aspect describes the possible tradeoff that exists between the circuit speed and the delay-fault test set size required to guarantee that a manufactured circuit operates at that speed. The notion of robust delay of a circuit is used to represent the minimum useful delay of a circuit under a given bounded delay model. The quantitative tradeoff between the robust delay and the testing effort has been formulated using a linear program, while also accounting for the presence of RD delay-faults.

The final aspect reduces the delay-fault test set size without decreasing the delay-fault coverage or changing the desired circuit speed.

Short path delay faults can also occur and cause problems. Examples are in wave pipelined circuits, clocking schemes using transparent latches, and asynchronous circuits. While the theory and algorithms in this paper have focussed on upper bounds on circuit performance, all the results can be easily extended to analysis of lower bounds.

Acknowledgements

William K. Lam is supported by the Hertz Foundation and a SRC grant under contract 92-DC-008. Alexander Saldanha is supported by Fujitsu Laboratories Ltd..

References

[1] H-C. Chen and D. Du. Path sensitization in critical path problem. In *Proceedings of the International Conference on Computer-Aided Design*, pages 208–211, November 1991.

[2] S. Devadas, K. Keutzer, and S. Malik. Delay computation in combinational logic circuits: Theory and algorithms. In *Proceedings of the International Conference on Computer-Aided Design*, pages 176–179, November 1991.

[3] P. McGeer and R. Brayton. Provably correct critical paths. In *The Proceedings of the Decennial Caltech VLSI Conference*, 1989.

[4] G. Smith. Model for delay faults based upon paths. In *Proceedings of the International Test Conference*, pages 342–349, August 1985.

[5] K. Wagner. The error latency of delay faults in combinational and sequential circuits. In *Proceedings of the International Test Conference*, pages 334–341, November 1985.

Programmable Active Memories: a Performance Assessment

Patrice Bertin, Didier Roncin
and Jean Vuillemin
Digital Equipment Corporation
Paris Research Laboratory
85, avenue Victor-Hugo
92563 Rueil Malmaison Cedex, France

Abstract

We present some quantitative performance measurements for the computing power of Programmable Active Memories (PAM), as introduced by [2]. Based on Field Programmable Gate Array (FPGA) technology, the PAM is a universal hardware co-processor closely coupled to a standard host computer. The PAM can speed up many critical software applications running on the host, by executing part of the computations through a specific hardware design. The performance measurements presented are based on two PAM architectures and ten specific applications, drawn from arithmetics, algebra, geometry, physics, biology, audio and video. Each of these PAM designs proves as fast as any reported hardware or super-computer for the corresponding application. In cases where we could bring some genuine algorithmic innovation into the design process, the PAM has proved an order of magnitude faster than any previously existing system (see [19] and [18]).

1 PAM concept

Like any RAM memory module, a PAM is attached to the system bus of a host computer. The processor can write into, and read from the PAM. Being an active hardware co-processor however, the PAM processes data between write and read instructions. The specific processing is determined by the content of its *configuration memory*. The host can change the PAM configuration by *downloading* a new design into it, within a few milliseconds.

We speed up a specific software application running on the host, by executing its *critical inner-loop* through an appropriate hardware design downloaded into the PAM. Ten examples of such designs and applications are presented below.

In selecting them, we have attempted to cover as many application areas as we could. In each, we picked basic and frequent problems, where a large inner-loop speedup through a specific PAM design results in a significant speedup of the application system, software and hardware combined.

For the sake of debugging, the functionality of each hardware design is matched, bit-wise at the I/O level, by a software implementation. Through successive refinements, both hardware and software implementations have been optimized for speed, while retaining the I/O behavior. This is our experimental data in this assessment of PAM computing power.

The traditional measurement unit here is the Mips[1], or its multiple the Gips ($1G = 10^9$) which is more appropriate to the levels of performance reported here. As explained in [9], quantitative performance comparisons between different computer architectures is a challenging art. For one, our Mips were measured in different units: for the 32b MIPS R3000 or the 64b Alpha AXP instruction sets? for which clock speed? cache size? bus bandwidth? For second, in a number of cases we had external data for both VLSI and super-computer implementations of systems close enough to some of ours that we could envision to make reliable speed comparisons.

To make quantitative comparisons significant across such a wide spectrum of feasible implementation technologies, we introduce a common unit for measuring all forms of computing power, the *Gbops* (billion of binary operations per second) where each useful boolean operation with up to 4 inputs, be it clocked or not, counts for one. For example the computing power of a 32b adder at 100 Mhz is 6.4 Gbops, accounting for both carries and sums. That of a n bits multiplier at 1 GHz is $2n^2$ Gbops. Any specific computer operation can be similarly decomposed at the bit level, so we can evaluate the corresponding power in Gbops.

2 Two PAM architectures

Our assessment is based on two PAM architectures realized at DEC's Paris Research Lab., DECPeRLe-0 (see [2]) and DECPeRLe-1 (see [3]) which we respectively refer to as P_0 and P_1 in the following.

Each PAM is built around a large array of bit-level configurable logic cells (hereafter called Programmable Active Bits or PABs) in which the application-specific hardware operator is programmed. This array is surrounded with local RAM banks used as a cache (wide and fast enough to match the PAM's external bandwidth), a programmable clock generator, and some additional non-configurable logic to manage the host bus interface and the download process.

[1]Million of instructions per second

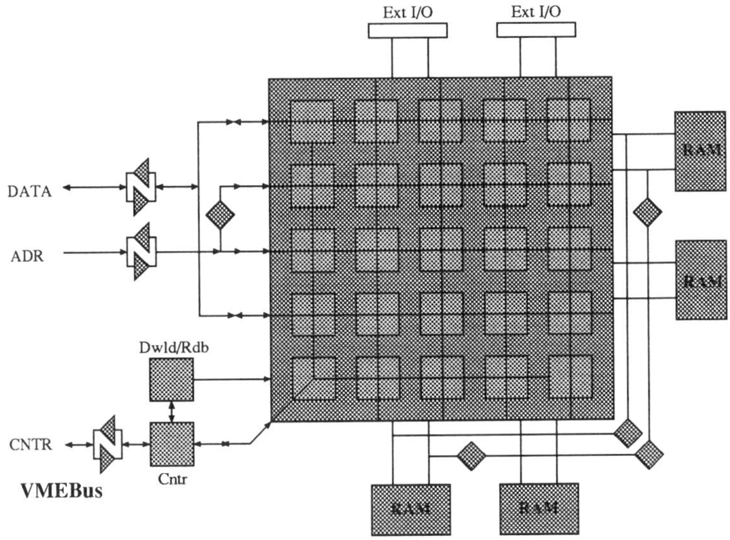

The above figure sketches the architecture of P_0 (1988). The central computational array is made of a 5×5 matrix of Xilinx XC3020 Programmable Gate Arrays [23]; it has two 32-bit wide RAM banks on the south and east sides, a VME bus interface on the west side and general-purpose interface connectors on the north side. The control and bootstrap logic is implemented in two extra XC3020 (non user-programmable). Finally, additional bus switching resources are provided for global data routing (represented here with diamond-shape boxes).

The architecture of P_1 was designed after two years of P_0 usage and is described in details in [3]. It features a 4-times-bigger central computational matrix with accordingly wider RAM, a faster host interface, and a much more flexible global interconnection network for data routing and switching.

For the purpose of evaluating Xilinx-based designs such as ours, it is convenient to define the Programmable Active Bit as in [2]: a PAB consists of a universal 4 inputs combinatorial gate and a synchronous flip-flop. Using this measurement unit, our two PAM architectures P_0 and P_1 have the following vital statistics:

PAM	part	nb	PABs	F_{max} (MHz)	Power (Gbops)	RAM (MB)	Host bus (MB/s)
P_0	XC3020	25	3.2K	25	80	0.5	8
P_1	XC3090	23	14K	40	588	4	100

This chart exhibits the three most important architectural parameters conditioning which application benefits most from a PAM speed-up:

- The number of PABs (3.2K for P_0 and 14K for P_1), together with the application-dependent maximal clock frequency (25 MHz and 40 MHz) at which we can reliably operate. The product of these two numbers is the maximum theoretical computing power of the PAM, expressed in Gbops (80 Gbops for P_0 and 588 Gbops for P_1).

- The host bus bandwidth: 8 MB/s for P_0 through a VME bus, and 100 MB/s for P_1 through a TURBOchannel interface [6].
- The size of the local (fast) RAM: 0.5 MB for P_0 and 4 MB for P_1.

3 PAM programming

Programming an algorithm on the PAM is similar to casting it in conventional hardware (gate-array or VLSI), with two very important differences however: first the target hardware provides a completely clean implementation of the synchronous logic model, so there is no need to worry about low-level electrical details; second the whole design process is purely software with a fast (5 to 30 mn) turnaround, so it can be approached with the same methodology of piecewise testing and successive refinements as a software design. For a given application, it involves:

1. identifying the critical computations which are best implemented in hardware; this is usually done by successive refinement, under constraints of communication bandwidth and load balancing between the software and hardware;
2. implementing the hardware part on the PAM and gradually optimizing it;
3. implementing the software part on the host processor and gradually optimizing it.

Step 3 above is done with conventional techniques. Step 2 consists in describing the logic design to implement in the PAM down to the individual bit level, as well as its geometric layout, much as is done for a conventional VLSI design.

For this, we have developed a tool suite in which the hardware design is described by the writing of a program in a conventional programming language (we use Lisp, C++ [21] and Esterel [1]) using a specialized library. This program will describe the various logic modules by their bit-level logic equations, or by using standard library modules (adders, registers, standard interfaces...). It will also contain layout (geometric) information to the level of details the designer wishes to specify: this usually starts with global floorplanning and detailed layout of the important datapath components only, to be enriched with more precise descriptions along the optimization process.

Executing this program produces a partially placed, hierarchical netlist which can either be simulated, or compiled to the final PAM configuration in a fully automatic way, through tools developed in-house (board-level routing, logic optimization) and standard Xilinx back-end software (chip-level placement and routing, bitstream generation).

The design can then be downloaded into the PAM for debugging and testing under real conditions; its maximum running speed can be characterized, and its critical paths identified (we have developed specialized interactive tools to help visualize the latter). The designer can then gradually optimize it, for example by adding extra levels of pipeline to increase the clock speed, or optimizing the geometric layout, or adding new functionalities to further unload the software. The most important point there is that, as compiling and trying a new version costs no money and only a few minutes to an hour of time, it is possible to incrementally design, test and optimize the algorithms and implementations involved, as is usually done with software, as opposed to having to have them correct and optimal on the first try.

4 Ten PAM applications

The following applications were chosen to span a wide range of current leading edge computational challenges. In each case, we provide a brief description of the design, the names of the implementors, and a performance comparison with similar reported work. In each case, the following simple paradigm has been applied:

> *compile the inner loop in PAM hardware, and let software handle the rest!*

In what follows, we let $(a \div b)$ represent the quotient and $(a \cdot\!|\!\cdot b)$ the remainder in the integer division.

4.1 Long multiplication

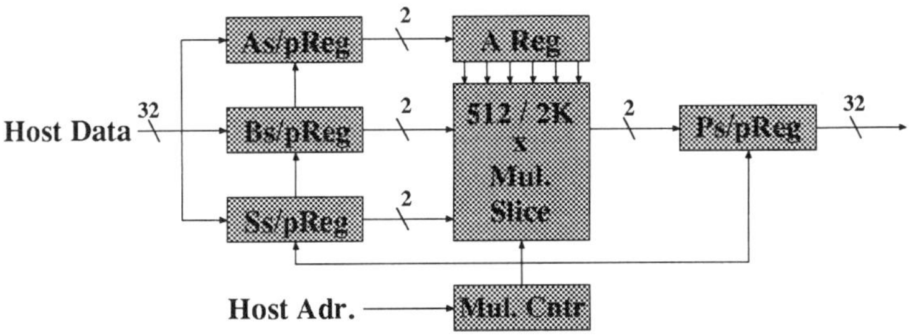

We have programmed both PAMs into long multipliers ($n = 512$ bits for P_0, and $n = 2K$ bits for P_1) computing $P = A \times B + S$, with A a n-bits multiplier, and B, S arbitrary size multiplicands and summands (see [10], [2]

and [19]). These multipliers are interfaced with an arbitrary-precision arithmetic package *BigNum* (see [16]) so that any program based on that software takes advantage of the PAM without modification, by simply relinking with a modified BigNum library. This respectively speeds up raw multiplication by a factor up to 24 (P_0) and 30 (P_1) for long operands, as compared to optimized assembly code running on the host workstation. For example, P_0 equipped with this design computes RSA encryption/decryption at 1500 bits per second for arbitrary 512-bit keys (see [14]); this is about 10 times faster than our best software version on the same host.

The P_1 implementation produces product bits at 66 Mbits/s, which makes it faster than *any* known machine for which we could obtain benchmark measures. It is at least 16 times faster than the best figures for a Cray II or a Cyber 170/750 reported by [5]. This multiplier can be used to compute a 50 coefficients (16-bits) polynomial convolution (FIR filter) at 16 times *audio real time* (2×24 bits samples at 48 kHz).

4.2 RSA cryptography

To further investigate the tradeoffs which are possible in our hybrid hardware/software system we focused on the RSA cryptosystem (see [14]), which can be cast entirely in terms of long multiplications. Starting from the above general-purpose multiplier, M. Shand from DEC-PRL implemented a series of hardware/software systems spanning two orders of magnitude in performance. The latest version is based on an original hardware design for computing modular products at the rate of two bits per cycle [20].

The system originally used three differently programmed P_0 boards, all operating in parallel with the host (see [19]). At 200 kbit/s decoding speed, it was faster than *any* currently existing 512 bits RSA implementation, in *any* technology, as of February 1990. A survey by [4] grants the previous speed record for 512 bits keys RSA decryption to a VLSI from AT&T, at 19 kbits/sec.

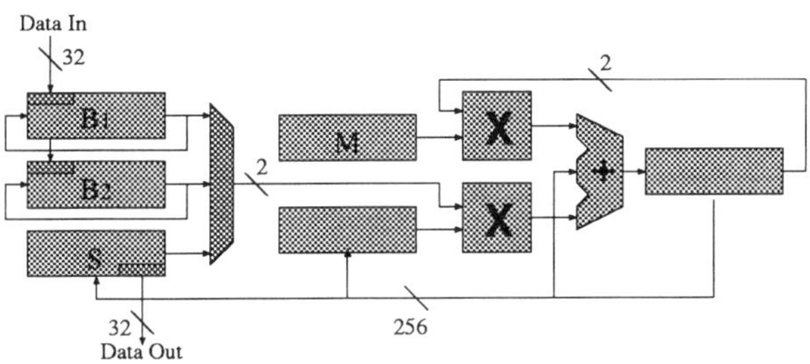

M. Shand recently ported this RSA system to a single P_1 board; at 40 MHz, this design provides either two independent 600 Kb/s RSA encryption channels for 480b keys, or one 175 Kb/s RSA encryption channel for 970b keys.

4.3 Data compression

M. Skubiszewski from DEC-PRL has implemented a P_0 design to speed up the algorithm of [24], which is well known to achieve an average data compression ratio varying from 2 for English (or French, or Polish...) plain text to 3 for C (or Lisp, or Pascal...) source code.

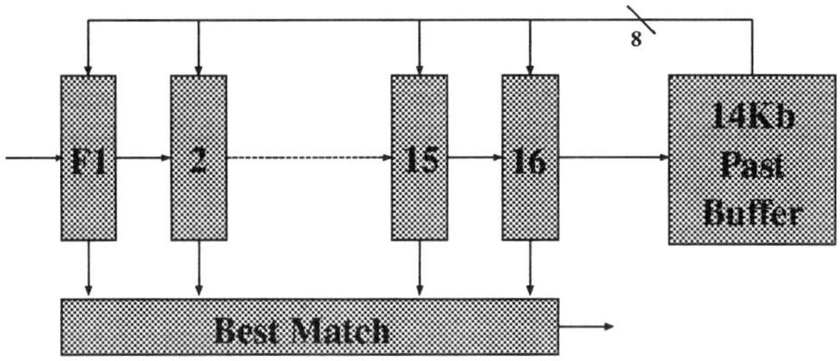

The design is a massively parallel method which computes 64 byte comparisons on each (70 ns) cycle; it matches the next 16 bytes in the file to be compressed against the last 4k bytes seen (stored in the local RAM), in order to detect the longest substring previously seen. While this design performs a respectable 1 Gops (8 bits integer comparison), it ends up in a disappointing factor two speed-up, when compared to optimized software such as Unix `compress`. Indeed, such optimized software avoids most of the comparisons performed in the hardware, by detecting early that they are irrelevant to the final output. A more elaborate hardware design is needed to genuinely speed up this particular algorithm.

4.4 String matching

Given an alphabet $\mathcal{A} = (a_1, \ldots, a_n)$, a probability $(S_{ij})_{i,j=1\ldots n}$ of substitution of a_i by a_j, and a probability $(I_i)_{i=1\ldots n}$ (resp. $(D_i)_{i=1\ldots n}$) of insertion (resp. deletion) of a_i, one can use a classical dynamic programming algorithm to compute a probability of transformation of w_1 into w_2; this defines a *distance* between any two words w_1 and w_2 over $\mathcal{A}$.

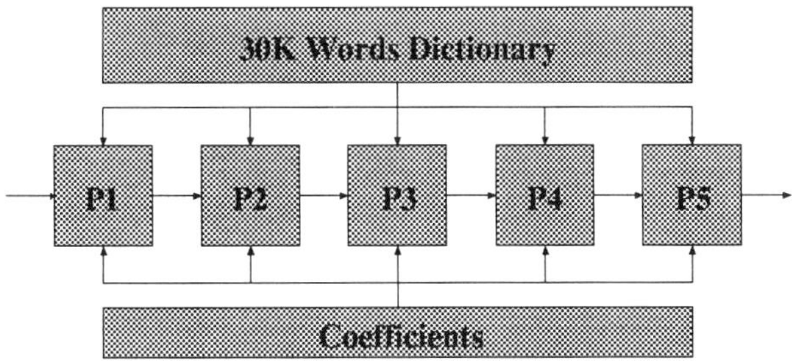

D. Lavenier from IRISA (Rennes, France) has implemented this algorithm with a P_0 design which computes the distance between an input word and all 30K words in a dictionary; it reports the k words found in the dictionary which are closest to the input. The system processes 200K words/sec: this is faster than a solution previously implemented at CNET using 12 Transputers; it has only half of the performance obtained by a system previously developed at IRISA, based on 28 custom VLSI chips and 2 PC boards.

Applications of this algorithm include automated mail sorting using OCR scanners, on-the-fly keyboard spelling corrections, and DNA sequence matching (see [11]).

4.5 Heat and Laplace equations

[22] shows how to adapt the classical finite difference method (see [7]) to compute solutions of the heat and Laplace equations in n dimensions with help from special purpose hardware. An implementation of the method on P_1 operates with a pipe-line depth of 128 operators:

Each operator computes:

$$\mathcal{O}(v_0, v_1) = \begin{cases} v_0, & \text{when } v_0 \cdot|\cdot 2 = 1, \\ 2(v_0 \div 4 + v_1 \div 4 + (v_0 \cdot|\cdot 4) \div 2), & \text{otherwise,} \end{cases} \quad (1)$$

all with 24b fixed-point data format. At 20 MHz, this amounts to 5 Gops (24b adds, tests and shifts); it is easy to show (see [22]) that fixed-point gives the same results as floating-point operations for this specific problems; the achieved performance thus exceeds those reported in [12] and [13] for solving the same problem with super-computers. A sequential computer needs to execute 25 billion instructions per second (25 Gips), to reproduce the same computation.

The heat and Laplace equations have many applications in mechanics, circuit technology, fluid dynamics, electrostatics, optics, finance...

4.6 Newton's mechanics

J. Vuillemin has specified a P_1 design for computing the evolution of a n-body system, using Newton's equations. The design computes the gravitational field acting on body k by summing the individual fields induced at k by each other body in the system. This amounts to the following 18 operations:

$(x_i - x_k)$	$\Rightarrow$	dx	$(dx2 + dz2)$	$\Rightarrow$	dxz	$fm \times dx$	$\Rightarrow$	fdx
$(y_i - y_k)$	$\Rightarrow$	dy	$(dxz + dy2)$	$\Rightarrow$	$d2$	$fm \times dy$	$\Rightarrow$	fdy
$(z_i - z_k)$	$\Rightarrow$	dz	$\sqrt{d2}$	$\Rightarrow$	d	$fm \times dz$	$\Rightarrow$	fdz
$(dx \times dx)$	$\Rightarrow$	$dx2$	$d \times d2$	$\Rightarrow$	$d3$	$fx + fdx$	$\Rightarrow$	fx
$(dy \times dy)$	$\Rightarrow$	$dy2$	$\frac{1}{d3}$	$\Rightarrow$	fd	$fy + fdy$	$\Rightarrow$	fy
$(dz \times dz)$	$\Rightarrow$	$dz2$	$fd \times m_i$	$\Rightarrow$	fm	$fz + fdz$	$\Rightarrow$	fz

Positions and forces are represented as 20 bits floating-point numbers. Assuming a 40 ns internal cycle (achievable through deep pipe-line) the expected throughput exceeds 2.5 GFlops (this design has not been tested by printing time).

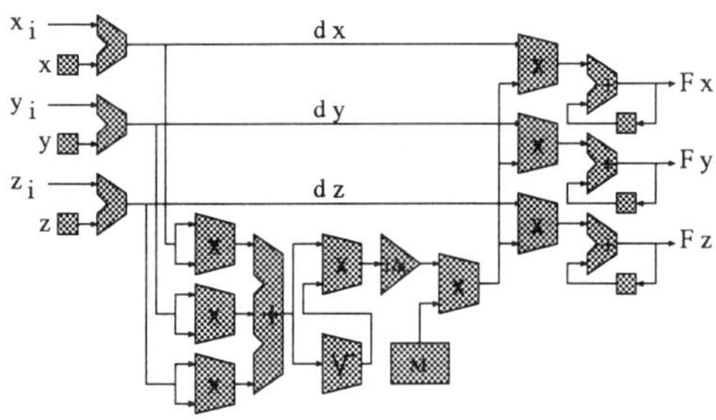

4.7 Binary 2D convolution

B. Chen and J. Vuillemin have implemented a 7×7 binary 2D convolver on P_0, for performing erosion, dilation and matching on black and white images, as defined in [15].

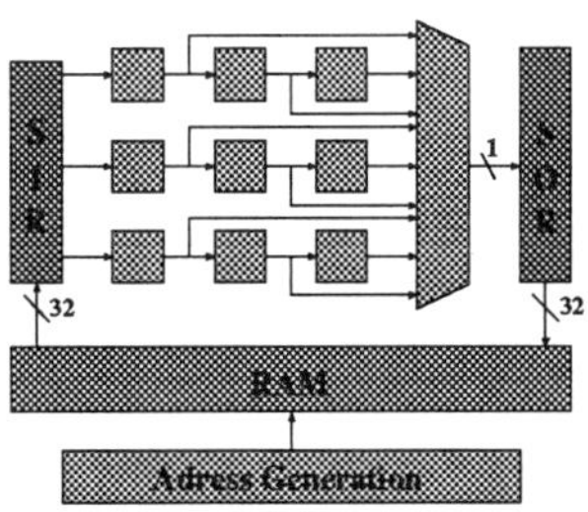

The convolver runs at 25 MHz, generating one pixel each 40 ns; it completes a single convolution pass over one 512×512 image in 10 milliseconds;

this allows for up to 4 successive operations (erosion, dilation, or matching) at video rate. Reproducing this performance through optimized software would require a 200 Mips computer.

4.8 Boltzmann machine

M. Skubiszewski has implemented two successive versions of a hardware emulator for binary neural networks, based on the *Boltzmann machine* model (see [17] and [18]).

The Boltzmann Machine is a probabilistic algorithm which minimizes quadratic forms over binary variables, i.e. expressions of the form

$$E(\vec{N}) = \sum_{i=0}^{n-1} \sum_{j=0}^{i} w_{i,j} N_i N_j \tag{2}$$

where $\vec{N} = (N_0, \ldots, N_{n-1})$ is a vector of binary variables and $(w_{i,j})_{0 \leq i,j < n}$ is a fixed matrix of *weights*. It is typically used to find approximate solutions to $\mathcal{NP}$-hard problems, such as graph partitioning or circuit placement.

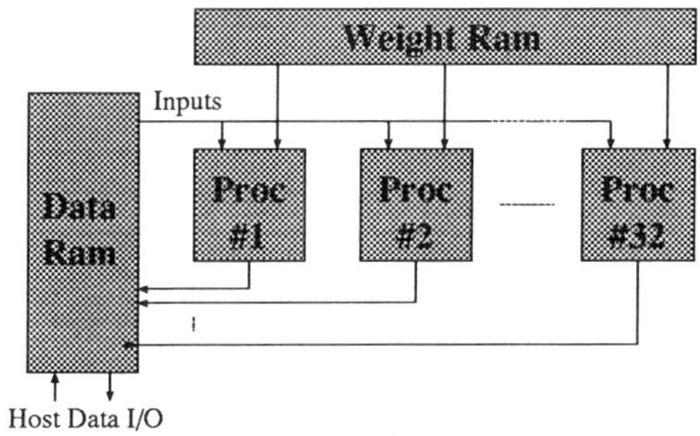

The latest realization, on P_1, can solve problems with up to 1400 variables, using 16-bit weights, for a total computing power of 500 *megasynapses per second* (the megasynapse is the traditional unit used in this field, it amounts to one million additions and multiplications, or one million terms of (2)).

4.9 3D Geometry

H. Touati from DEC-PRL has implemented a 3-D graphic accelerator for P_1, which supports translation, rotation, clipping and perspective projection, to directly compute the screen image of a cloud of points in 3-D space.

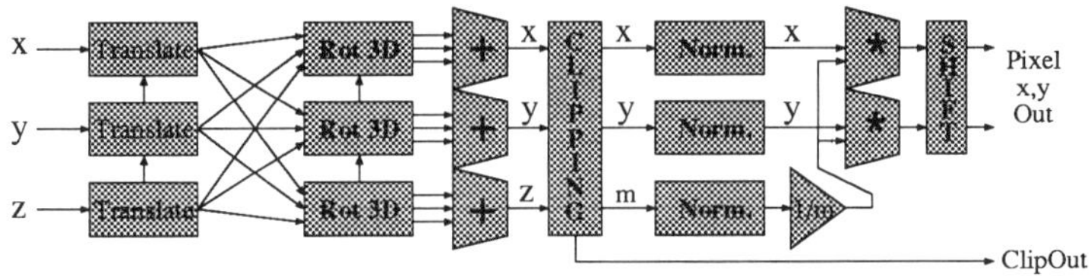

At 25MHz, it has a peak performance of 1.56 million points per second, using 16 bit fixed point coordinates for the input and output, and up to 32 bits for the intermediate results. One needs a 300 Mips processor to achieve the same throughput in software.

4.10 Discrete cosine transform

This design (by J. Vuillemin and D. Martineau on P_1) compresses a video stream in *real time* through multi-dimensional fast discrete cosine transform. The fDCT implements the following network:

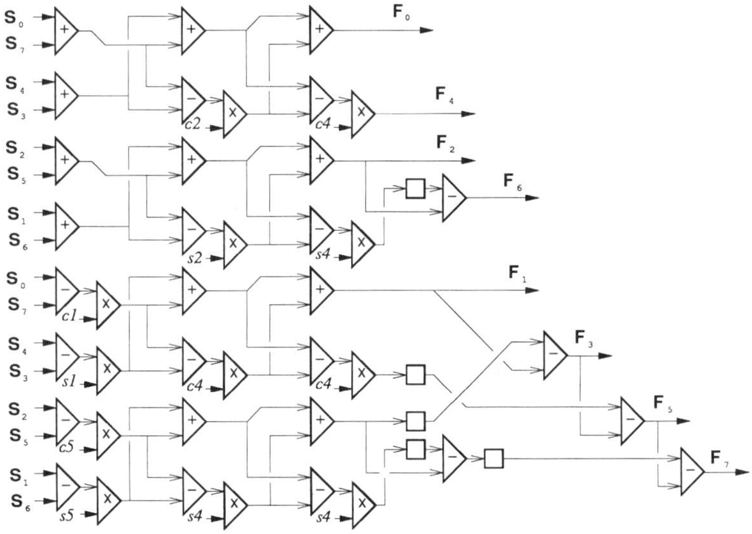

The overall design computes 48 fixed-point (32 bits outputs) operations (add, subtract, multiply and shift) on each 40 ns cycle, for a total of 1.4 Gops. To match this performance through software would require a 15 Gips processor.

5 Conclusion

The following chart summarizes the practical PAM performance achieved by each of our ten designs:

Design	PABs	MHz	Gbops	Gops	Gips	PAM
Multiplier	8k	33	264	0.8	2.6	P_1
RSA	8k	32	256	0.5	4	P_1
DCT	10k	25	250	1.4	15	P_1
Newton	10k	25	250	2.5 †	5	P_1
Laplace	10k	20	200	7.5	20	P_1
Boltzmann	8k	25	200	1	1.5	P_1
3D Geometry	3k	25	75	0.5	0.7	P_1
Ziv-Lempel	3k	15	45	1	2	P_0
String	3k	10	30	0.15	0.3	P_0
2D convolution	1k	25	25		1	P_0

†These are Gflops, with 20 bit floating-point numbers.

The applications are ranked according to the most reliable performance measure, namely the *Gbops*. As a comparative measure of resource utilisation in such systems, the following table charts the maximum theoretical performance of generic PAM hardware (in Gbops) obtained by multiplying the maximal clock frequency (in MHz) by the area (in PABs):

PAM	Area	1 MHz	20 MHz	50 MHz
XC3020	128	0.1	2.5	6.4
XC3090	640	0.6	12.8	32
P_0	3.2K	3.2	64	
P_1	14K	14	280	700

Three years of PAM design lead us to believe the following:

1. For each of the chosen application, we have shown that the level of performance achieved with the PAM is comparable to the best figures reported using super-computers or custom silicon circuits.

   Our applications have been carefully selected for having a clearly identified (PAM implementable) inner-loop, which accounts for a vast percentage of the software run-time. For such low level processing, the PAM proves more cost effective than any super-computer.

   Due to their software complexity, many current super-computers applications still remain outside the possibilities of current PAM technology.

2. Each mentioned PAM design was implemented and tested within one or two months, starting from the delivery of the specification software. This is roughly equivalent to the time it takes to implement a *highly optimized* software version of the same system with a super-computer; both are technically challenging, yet remain an order of magnitude faster than the time it takes to cast a system into silicon.

3. The cost of P_1 is comparable to that of a high-end workstation. This is orders of magnitude lower than the cost of a super-computer. Based on figures from [12], we find that the price (in $ per operation per second) of solving the heat and Laplace equations is 100 times higher with super-computers than with the PAM.

4. Another field of applications, not covered by any existing super-computer, is open to PAM technology: high-bandwidth interfaces to the external world, with *fully programmable real-time* capabilities. The P_1 PAM has a 256b wide connector, capable to deliver up to 6.4 Gb/s of external bandwidth. It is a "simple matter of hardware programming" to interface directly with any electrically-compatible external device, by programming its communication protocol into the PAM itself. Applications for this capability are numerous, including interfaces to high-bandwidth networks, audio and video input or output devices, on-the-fly data acquisition and filtering, etc...

References

[1] Gérard Berry. A Hardware Implementation of Pure Esterel. *PRL Report 15*, Digital Equipment Corp., Paris Research Lab., 85 av. Victor Hugo, 92563 Rueil Malmaison, France, 1991.

[2] Patrice Bertin, Didier Roncin and Jean Vuillemin. Introduction to programmable active memories. *Systolic Array Processors, J. McCanny, J.McWhirther, E. Swartslander Jr. editors*, Prentice Hall, 300–309, 1989. Also available as *PRL report 3*, Digital Equipment Corp., Paris Research Lab., 85, Av Victor Hugo. 92563 Rueil Malmaison Cedex, France, 1989.

[3] Patrice Bertin, Didier Roncin and Jean Vuillemin. Programmable Active memories: the Coming of Age. *PRL report in preparation*, Digital Equipment Corp., Paris Research Lab., 85, Av. Victor Hugo. 92563 Rueil-Malmaison Cedex, France, 1992.

[4] E.F. Brickell. A Survey of Hardware Implementations of RSA. *Proceedings of Crypto'89, Lecture Notes in Computer Science*, Springer Verlag, 1990.

[5] D.A. Buell and R.L. Ward. A multiprecise integer arithmetic package. *The Journal of Supercomputing, Kluwer Academic Publishers*, Boston, 3:89–107, 1989,

[6] Digital Equipment Corporation. TURBOchannel hardware specification. *DEC document EK-369AA-OD-007A*, Digital Equipment Corp., 1991.

[7] R.P. Feynman, R.B. Leighton and M. Sands. The Feynman lectures on PHYSICS. Addison-Wesley, 1963.

[8] J.P. Gray and T. Kean. Configurable hardware: two case studies of micro-grain computation. *Systolic Array Processors, J. McCanny, J.McWhirther, E. Swartslander Jr. editors*, Prentice Hall, 310–319, 1989.

[9] J.L. Hennessy and D.A. Patterson. Computer architecture: a quantitative approach. *Morgan Kaufmann Publishers, Inc*, 1990.

[10] R.F. Lyon. Two's complement pipeline multipliers. *IEEE Trans. Comm.*, COM-24:418–425, 1976.

[11] D.P. Lopresti. P-NAC: A systolic array for comparing nucleic acid sequences. *Computer Magazine*, 20(7):98–99, 1987.

[12] O.A. McBryan. Connection machine application performance. In *Scientific Applications of the Connection Machine, World Scientific*, 94–114, 1989.

[13] O.A. McBryan, P.O. Frederickson, J. Linden, A. Shüller, K. Solchenbach, K. Stüben, C.A Thole and U. Trottenberg. Multigrid Methods on Parallel Computers - A survey of recent developments. *Impact of Computing in Science Engineering, Academic Press*, 3(1):1–75, 1991.

[14] R.L. Rivest, A. Shamir and L. Adleman. Public key cryptography. *CACM 21*, 120–126, 1979.

[15] J.P. Serrat. Image Analysis and Mathematical Morphology. *Academic Press, N.Y.*, 1982.

[16] B. Serpette, J. Vuillemin and J.-C. Hervé. BigNum: a portable efficient package for arbitrary-precision arithmetic. *PRL Report 2*, Digital Equipment Corp., Paris Research Lab., 85, av. Victor Hugo. 92563 Rueil Malmaison Cedex, France, 1989.

[17] Marcin Skubiszewski. A hardware emulator for binary neural networks. *Proceedings of the International Neural Network Conference, Paris*, 2:555–558, 1990.

[18] Marcin Skubiszewski. An exact hardware implementation of the Boltzmann Machine. *PRL Report in preparation*, Digital Equipment Corp., Paris Research Lab., 85, av. Victor Hugo. 92563 Rueil Malmaison Cedex, France, 1992.

[19] Mark Shand, Patrice Bertin and Jean Vuillemin. Hardware speedups in long integer multiplication. *Computer Architecture News*, 19(1):106–114. 1991.

[20] Mark Shand and Jean Vuillemin. A hardware implementation for fast RSA Cryptography. *To appear*, 1993.

[21] Hervé Touati. Perle1DC: a C++ Library for the Simulation and Generation of DECPeRLe-1 Designs. *PRL Technical Note 4*, Digital Equipment Corp., Paris Research Lab., 85 av. Victor Hugo, 92563 Rueil Malmaison, France, 1992.

[22] Jean E. Vuillemin. Contribution à la résolution numérique des équations de Laplace et de la chaleur. *PRL Report 16*, Digital Equipment Corp., Paris Research Lab., 85 av. Victor Hugo, 92563 Rueil Malmaison, France, 1992.

[23] Xilinx, Inc. The Programmable Gate Array Data Book. *Product Briefs, Xilinx, Inc.*, 1987.

[24] J. Ziv and A. Lempel. A Universal algorithm for sequential data compression. *IEEE Trans. on Information Theory*, IT-23(3):337-343, 1977.

A High-Speed FPGA Using Programmable Mini-tiles

Paul Chow[1], Soon Ong Seo, Kevin Chung, Gerard Paez[2], and Jonathan Rose
Department of Electrical and Computer Engineering
University of Toronto
Toronto, Ontario, Canada M5S 1A4

Abstract

Field-Programmable Gate Arrays (FPGAs) are now a recognized technology for the implementation of digital systems, but they suffer from reduced speed and logic density compared to Mask-Programmed Gate Arrays. Many studies have been performed concerning the effect of an FPGA's architecture on its speed and density. In this paper we describe how these studies are used in an actual implementation of a high-performance FPGA.

The architecture of the FPGA logic block was determined through an experimental process using custom-built CAD tools. The result is a logic block that is an asymmetric tree of four-input lookup tables that are hardwired together. A segmented routing architecture, also tuned using experiments, is employed to improve the speed of the interconnect.

To address the problems of full-custom design, a novel layout style for FPGAs is proposed. It can be likened to the technique used in PLAs, in which a 'mini-tile' contains a portion of most components in the logic tile. The mini-tile is optimized for layout density and speed, and placed into a 4x4 array, where it is then customized by adding vias to obtain the desired hardwired connections. As well as providing ease of layout, this technique gives the capability to easily change the hardwired connections in the logic block architecture, and the segmentation length distribution in the routing architecture.

1 Introduction

The generality of Field-Programmable Gate Array (FPGA) technology permits the design of reasonably complex digital circuits using a commodity part. The price of this generality is a reduced maximum clock rate for systems, and a reduction in the amount of logic that can be integrated onto one chip. A Mask-Programmed Gate Array (MPGA) implementation is roughly three times faster and requires about ten times less silicon area when compared to an FPGA implementation [1]. In this paper we describe the design

[1]Email: pc@eecg.toronto.edu
[2]Now with CEMISID, Universidad de Los Andes, Venezuela.

and implementation of an FPGA where the primary goal is high performance with a secondary goal of maintaining a "reasonable" logic density.

FPGAs were first introduced in 1986 by Xilinx using a RAM-based programming technology [2]. Since then there have been many new commercial architectures and several new programming technologies [3, 4, 5, 6]. All of these papers describe the architectures and their various features but very little discussion is devoted towards the design decisions that resulted in the architectures. Even less is mentioned about how the choices were made in circuit design.

A few non-commercial FPGA architectures have been described, and details about those designs are more readily available. The Triptych FPGA [7] matches the physical structure of the routing architecture to the fanin/fanout nature of the structure of digital logic by using short connections to the nearest neighbours. Segmented channels are used between the columns to provide for fanouts greater than one. No discussion is given about how the segmentation length distribution was selected. This routing architecture does not allow arbitrary point-to-point routing as provided in general FPGA structures. The logic block implements logical functions using a multiplexer-based three-input lookup table followed by a master-slave D-latch, and can also be used for routing. Initial results show potential implementation efficiencies in terms of area using this structure. The Montage FPGA [8] is a version of the Triptych architecture modified to support asynchronous circuits and interfacing separately-clocked synchronous circuits. This is achieved by the addition of an arbiter unit and a clocking scheme that allows two possible clocks or makes the latches transparent.

Early work at the University of Toronto resulted in the implementation of an architecture (UTFPGA1) using three cascaded four-input logic blocks and segmented routing [9]. UTFPGA1 used information from our architectural studies but there was very little done to achieve a good design from the circuit and layout aspects. This was a first attempt that provided some insight into the problems faced in the design and layout of an FPGA.

Modern trends in the design of computer architecture have shown the importance of considering both the compiler technology and the hardware technology at the same time, when designing for high performance [10]. This is also true in the design of an FPGA, where it is important that the CAD tools match well with the architecture of the FPGA. At the University of Toronto, we have developed a number of custom CAD tools that have been used to perform "experimental" architectural studies. In this paper, we will describe two studies that influenced the architecture of our second-generation FPGA. We will also introduce a novel layout style for FPGAs that builds the basic tile using a set of identical mini-tiles and customization by the addition of vias. This allowed us to still achieve reasonable density while significantly reducing the time spent doing custom layout. These features have all been

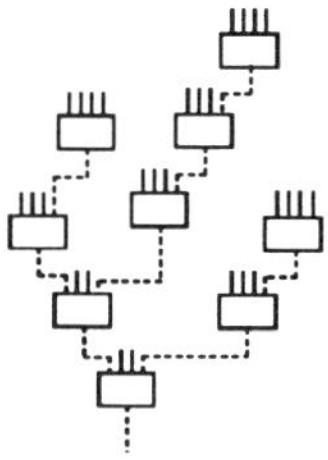

Figure 1: Circuit connections without hardwired connections.

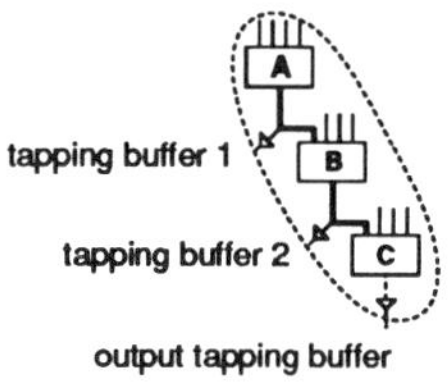

Figure 2: Hardwired logic block.

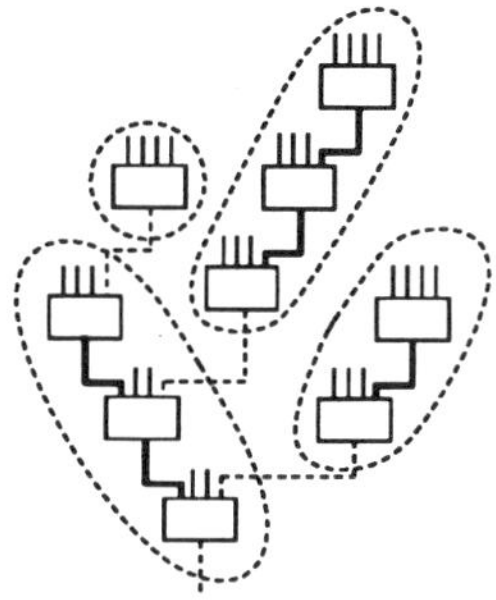

Figure 3: Circuit connections using hardwired logic block.

implemented in our test chip, called LEGO[3].

In the following section we summarize the results that led to the choice of our logic block architecture. Section 3 discusses the segmented routing architecture. The circuit design and layout issues of the chip are presented in Section 4. The results are summarized in Section 5.

2 Logic Block Architecture

The speed and density gap between FPGAs and MPGAs is mostly caused by the routing structures used to connect logic components in each technology. In MPGAs, logic blocks are connected with mask-programmed metal wires, whereas in FPGAs, logic block pins are connected using field-programmed connections, which are wire segments linked by programmable switches. Regardless of the type of programmable switches used in the FPGA (e.g. static RAM-controlled pass transistors [3] or anti-fuses [4]), the capacitance, resistance, and size of programmable connections makes them much slower and larger than a simple metal wire.

One way to improve the speed and density of FPGAs is to replace some of the programmable connections between basic logic blocks with hardwired connections, which are simple metal wires. By using hardwired links to construct coarser-grained logic blocks (called Hardwired Logic Blocks, or HLBs) from several basic blocks, the delay and size of circuits can be reduced. For example, Figure 1 shows a network of basic blocks with five programmable connections in the routing along the critical path and nine programmable connections in total. Suppose that three of the basic blocks are hardwired together to create a hardwired logic block with a fast three-block path as shown in Figure 2. Now the resulting circuit in Figure 3 has only two slow programmable links instead of five along the critical path. This is a sizable

[3]Logic that's Erasable and Greatly Optimized

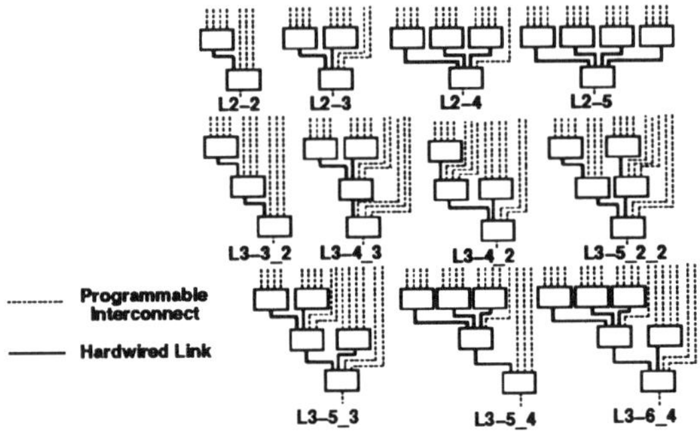

Figure 4: Example tree topologies.

reduction in routing delay. Also, the total number of programmable connections has been reduced from nine to four and this may lead to a significant reduction in routing area.

2.1 The Hardwired Logic Block Design Space

There are many possible hardwired logic blocks and each possibility offers different delay and area for a given circuit. In general, the more hardwired links in a hardwired logic block, the faster will be the circuits implemented using that logic block. However, more hardwired connections leads to less connection flexibility and this may lead to lower density. There are many possible topologies for connecting basic logic blocks using hardwired links. Some of the tree topologies that were investigated for speed and area-efficiency are illustrated in Figure 4[4]. The granularity of the basic block also affects the delay and area of the implemented circuit. In FPGAs without hardwired connected blocks, smaller basic blocks tend to be better for logic density while coarse-grained basic blocks are better for speed. The results of previous studies [11, 12] have shown that for FPGA architectures without hardwired connections, a 4-input lookup table (4-LUT) is a good choice for both density and speed. The goal of the experiments described below is to select from among the many possible hardwired 4-LUT architectures one that produces high system speeds, at a reasonable cost in area.

An important architectural assumption is that each HLB features tapping buffers on the output of each LUT basic block to make the output accessible to the routing. Tapping buffers lead to faster HLB circuits since the output of one basic block can be accessed directly instead of propagating it through another basic block. Tapping buffers also improve logic density since unrelated pieces of logic can be packed together in the same HLB, with each piece

[4]The topology name is the letter "L", then the height of the HLB, then a "-" followed by a listing of sizes of the subtrees from a pre-order traversal of the canonical HLB tree. Each subtree size is separated by "_", and leaf inputs and single-LUT subtrees are not listed.

using a separate tapped basic block output. In summary, with HLBs one gets a coarse-grained logic block, which leads to fast circuits, yet the HLB tapping buffers allow use of the relatively fine-grained basic blocks of an HLB, which leads to reasonable logic density.

To evaluate the various HLB topologies, a set of 15 MCNC combinational benchmark circuits are each implemented in a set of FPGAs with each HLB topology. The area and delay of each implemented circuit is then calculated using area and delay models, and the results over all circuits are summarized for each HLB topology.

The area of the circuit consists of the logic area and the routing area. The logic area consists of the area of the LUT bits, tapping buffers, and the D flip-flops. For routing area estimates, we assume a routing architecture similar to Brown [13]. The basic architecture is illustrated in Figure 6, consisting of switching from pins to channel wires and between channel wires. We perform global routing to find the maximum number of routing tracks over all channel segments. This maximum number is assumed for every channel segment. The routing area is approximated by the area of the programmable switches in the connection blocks (C) and the switch blocks (S). The number of programmable switches in each connection block is the product of the number of HLB I/O pins and the number of routing tracks to which each pin can connect. The number of programmable switches in each switch block is determined by the number of tracks each track can connect to at a switch block, as well as the number of tracks in each channel segment.

The delay of the circuit is the sum of the LUT delays and the programmable connection delays along the critical path. We assume zero delay for hardwired connections. The delay for a 4-LUT was determined by HSPICE[16] simulations. The programmable connection delay is difficult to determine since it depends upon several circuit and routing architecture properties, such as connection fanout, number of switching stages in the connection and number of switches connected to each channel. We chose the simplest possible model for average programmable connection delay, a constant value parameter, since our studies concerned only logic block architecture. Note that using constant delay per programmable connection penalizes larger HLBs. With a larger HLB, the average number of switching stages per connection decreases in the circuit, and thus the average programmable connection delay should also decrease.

In the following, two studies are described: one in which the circuits are optimized purely for delay, and then purely for area. By comparing how the blocks perform in each domain, we can deduce a choice that performs well in both.

2.2 Experimental Results: Speed Study

In the first experiment, the circuits were optimized for the greatest possible speed, sometimes at significant area expense. The HLB topologies investi-

# LUTs	HLB Name	Speed Factor		Area Factor	
1	L1	1	(35 MHz)	1	(15.3 x 10^6 μm^2)
2	L2-2	1.14		1.19	
3	L2-3	1.26		1.21	
4	L2-4	1.34		1.32	
4	L3-4_2	1.28		1.31	
5	L2-5	1.42		1.40	
5	L3-5_2	1.37		1.37	
6	L3-6_5	1.43		1.65	
7	L3-7_3	1.43		1.66	
8	L3-8_3_2	1.47		1.61	
9	L3-9_4_2	1.50		1.70	

Table 1: HLB speedups with respect to no hardwired links.

gated were restricted to all possible tree topologies with nine or fewer 4-LUTs and three or fewer LUT levels. HLBs with more than nine 4-LUTs would provide greater speedups but require too much area to be considered practical.

In Table 1 each row shows the HLBs with a given number of 4-LUTs that produce the greatest speedup. If there are two rows for a given number of LUTs, the second entry is an alternative HLB that offers a slower speed but at a lower area cost. The first column of Table 1 lists the number of 4-LUTs in the HLB, the second column shows the name of the HLB topology, the third column displays the normalized speedup of the HLB with respect to L1 (a 4-LUT FPGA with no hardwired links) and the last column shows the normalized area with respect to L1. The geometric mean of the areas[5] of the circuits for L1 is 15.3 x 10^6 μm^2 and the mean of the delays is 28.5 ns or a mean system speed of 35.0 MHz.

The results in Table 1 show that the HLBs investigated can increase system speed from 14% up to 50% compared to L1. However, the area cost of these speedups is an increase of 19% to 70% in circuit size. The area increase is caused by the HLB mapper, which replicates LUTs during delay optimization to place as many hardwired links as possible in the HLBs along the critical paths. The addition of only one hardwired LUT (L2-2 HLB) yields a speedup of 14% with an area cost of 19%. Adding a second hardwired LUT (L2-3) gives a further 12% speedup (system speed of 44 MHz) but only an extra 2% area cost. The fastest investigated HLB (L3-9_4_2) yields a system speed of 52.5 MHz but costs 70% more in area compared to L1.

2.3 Experimental Results: Area-efficiency Study

The goal of this study is to find the minimum area cost of using each HLB topology to implement the benchmark set. The same HLB topologies as in

[5]All measurements are for a 1.2 μm CMOS process. Absolute areas come from actual layouts of cells used in an area model. Absolute times come from a delay model, which incorporates delay data derived from the simulation of the cell layouts.

# LUTs	HLB Name	Area Factor		Speed Factor	
1	L1	1	(6.7×10^6 μm^2)	1	(22.1 MHz)
2	L2-2	0.93		1.17	
3	L2-3	0.96		1.17	
4	L2-4	1.00		1.09	
4	L3-4_2	0.93	(9.67×10^6 DSM^2)	1.27	(28 MHz)
5	L2-5	1.10		1.03	
5	L3-5_2	1.04		1.24	
6	L3-6_5	1.23		1.04	
7	L3-7_3	1.28		1.19	
8	L3-8_3_2	1.32		1.25	
9	L3-9_4_2	1.42		1.24	

Table 2: HLB areas with respect to no hardwired links.

the speed study are investigated. It should be noted that when minimizing area, the HLB mapper does not replicate LUTs.

Table 2 presents the same HLB topologies as in the previous table but for area-optimized circuits. Note that even though the circuits are optimized for area the presence of hardwired links in the HLBs will cause a speedup with respect to L1, a 4-LUT FPGA without hardwired links. The first column lists the number of 4-LUTs in the HLB, the second column lists the HLB name, the third column lists the normalized area with respect to L1 and the fourth column lists the normalized speed with respect to L1. The geometric mean of the area of the area-optimized L1 circuits is 6.7×10^6 μm^2 and the mean of the delays is 45.3 ns or a mean system speed of 22.1 MHz.

From Table 2, there are three HLBs, L2-2, L2-3 and L3-4_2, that offer better logic density and speed than the 4-LUT FPGA without hardwired connections. The L3-4_2 HLB offers the best logic density and the best speedup among the HLBs in Table 2. In fact, the speed of the area-optimized circuits using L3-4_2 is only 20% slower than the delay-optimized L1 circuits, while using only 40% of the area of the delay-optimized L1 circuits. From Table 1, when optimizing purely for delay, the L3-4_2 HLB offers a 28% speedup over L1 with only a 31% area increase. The area-efficiency of L3-4_2 when optimizing purely for area coupled with the good speedup and area-efficiency when optimizing purely for delay led us to conclude that the L3-4_2 HLB was the best overall HLB.

3 Segmented Routing Architecture Design

The effect of an FPGA's routing architecture on system speed and integration logic density can be very significant. As discussed above, the main reason is that the programmable switches, which are pass transistors driven by SRAM cells, have significant resistance and capacitance and require large area. The *routing architecture*, which is the manner in which wire segments and switches are organized, can be designed to mitigate these effects.

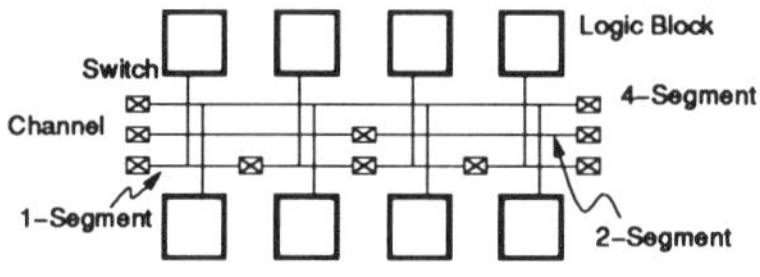

Figure 5: Segmented routing architecture example.

In the original Xilinx 2000 and 3000 architectures [2, 14], a very simple architecture was employed: most wire segments spanned only the length (or width) of one logic block, having switches at either end. One way to improve the speed of connections that travel long distances is to provide segments that span multiple logic blocks without being switched, as illustrated in Figure 5. In this way, a segment with appropriate length for a connection can be selected, which results in less resistance along the path.

This leads to the architectural question: How many segments of each length should there be? That is, what is the *segmentation length distribution*? There is a tradeoff in this design question: the more longer segments there are, the greater the likelihood that long and otherwise time-consuming connections will be routed with fewer series switches, resulting in a faster circuit. On the other hand, an excessive number of longer segments will mean that some of the long segments will have to be used for short connections that do not span the entire segment length, resulting in the waste of part of the segment. This waste leads to a decrease in logic density. Thus it is important to find a segmentation distribution that improves speed, but does not waste too much area.

There is a second important architectural feature of segmented routing architectures besides the distribution: a segment is called internally *populated* if it is possible to make connections from the middle of a segment either to logic blocks or other routing segments. The advantage of un-populated segments is that there is less parasitic switch capacitance connected to the segment, which makes it faster. The disadvantage is that the reduction in routing flexibility may result in the need for more tracks and so a loss of logic density.

In this section we will summarize a study on segmented routing architecture distribution and population [15] that was used as the basis for decisions on the routing architecture for LEGO. The basic approach was experimental, similar to the one described in the previous section: several benchmark circuits are "implemented" as FPGAs, each with a different segmentation distribution. These experiments required several CAD tools, including a detailed router specifically designed for FPGAs [13]. The number of tracks needed to complete the route using segments of length one is called the minimum. When the channel includes segments of other lengths, the number of tracks required above the minimum is measured.

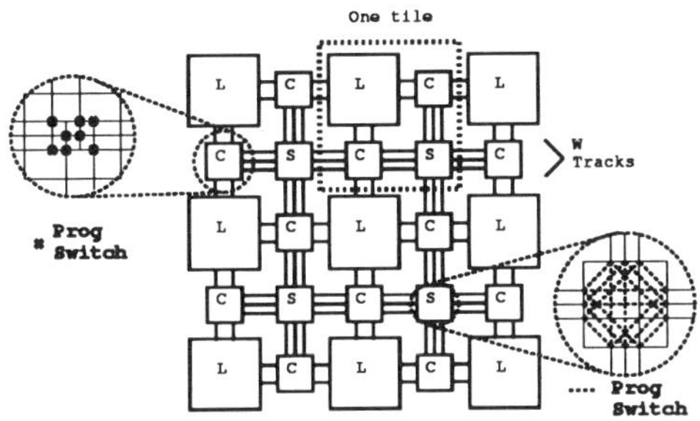

Figure 6: Architecture Definitions.

3.1 Definitions and Experiments

Figure 6 illustrates the broad architectural features of the LEGO FPGA that is being designed. The logic block (L) has pins that are connected to routing channels in a structure called a *connection block* (C). The *switch block* (S) provides connectivity among its attached channels. Thus a channel consists of C and S blocks. Each channel contains W routing tracks. We assume that each track consists of only one type of segment length. We will consider only segments of length one, two, and three, which will be referred to as 1-segments, 2-segments and 3-segments.

Let D_1 be the fraction of the W tracks that contain 1-segments. Similarly D_2 and D_3 are the fractions for 2- and 3-segments. Thus $D_1 + D_2 + D_3 = 1$ and the number of tracks containing segments of length i is $D_i \times W$. Note that the longer segments pass through switch blocks without actually traversing a programmable switch.

As mentioned above, it is possible to consider the removal (or de-population) of the switches internal to a segment. We divide this population question into two parts: whether or not the connection blocks internal to the segment (i.e., those not at its ends) should be populated, and whether or not the internal switch blocks should be populated. The following notation indicates the level of population: CB_u indicates that all unpopulatable connection blocks are un-populated, while CB_p means that they are populated. Similarly for SB_u and SB_p. Thus there are four combinations of population. Notice that for 2-segments, only the middle switch block can be de-populated, and for 3-segments, there are two switch blocks and one connection block that can be de-populated.

3.2 Results and Decisions

The following experiments were performed: for all possible values of D_1, D_2, and D_3, using all four population combinations, the number of excess tracks needed over the minimum (for $D_1 = 1$) were measured and averaged for five benchmark circuits. We addressed the questions of population and

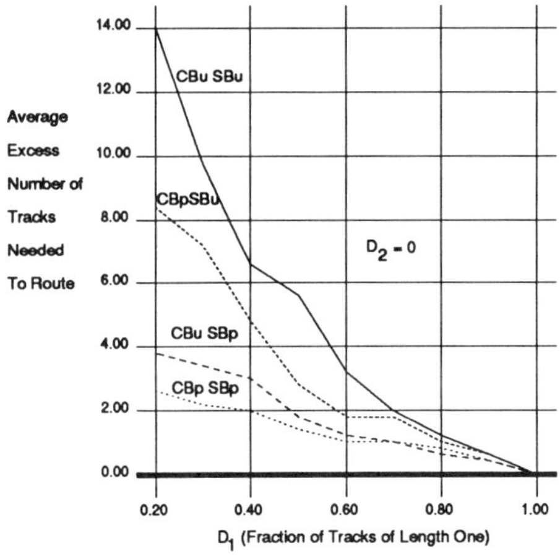

Figure 7: Population comparison.

distribution using these experiments. Note that the logic block used in these experiments was a single 4-input LUT, with a D flip-flop. While this was not the block chosen for LEGO, we believe the the general results discussed below will hold for most logic blocks.

3.2.1 Population of Segments

The first question we address is that of population of the S and C blocks. Figure 7 is a plot, for $D_2 = 0$, of the average number of excess tracks versus D_1 for the five circuits. The four curves are the different combinations of population for the switch and connection block. Notice that as D_1 decreases from 1, D_3 increases from 0. Since this provides a greater number of longer segments with lower flexibility, we expect the number of tracks needed to successfully route will increase, and this is borne out in the figure.

The data show that both cases in which the switch block is un-populated (the upper two curves) result in significantly greater excess tracks. For this reason we decided to populate the switch block in LEGO. On the other hand, the bottom two curves (with the switch block populated) illustrate that only a minor increase in tracks is experienced when the connection block is de-populated. Since there is an advantage in speed for de-population (less capacitance on the track) we decided to de-populate the connection block in LEGO.

3.2.2 Segmentation Distribution

Table 3 gives the average number of excess tracks for many possible values of D_1, D_2, and D_3. Note that since $D_1 + D_2 + D_3 = 1$ there are only two independent variables, which we chose as D_2 and D_3 in the table. The rows

D2/D3	0.0	0.1	0.2	0.3	0.4	0.5	0.6	0.7	0.8
0.0	0.0	0.4	0.6	1.0*	1.2*	1.8	3.0	3.4	3.8
0.1	0.2	0.6	1.0*	1.2*	1.6	2.2	2.8	3.4	
0.2	0.6	0.8	1.0*	1.0*	1.8	3.0	3.2		
0.3	0.6	0.6	1.2*	1.8	2.2	2.4			
0.4	0.6	1.0*	1.0*	2.0	2.6				
0.5	0.8	1.2*	1.4	2.0					
0.6	1.2*	1.6	2.0						
0.7	1.4	1.8							
0.8	1.8								

Table 3: Average number of excess tracks for each distribution CB_u, SB_p.

vary D_2 and the columns vary D_3. This table is for architectures with both the C blocks un-populated and the S blocks populated, as discussed above.

Table 3 illustrates that even with 80% length three segments, that only about four extra tracks on average are required. (The minimum number of absolute tracks required for the five circuits ranged from 9 to 15.) We decided that we were willing to tolerate only about one extra track per channel above the minimum, so that the area cost of the higher speed tracks would be small. Thus architectures with D_2 in the range of 0.0 to 0.6 and D_3 in the range of 0.1 to 0.4 (those marked with a * in Table 3) would reflect such a choice. The faster architectures are the ones with higher values of D_3.

In Section 4.2 we give the reasons that lead to the choice of 16 tracks in LEGO. For the segmentation distribution, there is an additional constraint that arises from the one-tile style of layout that was used: there must be an even number of 2-segment tracks, and the number of 3-segment tracks must be a multiple of three. Taking this constraint and the above distribution into account, the following segmentation distribution was used in LEGO: nine of the 16 tracks are 1-segments, four tracks are 2-segments, and three are 3-segments. This corresponds to a segmentation distribution in which $D_1 = 0.56$, $D_2 = 0.25$ and $D_3 = 0.19$, which fits within the architectural range indicated above.

4 Chip Design

Given that the major architectural choices have been made there are still many decisions in the circuit-level implementation. The determination of the hardwired logic block and the segmented routing were done independent of the circuit design issues. The only implementation consideration was the area and timing models used to select the logic block. However, there were a few lower-level architectural questions that could only be answered after we considered the circuit and layout design. We will address those here, as well as describe a novel layout methodology.

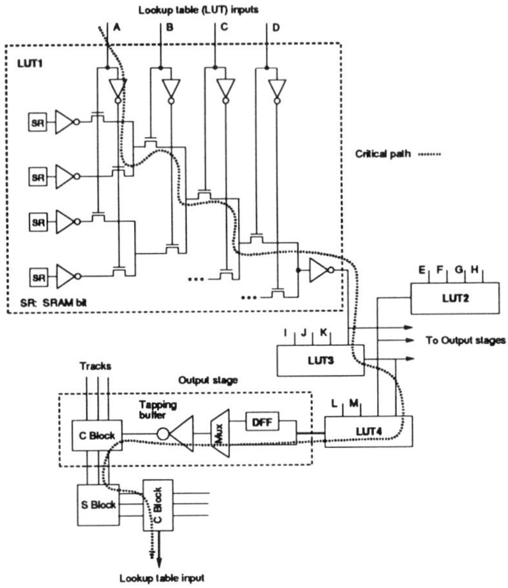

Figure 8: Critical Path

4.1 Circuit Design

The primary objective of this work is to achieve high speed by finding better architectures for FPGAs. Even with a good architecture, it is still necessary to carefully tune the circuit design and we address some of the important circuit issues here. The technology that we used is a 1.2μm two-level metal n-well CMOS process.

Recall that the logic block chosen is the one labelled L3-4_2 in Figure 4. Each 4-input LUT is implemented with a multiplexer tree, attached to static RAM cells. The critical path is shown in Figure 8. It starts at the most heavily loaded input of the first LUT multiplexer tree. Note that the outputs of the cascaded LUTs connect to the least loaded input of the next LUT. (Not only is this the least loaded input, but it also has the shortest distance that the signal can travel, which is a significant saving in delay compared to the other inputs). From the LUTs the path goes through a multiplexer that selects the combinational output or a latched output before going to the C block. In this case, the combinational path is used. The C block connects the signal to a track, which then goes through an S block (or number of S blocks) to another C block and then the input of another lookup table.

Extensive simulation with HSPICE[16] was used to do the circuit design. Here we address the individual components separately.

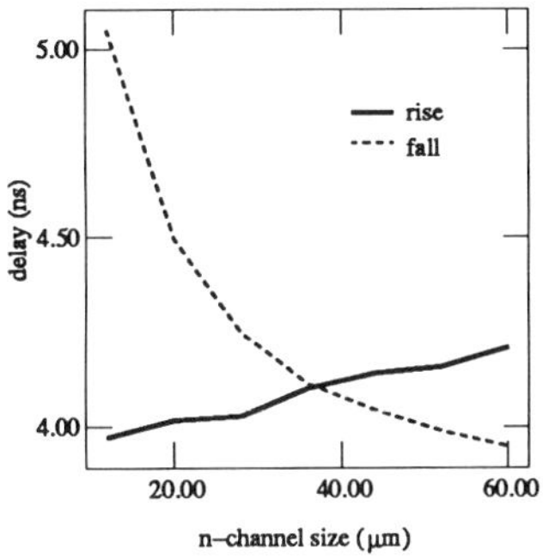

Figure 9: Transistor sizing for the logic block drivers.

4.1.1 Switch and Driver Transistor Sizing

Transistor sizing is a major concern for achieving high performance. In Triptych [7], the method of Logical Effort [17] was used to assist in transistor sizing and buffer insertion for speed. We chose to use actual simulation because we felt it was easier to balance the rise and fall times. We began by simulating the components of the critical path individually but later tuned the design by simulating the entire critical path at once. Our initial assumption was that the n-channel pass transistors, which were used for the switches, had to be at least as large as the transistors of the drivers, so as to reduce the effects of the switch sizes on the critical path delay. This allowed us to find an initial driver size. We then tuned the design by simulating with different switch sizes. An example result is shown in Figure 9, which shows the effect of varying the logic block output driver size on the total delay of the critical path. It can be seen that the rise and fall times are equal at about 37μm for the size of the n-channel driver pulldown. In LEGO we actually used 34μm because in our initial simulations we used nine tracks in the channels, which was later changed to 16 tracks, and thus a higher load.

4.1.2 Muxes versus Switches

The C block, as illustrated in Figure 6, connects the track to the logic block inputs and the logic block outputs to the track. Consider the track to logic block connection. The connection can be done by means of switches, where there is a single switch for each track to pin connection, or by multiplexing N tracks to each input pin. The output pins can be treated in a similar fashion using switches or de-multiplexers. In UTFPGA1 [9] we chose to use multiplexers for the inputs and switches for the outputs. The reasoning was that since only one wire in the channel can be connected to an input pin, there is no need for the flexibility provided by switches. This saves in the number of bits needed to select the input connection and the total number of transistors needed. For the output, having a switch from a pin to each

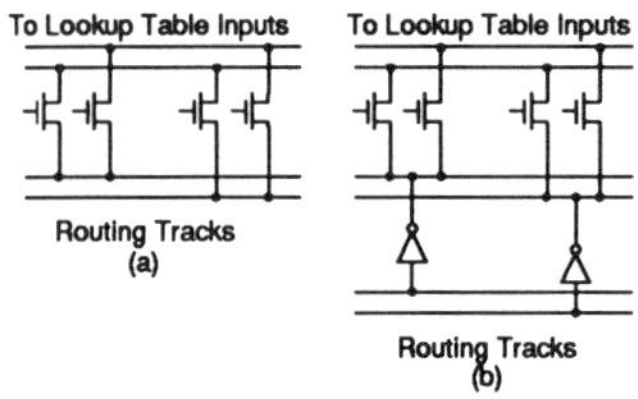

Figure 10: (a) C-block inputs with no receivers. (b) C-block inputs with isolation.

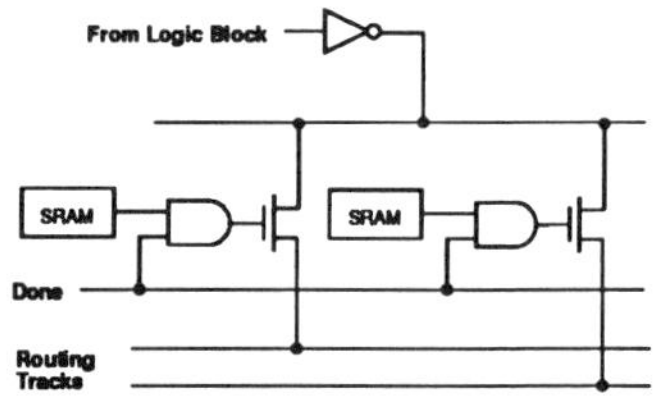

Figure 11: Power-up and programming protection without affecting the critical path.

track provides some fanout capability. Although using a multiplexer should result in less area, it has a significant performance impact because the signal must go through a number of series transistors. In LEGO we chose to use switches at both the inputs and the outputs of the logic block. The price is that a significant fraction of the area is devoted to the C blocks.

4.1.3 Routing Track Isolation

Another important issue is the capacitance at the C block inputs. With the switches connected as shown in Figure 10(a) the capacitance depends on whether the switch is on or off. If a switch is off the contribution to the capacitance is due to the switch transistor itself. If the switch is on, then any of the circuitry on the other side of switch will contribute to the load seen by the driving signal. This means that the loading will be a function of the number of C-block switches that are turned on. This is related to the problem noted in the design of the Triptych [7] logic block, and their solution was to use gate logic instead of switch logic. The problems in LEGO can be reduced by adding a driver between the track and each switch input, which makes the load independent of how many switches are turned on. However, this still has one load per switch input. Further isolation can be provided by using the configuration shown in Figure 10(b). This is the configuration used in LEGO.

4.1.4 Power-up and Programming Protection

FPGAs face a difficult problem when the chip first powers up: if the programming bits are unfavorable, then several buffers may be driving the same line in opposite directions, causing a current that may be enough to physically damage the chip. This problem has been previously addressed [9, 7]. One possibility is to make the driver a tri-state driver, but this has the disadvantage that the transistors that are used to disable the driver are in the critical path, either directly in series with the driver transistors or in the path that turns on the drive transistors. The solution used in LEGO is shown in Figure 11. We use the switch transistor to isolate the driver from the tracks.

4.2 Layout Style

Besides the desire to achieve a high-performance circuit design we wanted to find a better layout style than just trying to attack the problem purely as a full-custom design. When doing VLSI layout, it is desirable to find some regularity in the topology to make the task easier. In an FPGA the first level of regularity is the arraying of the tiles as shown in Figure 6, where a tile includes an L block, an S block, and two C blocks. This partitioning was used for UTFPGA1 and in LEGO. However, even the full custom layout of a tile is a significant task, and it would be desirable to find even more regularity. In UTFPGA1 we attempted to use the same C block layout for both the vertical and horizontal channels but this resulted in a very poor tile floorplan because it was difficult to pack the pieces together. For LEGO we wanted a solution that had smaller granularity.

The technique that we used was to divide the tile into 16 mini-tiles. Each mini-tile contains one sixteenth of the L block, S block, and C blocks. The tile[6] is implemented by creating a 4x4 array of mini-tiles, and adding vias at the appropriate locations to customize the local function of each mini-tile. The flip-flops, muxes to select the latched or combinational outputs, and output tapping buffers are at the edge of the array, since there are only four of each. These are shown as part of the output stage in Figure 8. The motivation for this topology comes from the topology used for PLAs [18], where the goal is to make the overall structure regular and have irregularity restricted to the final programming. Figure 12 shows a schematic of the mini-tile without details of the lookup table (LUT) logic included.

The major problem with dividing the functionality of the tile across 16 mini-tiles is the routing needed to interconnect the mini-tiles. It is difficult to organize the routes correctly, and it is likely that the extra wires will cost more in area than a non-mini-tile approach. The area issue is addressed below. Figure 12 shows all the necessary wiring. Examples of via placement are shown by the Xs. The horizontal and vertical tracks of Figure 6 are shown in heavier lines intersecting at the switch block (S block).

The choice of how many tracks to use depends on the actual size of the overall array of tiles in a chip and on the number of mini-tiles. For larger FPGAs more tracks would be needed, and in the case of LEGO, the number will be a multiple of four because of the mini-tile array size.

Each mini-tile contains a $\frac{1}{4}$ Mux-tree, which is really a 4-to-1 mux that has the SRAM bits at its inputs and a separate 2-to-1 mux. The three mux interconnect tracks going through the $\frac{1}{4}$ Mux-tree are used to connect between the mini-tiles to form the full lookup tables. These tracks are segmented in the mini-tile to allow sharing of the tracks and can be connected with a piece of metal as needed during final configuration. One of the 2-to-1 muxes is not required to make the full lookup table.

---

[6]We will use *tile* to denote the complete tile, and *mini-tile* to denote the smaller building block piece.

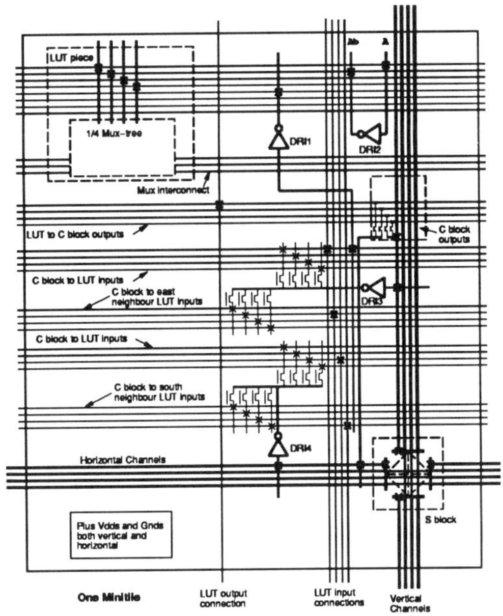

Figure 12: Mini-tile schematic.

In addition, the outputs of the LUTs are routed on the *Mux interconnect* tracks to the edge of the tile for connection to the D flip-flops and the driver to the tracks. The output of the driver is routed back into the mini-tiles along a *LUT to C block outputs* track for connection to the C block outputs. The remaining vertical wires are to distribute signals to the mini-tiles along the vertical dimension of the array.

Driver *DRI2* is the inverter shown at the inputs to the LUTs in Figure 8. Driver *DRI1* drives the input from the C block to the LUT inputs.

The drivers marked *DRI3* and *DRI4* are the track isolation drivers shown in Figure 10(b). The transistors at the outputs of these drivers are the switches described in Section 4.1.2. In our partitioning, the C block includes connections to the south- and east-neighbour tiles so some of the wires in each mini-tile are used to route signals to the LUT inputs in those neighbours.

Using the mini-tile methodology we sacrifice the potential for global layout optimization across the whole tile in exchange for a more simple full-custom layout task. However, we could afford to spend some time doing local optimization of the mini-tile. A plot of the mini-tile is shown in Figure 13 and the floorplan is shown in Figure 14. It can be seen that the vertical dimension is limited in pitch by the number of horizontal metal 2 tracks. In the horizontal dimension, there are still some visible internal vertical routing channels that do not go over any active area. If we had a third metal routing layer so that the vertical routing could go over active area, then the hori-

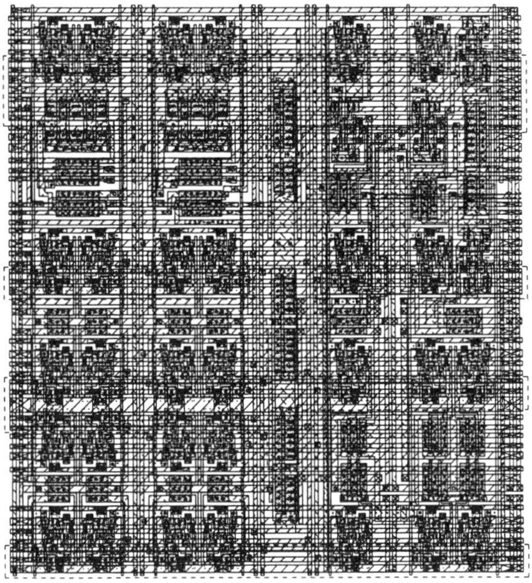

Figure 13: Mini-tile plot.

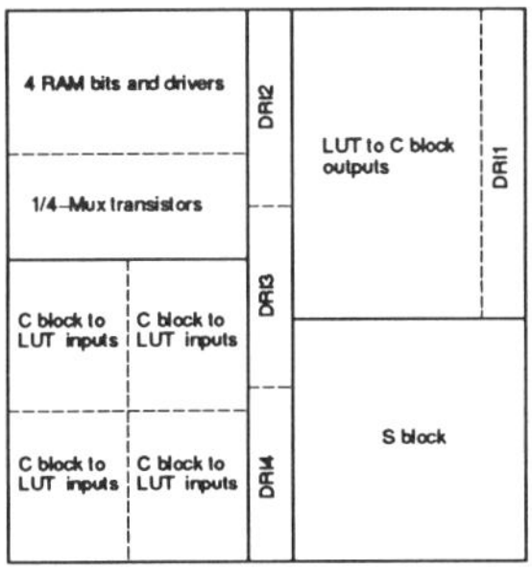

Figure 14: Mini-tile floorplan.

zontal dimension could be reduced by about 25%. With this area removed it can be seen that the layout efficiency would be quite good as most of the metal routing would be on top of active area, which cannot be compacted much. Since we did not attempt a full custom layout of the tile we do not have a quantitative measure of our layout efficiency. However, we feel that by using this methodology there was not a great sacrifice in overall density, and shortening the time to do the layout was more important[7].

An added benefit that has not been explored at this point is that with this ability to configure the tile after the hard layout work is done, it is possible to try other hardwired interconnections for the LUTs. This may be of benefit if our work in non-homogeneous logic blocks suggests that a mix of tiles with different types of hardwired connections is beneficial.

4.2.1 Other Layout Issues

The use of the mini-tiles made it easier to consider other issues such as access of the programming bits and the power distribution.

For programming, the top of the chip contains a shift register into which words of bits are serially loaded. Each mini-tile contains five word lines and seven bit lines. When the shift register is loaded, one of the word lines in the array is selected and data in the shift register are transferred into the array along the bit lines.

[7]Especially in a rapidly evolving industry.

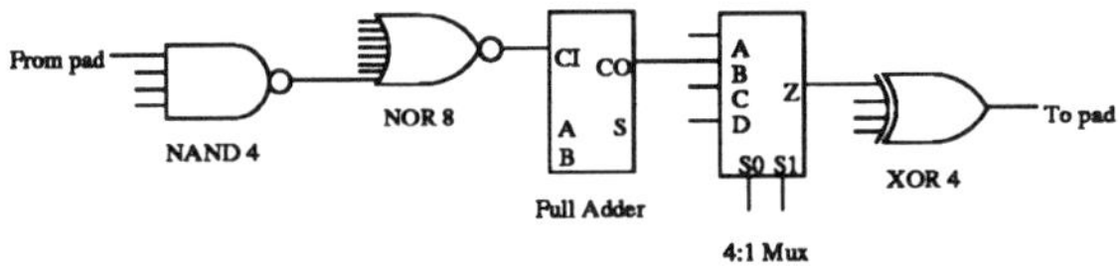

Figure 15: Benchmark combinational circuit.

With the mini-tile methodology, it is easy to have a rectangular floorplan for the tile, and this makes it easier to distribute power throughout the array.

5 Results

The size of each tile, which contains four four-input lookup tables, four D flip-flops, and connections for 16 tracks, is 1240μm by 1184μm. By way of comparison, the Xilinx 3000 tile in the same feature size technology (1.2μm CMOS), which contains the equivalent of two four-input lookup tables, two D flip-flops and connections for eight tracks (about half the functionality of the LEGO tile) is roughly 500μm by 400μm. This is about a factor of four smaller, which comes from the fact that there is twice as much logic and routing in the LEGO tile, the mini-tile strategy, and design decisions that sacrificed area for speed. In particular, the large size drivers and the switches (rather than a multiplexer) used in the C block take up significant area. Each LEGO tile requires 492 programming bits.

From simulation, we measure a delay of 4.1ns for the critical path shown in Figure 8. The equivalent path in a Xilinx 3000 FPGA would likely have to traverse at least two Configurable Logic Blocks (CLBs) with a delay of at least 11ns (in the -125 rated part). While this does not include inter-CLB routing time, the delay for LEGO is simulated time for nominal conditions, not worst-case.

An attempt to compare LEGO with commercial FPGAs is shown in Table 4. The circuit used is shown in Figure 15. The figure and some of the numbers in Table 4 are taken from some performance benchmarks published by Crosspoint Solutions [19]. The numbers in the *Published* column are the reported numbers assuming that the output goes to a pad driving a 50 *pf* load. The *adjusted* numbers subtract 25 *ns* as an attempt to take out the effect of the output pad[8]. This is a gross approximation as we do not have the numbers for the delays due to only the logic. However, it can be seen that LEGO compares well even though the others are implemented in higher performance technologies. Much of this is because of the larger functionality of the LEGO logic block. The LEGO number is from an HSPICE simulation under nominal conditions. The other numbers are for worst case process conditions.

[8]The Crosspoint benchmarks erroneously used data for a Xilinx 4005-7 part, which does not exist. The Xilinx data has been multiplied by 5/7 to reflect the speed of a 4005-5.

FPGA	Published *ns*	Adjusted *ns*
Crosspoint CP20420-1	37.4	12.4
Crosspoint CP20420-0	54.2	29.2
Actel A1280-1	70.0	45.0
Xilinx XC4005-5	41.4	16.4
LEGO	N/A	11.5

Table 4: Performance comparisons.

6 Conclusions

We have described the design of a high-performance FPGA from the architectural design decisions used to select the logic block architecture and the segmentation of the routing tracks, to the circuit design and layout methodology. The focus on high-performance illustrated important issues in circuit design that affect the critical path delay. We also presented a layout methodology for FPGAs that reduces the amount of custom layout required without sacrificing too much density.

A prototype of the LEGO architecture, containing two tiles, went to fabrication in September 1992.

Two areas where further work is definitely merited are in clocking and the use of the D flip-flops. All of the architectural studies to date focus on the combinational logic aspects. Questions about how many flip-flops are required, and how they should be clocked have not been addressed. In the case of LEGO, D flip-flops were added at the output of each tapping buffer in the hardwired logic block. The flip-flops are clocked from the system clock with an enable that can be selected from the routing channel.

Acknowledgements

This work was supported by funding from MICRONET, a Network of Centres of Excellence on Microelectronic Devices, Circuits and Systems for Ultra Large Scale Integration. Fabrication is provided by the Canadian Microelectronics Corporation and Northern Telecom.

We would like to thank the referees for their comments, which helped to improve the presentation of this work.

References

[1] Stephen D. Brown, Robert J. Francis, Jonathan Rose, and Zvonko G. Vranesic. *Field-Programmable Gate Arrays*. Kluwer Academic Publishers, 1992.

[2] William S. Carter, Khue Duong, Ross H. Freeman, Hung-Cheng Hsieh, Jason Y. Ja, John E. Mahoney, Luan T. Ngo, and Shelly L. Sze. A User Programmable Reconfigurable Logic Array. In *Custom Integrated Circuits Conference*, pages 233–235. IEEE, 1986.

[3] Hung-Cheng Hsieh, William S. Carter, Jason Ja, Ed Cheung, Steve Schreifels, Chuck Erickson, Philip Freidin, and Liane Tinkey. Third-Generation Architecture Boosts

Speed and Density of Field-Programmable Gate Arrays. In *Custom Integrated Circuits Conference*, pages 31.2.1–31.2.7. IEEE, 1990.

[4] Mike Ahrens, Abbas El Gamal, Doug Galbraith, Jonathan Greene, Sinan Kaptanoglu, Lynn Hutchings K.R. Dharmarajan, Sifuei Ku, Phil McGibney, John McGowan, Amer Samie, Kitty Shaw, Norma Stiawalt, Telle Whitney, Tom Wong, Wayne Wong, and Bortay Wu. An FPGA Family Optimized for High Densities and Reduced Routing Delay. In *Custom Integrated Circuits Conference*, pages 31.5.1–31.5.4. IEEE, 1990.

[5] David Marple and Larry Cooke. An MPGA Compatible FPGA Architecture. In *Custom Integrated Circuits Conference*, pages 4.2.1–4.2.4. IEEE, 1992.

[6] Algotronix Ltd., Edinburgh, Scotland. *CAL 1024 Datasheet*, 1989.

[7] Carl Ebeling, Gaetano Borriello, Scott A. Hauck, David Song, and Elizabeth A. Walkup. TRIPTYCH: a New FPGA Architecture. In Will Moore and Wayne Luk, editors, *FPGAs*, chapter 3.1, pages 75–90. Abingdon EE&CS Books, 15 Harcourt Way, Abingdon OX14 1NV, England, 1991. Presented at the Oxford 1991 International Workshop on Field Programmable Logic and Applications.

[8] Scott Hauck, Gaetano Borriello, Steven Burns, and Carl Ebeling. MONTAGE: An FPGA for Synchronous and Asynchronous Circuits. In *Proceedings of the Second International Workshop on Field-Programmable Logic and Applications*, Vienna, Austria, September 1992.

[9] Paul Chow, Soon Ong Seo, Dennis Au, Terrence Choy, Bahram Fallah, David Lewis, Cherry Li, and Jonathan Rose. A 1.2μm CMOS FPGA Using Cascaded Logic Blocks and Segmented Routing. In Will Moore and Wayne Luk, editors, *FPGAs*, chapter 3.2, pages 91–102. Abingdon EE&CS Books, 15 Harcourt Way, Abingdon OX14 1NV, England, 1991. Presented at the Oxford 1991 International Workshop on Field Programmable Logic and Applications.

[10] John L. Hennessy and David A. Patterson. *Computer Architecture: A Quantitative Approach*. Morgan Kaufmann Publishers, Inc., 1990.

[11] Jonathan Rose, Robert J. Francis, David Lewis, and Paul Chow. Architecture of Field Programmable Gate Arrays: The Effect of Logic Block Functionality on Area Efficiency. *IEEE Journal of Solid-State Circuits*, 25(5):1217–1225, October 1990.

[12] Satwant Singh, Jonathan Rose, Paul Chow, and David Lewis. The Effect of Logic Block Architecture on FPGA Performance. *IEEE Journal of Solid-State Circuits*, 27(3):281–287, March 1992.

[13] Stephen Brown, Jonathan Rose, and Zvonko G. Vranesic. A Detailed Router for Field-Programmable Gate Arrays. *IEEE Transactions on Computer-Aided Design of Integrated Circuits and Systems*, 11(5):620–628, May 1992.

[14] Hung-Cheng Hsieh, Khue Dong, Jason Y. Ja Roy Kanazawa, Luan T. Ngo, Liane G. Tinkey, William S. Carter, and Ross H. Freeman. A 9000-Gate User-Programmable Gate Array. In *Custom Integrated Circuits Conference*, pages 15.3.1–15.3.7. IEEE, 1988.

[15] Gerad Paez, Immanuel Rahardja, and Jonathan Rose. On the Design of Area-Efficient Segmented Routing Architectures for Field-Programmable Gate Arrays. In preparation, 1992.

[16] Meta-Software, Inc., 1300 White Oaks Road, Campbell, CA 95008. *HSPICE User's Manual*.

[17] Ivan E. Sutherland and Robert F. Sproull. Logical Effort: Designing for Speed on the Back of an Envelope. In *Proceedings of the 1991 University of California at Santa Cruz Conference on Advanced Research in VLSI*, pages 1–16, March 1991.

[18] Carver Mead and Lynn Conway. *Introduction to VLSI Systems*. Addison-Wesley, 1980.

[19] Crosspoint Solutions. Performance benchmarks, rev 1.01 8/92, 1992.

Spectral K-Way Ratio-Cut Partitioning and Clustering

Pak K. Chan, Martine Schlag, Jason Y. Zien
Computer Engineering
University of California, Santa Cruz, USA
Santa Cruz, CA 95064
pak@cse.ucsc.edu martine@cse.ucsc.edu zien@cse.ucsc.edu

Abstract

Recent research on partitioning has focussed on the ratio-cut cost metric which maintains a balance between the sizes of the edges cut and the sizes of the partitions without fixing the size of the partitions a priori. Iterative approaches and spectral approaches to two-way ratio-cut partitioning have yielded higher quality partitioning results. In this paper we develop a spectral approach to multi-way ratio-cut partitioning which provides a generalization of the ratio-cut cost metric to k-way partitioning and a lower bound on this cost metric. Our approach uses Lanczos algorithm to find the k smallest eigenvalue/eigenvector pairs of the Laplacian of the graph. The eigenvectors are used to construct an orthogonal projection to map a vertex (of the graph) in an n-dimensional space into a k-dimensional subspace. We exploit the near-orthogonality of the projected points to effect high quality clustering of points in a k-dimensional subspace. An efficient algorithm is presented for coercing the points in the k-dimensional subspace into k-partitions. Advancement over the current work is evidenced by the results of experiments on the standard MCNC benchmarks.

1 Introduction

Circuit partitioning is the process of dividing a large circuit into subcircuits. As the underlying algorithmic technique or merely as a means of handling problem complexity, circuit partitioning has applications in many aspects of systems design and design automation [23, chp.6] including placement, routing, logic synthesis, technology mapping, hardware simulation and emulation [31, 27]. Circuit partitioning techniques can be categorized as

1. 2-way partitioning or k-way partitioning,
2. iterative or spectral-based, and
3. graph or hypergraph-based.

To avoid unbalanced partition sizes while minimizing the connections between partitions, a common practice is to constrain the desired partition sizes. Cheng and Wei [6] have shown that the quality of 2-way partitions can be improved by incorporating the balancing of partition sizes in the cost metric, rather than imposing constraints on the partition sizes. They define the ratio-cut cost metric for a 2-way partition to be

$$\frac{E_c}{|P_1||P_2|}$$

where E_c is sum of the weights of edges cut, and $|P_1|$ and $|P_2|$ are the sizes of the two partitions [5, 22]. The relation between the second smallest eigenvalue/eigenvector pair of the Laplacian of a graph and the 2-way ratio-cut cost metric is the basis of the recent spectral 2-way partitioning algorithm of Hagen and Kahng [15]. Hagen and Kahng's basic approach handles two partitions at a time. Their method must be applied recursively to achieve k-way partitioning. Our new k-way method generalizes Hagen and Kahng 2-way spectral ratio-cut method to a completely global approach, and our experimental results compare favorably. In our approach, a network of n nodes is modeled by a weighted-edge graph of n vertices. A numerical algorithm based on the Scott-Parlett's implementation of the Lanczos algorithm is used to find the k smallest eigenvalue/eigenvector pairs of the Laplacian of the graph [25]. The eigenvectors are used to construct an orthogonal projection to map a point in an n-dimensional space into a k-dimensional subspace. The total cost of the projection, in the sense of sum of squared error, is minimized by the Lanczos algorithm. The near orthogonality of the projected points is used to effect clustering of points in a k-dimensional subspace. An efficient algorithm is presented for clustering the projected points into k-partitions. This paper reports three main results on partitioning which reveal the true nature of the relation between ratio-cut partitioning and spectral-based partitioning:

1. a generalization of 2-way ratio-cut cost metric to a k-way ratio-cut cost metric,

2. an efficient heuristic to effect a k-way partitioning based on the k-dimensional spectral embedding of a graph, and

3. a lower bound on the optimal k-way ratio-cut partitioning.

We define the partitioning problem and review previous approaches in Section 2. In Section 3, we generalize Hall's result to k dimensions using Barnes' assignment matrix to show that the sum of the smallest k eigenvalues of the Laplacian is a lower bound on $\sum_{h=1}^{k} \frac{E_h}{|P_h|}$, where E_h denotes the weight of the edges between nodes inside partition P_h and nodes outside the partition. This is the basis for a new method for extracting k-way partitions from the eigenvectors described in Section 4.

2 Formulation and Previous Approaches

The standard formulation of circuit partitioning is in terms of a hypergraph, $G = (\mathcal{V}, \mathcal{H})$ where the vertices, $\mathcal{V} = \{\nu_1, \nu_2, \ldots, \nu_n\}$, represent the modules and the hyperedges, H_j, correspond to the nets (notice that we use ν_i for vertices and v_i for eigenvectors). The task is to find a partition of the set of vertices $\mathcal{P} = \{P_1, P_2, P_3, \ldots P_k\}$ for a given k, which optimizes some cost criterion based on the hyperedges cut and/or the sizes of the partitions.

If all of the nets are two-terminal nets then G is a graph, and the problem reduces to graph partitioning. However, this is not generally the case, and the hypergraph is usually converted to a graph by representing each net by a set of edges. (A few iterative methods operate directly on the hypergraph.) A common practice is to expand a net of degree d (a hyperedge connecting d vertices) into a clique of $d(d-1)/2$ edges and assign weights to these edges so that a net with many vertices will not dominate the cost function. Researchers have chosen $w = \frac{1}{d}$ or $w = \frac{1}{d-1}$ with the rationale that the total number of connections needed in the worst case is $d-1$ and d respectively. Frankle proposed using $w = (\frac{2}{d})^{\frac{3}{2}}$ for the purpose of minimizing the worst-case deviation of cost from span^2 [12]. Vanelli et al, calculate edge weights based on the worst case scenario in which the vertices on a net are distributed evenly among the k equal sized partitions. A more complicated but promising technique involves partitioning the intersection graph of the nets [17, 7]. In this case an additional phase is required to resolve the net partitions into module partitions, however the additional potential sparsity in the intersection graph offers a speedup in the partitioning phase.

This paper focuses on the graph partitioning aspect of circuit partitioning. We assume that the circuit hypergraph has been reduced by the standard net clique expansion to a weighted graph, $G(\mathcal{V}, \mathcal{E})$ with n nodes, in which $w(e)$ is the weight of the edge e. More sophisticated net-models could easily be incorporated.

Iterative-based partitioning heuristics are local in nature. Given an initial 2-way partition of a graph, the primary technique used in iterative-based methods involves finding a sequence of moves or pairwise swaps of nodes between the two partitions. The primary representatives are the Kernighan-Lin, (Fiduccia-Mattheyses) algorithm [18, 10] and the simulated-annealing based approach [19, 29]. These methods are highly sensitive to the initial (randowm) 2-way partition; improvements have been made [6], which generally involve multiple trials with different initial 2-way partitions. An extension of the technique to multi-pin nets (hypergraphs) is in [21]. Sanchis adapted Krishnamurthy's method to k-way partitioning [28].

Spectral-based graph partitioning uses the eigenvalues/eigenvectors of graph matrices to extract global information about the structure of a graph. A numerically stable algorithm to solve a symmetric eigenvalue problem for large sparse matrices is required; the Lanczos algorithm is the method of choice. With the advent of high performance workstations with floating-point accelerators, this avenue of partitioning has become practical.

The relation between the properties of a graph and its *spectrum* (the eigenvalues/eigenvectors of its associated matrices) has been an area of active research for several years [8, 3, 11]. The spectrums of the adjacency matrix and Laplacian of a graph are the basis for both partitioning [1, 9, 14] and placement techniques [12, 30]. The *adjacency matrix* of G is the $n \times n$ matrix $A(G) = [a_{ij}]$ defined by,

$$a_{ij} = \text{the number of edges between } \nu_i \text{ and } \nu_j.$$

If G has no loops nor parallel edges, then all of the entries in $A(G)$ are 1's or 0's and there are 0's along the diagonal.
The *degree matrix* of G is the $n \times n$ matrix $D(G) = [d_{ij}]$ defined by,

$$d_{ij} = \begin{cases} degree(\nu_i) & \text{if } i = j \\ 0 & \text{if } i \neq j \end{cases}$$

For weighted graphs, the adjacency and degree matrices are generalized by:

$$a_{ij} = \text{the sum of the weights of all edges between } \nu_i \text{ and } \nu_j$$

and

$$d_{ij} = \begin{cases} \sum_{k=1}^{n} a_{ik} & \text{if } i = j \\ 0 & \text{if } i \neq j \end{cases}$$

The *Laplacian* (aka *Kirchoff matrix* aka *admittance matrix*) of G is the $n \times n$ symmetric matrix $Q(G) = D(G) - A(G)$. Since the rows (and columns) sum to 0, the Laplacian is singular; it has rank at most $n - 1$ and 0 as an eigenvalue. In fact, the multiplicity of the 0 eigenvalue is the number of connected components of G [8].

Donath and Hoffman [9] derived a lower bound on the weight of the edges cut (E_c) by a partition satisfying pre-determined partition sizes. If $m_1 \geq m_2 \geq \ldots \geq m_k$ are the given partition sizes and $\lambda_1 \leq \lambda_2 \leq \ldots \leq \lambda_k$ are the smallest k eigenvalues of the Laplacian, then

$$E_c \geq \frac{1}{2} \sum_{i=1}^{k} \lambda_i m_i.$$

Barnes [1] formulates the k-way partitioning problem in terms of an *assignment* matrix. A k-way partition of the vertices of G, $\mathcal{P} = \{P_1, P_2, \ldots P_k\}$ is represented by an $n \times k$ *assignment matrix* $Y(\mathcal{P}) = [y_{ih}]$ where

$$y_{ih} = \begin{cases} 1 & \text{if } \nu_i \in P_h \\ 0 & \text{if } \nu_i \notin P_h \end{cases} \tag{1}$$

A k-partition of G, $\mathcal{P} = \{P_1, P_2, \ldots P_k\}$ can also be represented by the $n \times n$ *partition matrix* [1]. $P = [p_{ij}]$ where

$$p_{ij} = \begin{cases} 1 & \text{if } \nu_i \text{ and } \nu_j \text{ are in the same partition} \\ 0 & \text{otherwise} \end{cases}$$

For a given partition $\mathcal{P} = \{P_1, P_2, \ldots P_k\}$ we have that $P = YY^T$.

Given pre-determined partition sizes, $m_1, m_2, \ldots, m_k$ ($|P_i| = m_i$), Barnes solves the partitioning problem by finding an approximation of the adjacency matrix A by a partition matrix P with respect to the Frobenius norm. The problem involves finding the first k eigenvectors of A followed by solving a *transportation problem* [1, 2].

Researchers have incorporated these approaches into partitioners [14, 9, 26]. For example, [14] uses Barnes' method to find initial partitions (of pre-determined sizes) and then an iterative method to improve these partitions. Recently, Hagen and Kahng [15] have used Hall's weighted quadratic placement formulation to derive a lower bound on ratio-cut partitioning from the second eigenvalue of the Laplacian.

In Hall's one-dimensional weighted quadratic placement formulation [16], the total of weighted squared distance between n points is expressed as:

$$z = \frac{1}{2}\sum_{i=1}^{n}\sum_{j=1}^{n} a_{ij}(x_i - x_j)^2. \tag{2}$$

Here, x_i is the coordinate assigned to ν_i in a one-dimensional space. The constraint $\sum_{i=1}^{n} {x_i}^2 = 1$ is imposed to avoid the trivial solution in which all x_i's are equal (the vertices are all assigned the same coordinate). If the solution is expressed as a column vector $x = [x_i]$, then the one-dimensional minimization problem (2) can be rewritten in matrix notation (quadratic form) as,

$$\text{minimize } z = x^T Q x \text{ subject to } x^T \cdot x = 1.$$

This constrained minimization problem can be solved by the Lagrange multiplier method with multiplier β and the Lagrangian $x^T Q x - \beta(x^T x - 1)$. Solving for x yields,

$$Qx = \beta x \tag{3}$$

which is the symmetric eigenvalue problem whose solutions are the eigenvectors/eigenvalues of Q. A fundamental result in matrix theory is that solving (3) yields real eigenvalues and the corresponding eigenvectors are orthonormal since the Laplacian Q is symmetric and real [13, pg. 410]. That is

$${v_i}^T \cdot v_j = 0 \text{ for } 1 \le i < j \le n \text{ and } {v_i}^T \cdot v_i = 1 \text{ for } 1 \le i \le n. \tag{4}$$

For any eigenvector v, of Q with corresponding eigenvalue, λ, we have

$$v^T Q v = v^T \lambda v = v^T v \lambda = \lambda. \tag{5}$$

The quality of the solution v is dictated by the eigenvalue, λ. There are n possible eigenvalues for a system of n linear equations, $\lambda_1 \le \lambda_2 \le \ldots\lambda_n$. For a connected graph, the Laplacian has rank of $n - 1$, and the minimum eigenvalue $\lambda_1 = 0$ which yields the trivial solution $v_1 = (\frac{1}{\sqrt{n}}, \frac{1}{\sqrt{n}}, \ldots, \frac{1}{\sqrt{n}})$. The second smallest eigenvalue, λ_2, is a lower bound on a non-trivial solution to (2). In his paper, Hall suggested a k-dimensional generalization in

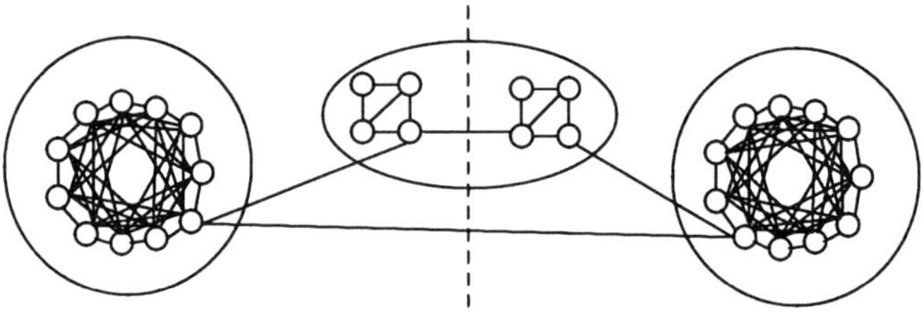

Figure 1: The optimal minimum-width bisection or ratio-cut separates the two four-node subgraphs, preventing the optimal 3-way partition.

which the eigenvectors are used as the basis for geometric clustering [16]. In the next section we will re-examine the generalization to k dimensions to motivate the use of the cosine of the angle between k-dimensional embedding of the vertices rather than their proximity. This is the basis of the method presented in Section 4.

Recently, Hagen and Kahng [15] established a connection between Hall's formulation and 2-way ratio-cut partitioning. Given a 2-way partition, $\mathcal{P} = \{P_1, P_2\}$, they construct a vector (The actual construction in the Hagen and Kahng's paper uses Rayleigh's Principle). $x(\mathcal{P})$

$$x_i(\mathcal{P}) = \begin{cases} \sqrt{\frac{|P_2|}{n|P_1|}} & \text{if } \nu_i \in P_1 \\ -\sqrt{\frac{|P_1|}{n|P_2|}} & \text{if } \nu_i \in P_2 \end{cases}$$

which satisfies the constraint: $x(\mathcal{P})^T x(\mathcal{P}) = 1$. The term $a_{ij}(x_i(\mathcal{P}) - x_j(\mathcal{P}))^2$ of (2) is zero when ν_i and ν_j are in the same partition and $\frac{n \cdot a_{ij}}{|P_1| \cdot |P_2|}$ when they are in different partitions. Hence

$$x(\mathcal{P})^T Q\, x(\mathcal{P}) = \frac{n \cdot E_c}{|P_1| \cdot |P_2|} \tag{6}$$

from which a lower bound on the ratio-cut cost metric is obtained[15, 26]:

$$\frac{\lambda_2}{n} \leq \frac{E_c}{|P_1| \cdot |P_2|} \tag{7}$$

Hagen and Kahng construct a 2-way partition from the corresponding eigenvector of λ_2 by sorting the eigenvector and identifying a cut in the sorted eigenvector which yields the best ratio-cut metric. This heuristic achieves 17% improvement over the iterative-based ratio-cut 2-way partitioning on the MCNC benchmarks [20].

Any spectral 2-way algorithm can be recursively applied to achieve k-way partitioning, however this requires the reconstruction of the Laplacian and an eigenvalue/eigenvector computation for each round of 2-way partitioning. Moreover, this can entail a loss of solution quality, as noted in [23]. For example in the graph shown in Figure 1, the optimal 2-way partition (dashed line) precludes the optimal 3-way partition (ellipses). In some sense, the algorithm only uses a fraction of the global information embedded in the eigenvectors.

3 Spectral k-way ratio-cut partitioning

This section presents a *new* spectral k-way ratio-cut partitioning formulation which generalizes Hagen and Kahng's [15] spectral ratio-cut 2-way partitioning to k-way. We begin by generalizing Hall's one-dimensional weighted quadratic placement optimization to k dimensions. The k-way partitioning problem is modeled as the problem of finding an orthogonal projection of n-dimensional space onto a k-dimensional subspace, which maps the n canonical-vector basis (the n nodes of a graph) into k points (the partitions) to minimize the weighted quadratic placement. A connection is established between the new ratio-cut metric and the generalized Hall's weighted quadratic placement formulation to provide a lower bound on the cost metric.

Our formulation is an advancement over the spectral 2-way partitioning for several reasons [15, 26]. First, all k partitions are found simultaneously; it is therefore using all the global information available in the eigenvectors. Second, the orthonormality among the eigenvectors forms the basis of our heuristics to coerce the placement into partitions. This method is therefore taking the full advantage of the symmetric eigenvalue solver: the Lanczos algorithm.

3.1 Generalized Weighted Quadratic Placement

Let X be an $n \times k$ $(1 \leq k \leq n)$ matrix $X = [x_{ij}] = [x_1, ..., x_k]$. The sum of the weighted quadratic placement cost, z, is:

$$z = \frac{1}{2} \sum_{i=1}^{n} \sum_{j=1}^{n} \sum_{h=1}^{k} a_{ij}(x_{ih} - x_{jh})^2. \tag{8}$$

The following lemma establishes the connection between the weighted quadratic placement, z, and the quadratic form $X^T Q(G) X$.

Lemma 1 *If X is an $n \times k$ matrix then $X^T Q(G) X$ is a $k \times k$ matrix whose gh^{th} component is*

$$\frac{1}{2} \sum_{i=1}^{n} \sum_{j=1}^{n} a_{ij}(x_{ig} - x_{jg})(x_{ih} - x_{jh}). \tag{9}$$

Proof: Since $X^T Q(G) X = X^T (D(G) - A(G)) X = X^T D(G) X - X^T A(G) X$ the gh^{th} component of $X^T Q(G) X$ is

$$\sum_{i=1}^{n} x_{ig} d_{ii} x_{ih} \quad - \quad \sum_{i=1}^{n} x_{ig} \sum_{j=1}^{n} a_{ij} x_{jh} = \sum_{i=1}^{n} x_{ig} x_{ih} (\sum_{j=1}^{n} a_{ij}) - \sum_{i=1}^{n} x_{ig} (\sum_{j=1}^{n} a_{ij} x_{jh})$$

$$= \frac{1}{2}\sum_{i=1}^{n}\sum_{j=1}^{n} a_{ij}(x_{ig}-x_{jg})(x_{ih}-x_{jh})$$

Hence we have $z = \frac{1}{2}\sum_{i=1}^{n}\sum_{j=1}^{n}\sum_{h=1}^{k} a_{ij}(x_{ih}-x_{jh})^2 = trace(X^TQX)$. (The $trace(M)$ of a square matrix M is the sum of its diagonal entries.)

Theorem 1 *The $n \times k$ matrix which minimizes the $trace(X^TQ(G)X)$ subject to the constraint $X^TX = I_k$, is the $n\times k$ matrix whose k columns consist of the k eigenvectors of Q corresponding to the k smallest eigenvalues of Q.*

Proof: If $V = [v_1, ..., v_k]$ is formed with any k eigenvectors of Q, then $QV = V\Lambda_k$ where Λ_k is the diagonal matrix formed with the k eigenvalues corresponding to the k eigenvectors in V. This means that

$$V^TQV = V^TV\Lambda_k = I_k\Lambda_k = \Lambda_k$$

and $trace(V^TQV) = trace(\Lambda_k)$ which is the sum of the eigenvalues. To minimize $trace(V^TQV)$, the k eigenvectors corresponding to the k smallest eigenvalues should be selected. It remains to show that there exists no matrix X satisfying $X^TX = I_k$ with $trace(X^TQX) < trace(\Lambda_k)$, where Λ_k is formed with the k smallest eigenvalues. Suppose there is such an X. Let k be the smallest integer for which such an X exists. Since Q is positive semi-definite, i.e., $x^TQx \geq 0 = \lambda_1$ for all x, k must be greater than 1. Let y be the vector with $||y|| = 1$ in the subspace spanned by X which maximizes y^TQy. We can find an orthonormal basis for this subspace, B, which has y as one of its vectors. Since X and B span the same subspace and $B^TB = I_k$, there exists a $k \times k$ matrix S satisfying $B = XS$ and $S^{-1} = S^T$. Since similar matrices have the same eigenvalues and $B^TQB = S^{-1}(X^TQX)S$ and X^TQX are similar matrices, we have

$$trace(B^TQB) = trace(X^TQX) < trace(\Lambda_k).$$

Let B_* be the $n \times (k-1)$ matrix obtained by removing the column of B corresponding to y. Then, we must have $trace(B_*^TQB_*) \geq trace(\Lambda_{k-1})$ since k was the smallest integer for which this inequality did not hold. Then

$$y^TQy = trace(B^TQB) - trace(B_*^TQB_*) < trace(\Lambda_k) - trace(\Lambda_{k-1}) = \lambda_k.$$

But this implies that

$$\max_{\substack{x\in X\\ ||x||=1}} x^TQx = \max_{\substack{x\in X\\ x\neq 0}} \frac{x^TQx}{x^Tx} = y^TQy < \lambda_k$$

which violates the Courant-Fisher Theorem [32, pg. 101] which states that the kth smallest eigenvalue,

$$\lambda_k = \min_{dim(E)=k}\ \max_{0\neq y\in E} \frac{y^TQy}{y^Ty}.$$

This implies that all matrices X with $X^T X = I_k$ have $trace(X^T Q X) \geq trace(\Lambda_k)$. □

Hence the smallest k eigenvectors of Q, V, provide the solution to:

$$\text{minimize } trace(X^T Q X) \quad \text{subject to } X^T \cdot X = I_k. \tag{10}$$

3.2 Relation to Spectral k-Way Partitioning

If we use the assignment matrix Y defined by Barnes in (1) for X in the generalized Hall quadratic placement formula we obtain the Laplacian of the graph in which the partitions are collapsed into vertices. That is, given a weighted graph G and a partition $\mathcal{P}$, the *partition graph* $G_{\mathcal{P}}$ is the graph with k vertices $u_1, u_2, \ldots, u_k$ where the weight of the edge between u_g and u_h for $g \neq h$ is

$$\sum_{\nu_i \in P_g} \sum_{\nu_j \in P_h} a_{ij} = \sum_{i=1}^{n} \sum_{j=1}^{n} a_{ij} y_{ig} y_{jh}, \tag{11}$$

The partition graph $G_{\mathcal{P}}$ is the graph whose nodes are the partitions and whose interconnections are inherited from G.

That $Y^T Q(G) Y = Q(G_{\mathcal{P}})$ is the Laplacian of $G_{\mathcal{P}}$ can be seen by evaluating the gh^{th} entry of $Y^T Q Y$ whose formula is given in (9). The terms in the gth diagonal entry of $Y^T Q(G) Y$:

$$\frac{1}{2} \sum_{i=1}^{n} \sum_{j=1}^{n} a_{ij} (y_{ig} - y_{jg})^2.$$

are a_{ij} when exactly one of ν_i and ν_j is in P_g and 0 otherwise. Hence the sum is twice the total weight of edges from nodes in P_g to nodes not in P_g. The terms of gh^{th} entry of $Y^T Q(G) Y$:

$$\frac{1}{2} \sum_{i=1}^{n} \sum_{j=1}^{n} a_{ij} (y_{ig} - y_{jg})(y_{ih} - y_{jh}).$$

are $-a_{ij}$ exactly when one of ν_i or ν_j is in P_g and the other is in P_h, and 0 otherwise. Thus summing over all i and j gives minus twice the number of edges between vertices in partition P_g and vertices in partition P_h.

Because the Laplacian is a symmetric matrix, the sum of its eigenvalues is its trace and so by the handshaking lemma we obtain as in [24]:

$$\sum_{i=1}^{n} \lambda_i = trace(Q) = \sum_{i=1} d_{ii} = 2 \sum_{e \in E(G)} w(e).$$

We have established that $trace(Y^T Q Y)$ is *twice* the total weight of the edges in the partition graph $G_{\mathcal{P}}$. Thus finding an assignment matrix Y which minimizes $trace(Y^T Q Y)$ will minimize the weight of the edges cut in $\mathcal{P}$. However, we still require a constraint on Y, not only to ensure that it represents a partition, but also to balance the partition sizes. The approach taken by Barnes, Donath and Hoffman was to solve for Y given k and given $|P_1|, |P_2|, \ldots, |P_k|$, the sizes of the partitions [9, 1, 2]. Instead we reformulate the assignment matrix so that it meets the orthonormality constraint in (10). This allows us not only to use Hall's quadratic placement optimization, but also incorporate the size of the partitions into the cost metric instead of constraining the partition sizes.

Specifically, a k-way partition of the nodes of G, $\mathcal{P} = \{P_1, P_2, \ldots P_k\}$ is represented by an $n \times k$ *ratioed assignment matrix* $R = [r_{ih}]$ where

$$r_{ih} = \begin{cases} \frac{1}{\sqrt{|P_h|}} & \text{if } \nu_i \in P_h \\ 0 & \text{if } \nu_i \notin P_h \end{cases} \tag{12}$$

The rows no longer necessarily sum to 1, and column h now sums to $\sqrt{|P_h|}$. The related $n \times n$ ratioed partition matrix $P_R = [rp_{ij}]$ is defined by

$$rp_{ij} = \begin{cases} \frac{1}{|P_g|} & \text{if } \nu_i \text{ and } \nu_j \text{ both belong to } P_g \\ 0 & \text{otherwise} \end{cases}$$

The following properties of R and P_R will be needed both to derive the lower bound on k-way ratio-cut cost metric, and to extract the partitions from the k eigenvectors. They follow directly from the definitions of R and P_R. Proofs are in [4].

Lemma 2 *If R is the ratioed assignment matrix for $\mathcal{P} = \{P_1, P_2, \ldots, P_k\}$ then*

1. *The gth diagonal entry of $R^T Q(G) R$ is $\frac{E_g}{|P_g|}$.*

2. *The gh^{th} off-diagonal entry is minus the total weight of the edges between P_g and P_h divided by $\sqrt{|P_g| \cdot |P_h|}$.*

3. *$trace(R^T Q R) = \sum_{h=1}^{k} \frac{E_h}{|P_h|}$.*

4. *If P_R is the ratioed partition matrix, then $RR^T = P_R$.*

5. *$R^T R = I_k$.*

Since the ratioed partition matrix meets the constraint $R^T R = I_k$ in (10), we are now in the position to establish a connection between spectral k-way ratio-cut partitioning and (Hall's) generalized weighted quadratic placement formulation.

Theorem 2 *The sum of the smallest k eigenvalues of the Laplacian Q of a weighted graph G is a lower bound on $\sum_{h=1}^{k} \frac{E_h}{|P_h|}$ for any k-way partition of G.*

Proof: Let Π be the set of all k-way partitions of a graph G and let $R_{\mathcal{P}}$ denote the ratioed assignment matrix associated with a particular $\mathcal{P} \in \Pi$. Since by Lemma 2 we know that $R_{\mathcal{P}}$ satisfies the constraint on X in (10),

$$\sum_{i=1}^{k} \lambda_i = \min_{X^T X = I} trace(X^T Q X) \leq \min_{\mathcal{P} \in \Pi} trace(R_{\mathcal{P}}^T Q R_{\mathcal{P}}) = \min_{\mathcal{P} \in \Pi} \sum_{h=1}^{k} \frac{E_h}{|P_h|}. \tag{13}$$

3.3 A k-way ratio-cut cost metric

We define a generalized k-way ratio-cut cost metric of a k-way partition $\mathcal{P}$ of graph G to be

$$cost(\mathcal{P}) = \sum_{h=1}^{k} \frac{E_h}{|P_h|}. \tag{14}$$

(Recall that E_h is the total weight of the edges in G having exactly one endpoint in P_h.) For $k = 2$, if E_c is the weight of the edges cut in $\mathcal{P} = \{P_1, P_2\}$, then we have $E_c = E_1 = E_2$ and

$$cost(\mathcal{P}) = \frac{E_1}{|P_1|} + \frac{E_2}{|P_2|} = \frac{E_c}{|P_1|} + \frac{E_c}{|P_2|} = E_c \frac{|P_1| + |P_2|}{|P_1| \cdot |P_2|} = n \frac{E_c}{|P_1| \cdot |P_2|}. \tag{15}$$

Thus $cost(\mathcal{P})$ is a generalization from 2 to k partitions of the ratio-cut cost metric defined by Wei and Cheng after scaling by n (the number of nodes in the graph)[6].
For $k = 2$, since $\lambda_1 + \lambda_2 = 0 + \lambda_2 = \lambda_2$, from (13) we obtain

$$\lambda_2 \leq cost(\mathcal{P}) = \frac{E_c}{|P_1|} + \frac{E_c}{|P_2|} = n \frac{E_c}{|P_1| \cdot |P_2|}$$

which is equivalent to the lower bound (7) derived by Hagen and Kahng[15] $\frac{\lambda_2}{n} \leq \frac{E_c}{|P_1| \cdot |P_2|}$. We define a scaled cost of (14) to conform with the classical ratio-cut metric

$$scost(\mathcal{P}) = \frac{1}{n(k-1)} cost(\mathcal{P}) \tag{16}$$

The analysis of [6] suggested the merit of having a cost function which is constant for random graphs; all cuts should be invariant for random graph. For a random graph with edge probability f, and any k-way partition $\mathcal{P}$, our scaled cost metric yields

$$\begin{aligned} scost(\mathcal{P}) &= \frac{1}{n(k-1)} cost(\mathcal{P}) = \frac{1}{n(k-1)} \sum_{h=1}^{k} \frac{f|P_h|(n - |P_h|)}{|P_h|} \\ &= \frac{1}{n(k-1)} (fkn - fn) = f. \end{aligned}$$

Recently Yeh, Cheng and Ling proposed a different generalization of the ratio-cut cost metric from 2 to k partitions [33]:

$$\frac{\frac{1}{2}\sum_{h=1}^{k} E_h}{\sum_{h=1}^{k-1}\sum_{g=h+1}^{k} |P_h| \cdot |P_g|}. \tag{17}$$

Both generalizations (16) and (17) are constant for complete graphs with random edge weights. Both reward partitions with reduced cut weights and balanced partition sizes. However, they differ according to the distribution of the cut edges among the partitions. Cost metric (16) favors partitions in which the smaller sized partitions have smaller weight cuts, while cost metric (17) only varies according to the total weight of the edges cut, regardless of their distribution.

Another important issue is the selection of the proper (natural) value of k. Cost metric (16) increases with k, and therefore is not a good measure to indicate the proper choice of k. The lower bound in (13) predicts the proper choice of k based on the lowest value of

$$\frac{1}{n \cdot k \cdot (k-1)} \sum_{i=1}^{k} \lambda_i. \tag{18}$$

for "structured" graphs.

4 Extracting the partitions from a geometrical interpretation of the eigenvectors

Theorem 2 justifies the approximation of R by the first k eigenvectors $V = [v_1, ...v_k]$. Our approach to k-way partitioning is to "reverse engineer" the partitions R from these eigenvectors V. For this purpose a geometrical interpretation of R and V will provide some insight.

We associate the n vertices of the graph G with the n canonical vectors of $\mathbf{R}^n$, i.e., node ν_i is represented as a vector having a '1' in the ith position and '0's elsewhere. An ordered collection of these canonical vectors makes up the $n \times n$ identity matrix I_n. The $n \times n$ matrix $P_R = RR^T$, the ratioed partition matrix is an *orthogonal projector*, since $P_R{}^2 = P_R$ [13, pg. 75]. The significance of the orthogonal projector is that nodes belonging to the same partition are mapped to the same point by P_R. The orthogonal projector P_R completely characterizes the k-way partitioning.

Each row of R may be thought of as a vector in a k-dimensional subspace $\mathbf{R}^k$ representing a graph vertex (see Fig. 2). If V is an approximation to R, then an approximation to P_R is

$$Z = VV^T \tag{19}$$

since by Lemma 2 $P_R = RR^T$. Each z_{ij} element in Z is the dot product of the ith and j^{th} *rows* of V. If $V^T = [v'_1, ..., v'_n]$, then $z_{ij} = v'^T_i \cdot v'_j$. Z is the *orthogonal projection* onto the k-dimensional subspace spanned by V.

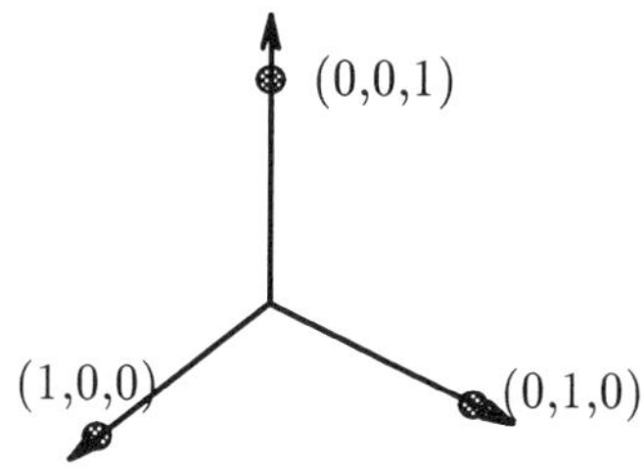

Figure 2: Representation of 3 nodes as canonical vectors

When ν_i and ν_j are in the same partition, P_h, the ij^{th} entry of P_R is $\frac{1}{|P_h|}$ while it is 0 when they are in different partitions. In contrast the ij^{th} entry of P is either 1 or 0. For this reason recognizing the partitions from an approximation to P rather than P_R is easier. To this end we exploit the following relation between P and P_R.

Given an $n \times n$ matrix M, let $N(M)$ be the $n \times n$ diagonal matrix where n_{ii} is the reciprocal of the norm of row i of M. Then we have $P = N(R) \cdot RR^T \cdot N(R) = N(P_R) \cdot P_R$. Performing the same normalization on Z gives us: $\hat{P} = N(V) \cdot VV^T \cdot N(V)$. In this case $N(V)$ is a diagonal matrix with

$$n_{ii} = \frac{1}{\sqrt{\sum_{h=1}^{k} {v'_{ih}}^2}} \text{ and the } ij^{th} \text{ entry of } \hat{P} \text{ is } \frac{{v'_i}^T . v'_j}{||v'_i|| \cdot ||v'_j||}.$$

which is the cosine of the angle between the two row vectors i and j of V (or the column vectors of V^T). Hence, the elements in $\hat{P}$ are the cosines of the angles between vectors v'_i and v'_j. These directional cosines provide a measure of how far each vertex is from other vertices. The geometric interpretation of $\hat{P}$ is that two vertices belong to the same partition, when the rows in Y which represent the k-space coordinates of the vertices are the same, and so the cosine of the angle between their rows in R is close to one. When two vertices belong to different partitions, the rows of the Y matrix are orthogonal, so the cosine of the angle between the two vectors tends towards zero.

Since it is expensive to compute all n^2 entries of $\hat{P}$ for graphs with $\approx 10^3$ nodes, a heuristic will be presented in the next section which computes the directional cosines on an as-needed basis; only a small fraction of the elements in $\hat{P}$ are actually computed.

4.1 A time and space efficient clustering heuristic

In this section we develop a clustering heuristic to form partitions using the directional cosines without calculating the entire $\hat{P}$ matrix. The directional cosines, the cosine of the angle between rows i and j of the V matrix $\cos(v'_i, v'_j)$ are computed only when needed. The goal is to "reverse engineer"

```
KP
 Repeat 4 times {
   Step 1.  Generate k prototypes.
            First iteration:  1st prototype = largest magnitude,
            other prototypes = most orthogonal wrt existing prototypes.
            Subsequent iterations:  prototypes = average of each partition.
   Step 2.  Examine each vertex and each prototype p_k, and assign it
            to p_k's partition if cos(v,p_k) ≤ cos(π/8)
   Step 3.  Build heap H with remaining unassigned vertices
            ordered by decreasing target cuts / degree
            while (H ≠ ∅)
                 Remove vertex at top of H and assign it to the partition
                      with which it has the most hyperedges (target).
                 For each partition, compute the cost if H
                      were merged with it and record if best so far.
            endwhile
 }
```

Figure 3: Pseudo-code for Clustering in **KP**.

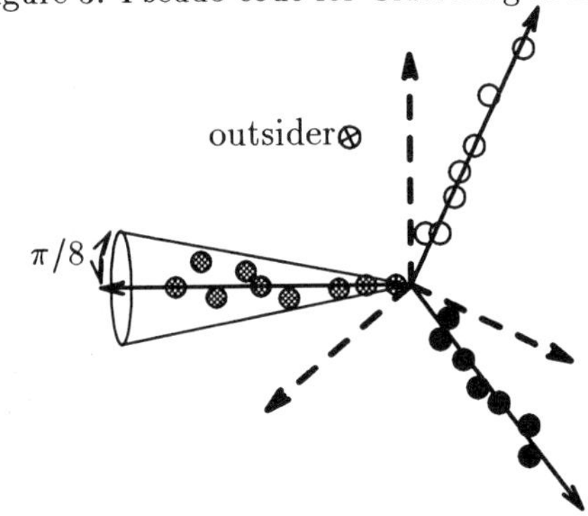

Figure 4: Representation of an outsider, a framework, and base vectors

the partitions under the premise that $\cos(v_i', v_j')$ is close to 1 when vertices ν_i and ν_j are in the same partition and 0 otherwise. The pseudo-code for the clustering heuristic in **KP** is given in Figure 3.

The first step is to generate a framework of k vertices whose row vectors are as orthogonal with respect to each other as possible. These vertices are the *prototypes* of their respective partitions. In Fig. 4 the framework is depicted by the solid lines. In the first iteration, the k prototypes are selected by examining the vertices in decreasing magnitude of their row vectors and selecting vertices if their row vector is (near) orthogonal to the row vectors of all previously selected vertices, until k vertices are obtained. In successive iterations, the average of all vertices in a partition is computed and the vertex of the partition closest to this average is selected as the partition prototype. Each cosine computation takes $O(k)$ time and so the prototypes are found in $O(nk^2)$ time in either case.

In the second step all vertices whose directional cosine is within $\cos(\pi/8)$ of a prototype are added to the partition corresponding to that prototype. This step is also accomplished in $O(nk^2)$ time.

In the third step, the remaining vertices (outsiders) are considered one at a time and merged into the partitions. When the graph does not have a "natural" k-way partition, the cost of the result depends largely on how the outsiders are merged into existing partitions. The number of hyperedge cuts from each outsider to each existing partition is calculated, and the partition with the largest number of cuts is chosen as the target partition for that vertex. All outsiders are placed on a heap and the number of hyperedge cuts with its target divided by hyperedge degree of the vertex is used as the key for the heap. The vertex at the top of the heap is removed, and placed into its target partition, and all of its neighbors still in the heap have their hyperedge cuts recalculated, and their keys adjusted. This is repeated until all outsiders have been assigned to a partition.

One problem with this method is that the outsiders may share more edges among themselves than with any of the existing partitions. To allow for this possibility, after each assignment of an outsider, we also consider merging *en masse* the entire set of remaining outsiders with each of the existing partitions. By incrementally tabulating the necessary information, the ratio-cut costs of these *en masse* mergings may be efficiently computed in $O(bk)$ time without actually performing them, where b is the maximum degree of a vertex. The result with the best ratio-cut cost is recorded, and then the algorithm proceeds to consider the next outsider at the top of the heap. The overall time for this entire third step is $O(n(bk^2 + bk\log(n)))$. This clustering algorithm uses $O(1)$ *additional* space per vertex, so summing over all vertices, we need only $O(n)$ additional space. The total space is dominated by the size of the eigenvector V matrix, which is $O(nk)$ space.

5 Experimental results

To evaluate the effectiveness of our k-way partitioning algorithm. We compare our results with successive spectral 2-way partitioning. We reimplemented Hagen and Kahng's spectral 2-way partitioning algorithm with the EIG1 heuristic (which evaluates and retains the best partitioning of the vertices at *every* point in their ordering by the second eigenvector) [15]. We refer to this algorithm as **SB**. We call our k-way spectral ratio-cut partitioning algorithm **KP**. Our implementation of the **SB** method first performs a 2-way partition. As in [15], if the desired number of partitions has not been obtained, it finds the largest existing partition and 2-way partitions the subgraph formed by that partition. This continues until the desired number of partitions is reached.

We ran both algorithms on the graphs derived from the MCNC layout synthesis benchmarks. The hypergraphs of the benchmarks were preprocessed to remove high fanout nets of degree greater than 99, and the pruned hypergraphs were then clique-expanded into graphs with weight of $\frac{1}{d}$ assigned to edges. However, these high-fanout nets are considered in the final ratio-cut metric. The SB and KP programs were tested on the MCNC benchmarks (`p1, p2_ga, t2, t3, t4, t5`) using k ranging from 2 to 25. We compare the hyperedge cuts, the scaled ratio-cut cost (16), and the run time in seconds of SB and KP on `p1` in Table 1. Due to space limitation, only a summary of the experimental results is given in Table 2, please refer to [34] for details. Overall, KP improves the number of hyperedges cut by 6%, and the ratio-cut metric by 12%. K-way partitioning algorithm (KP) appears to be very slow for $k = 2$ because it computes all 25 eigenvalue/eigenvector pairs when it starts up. Subsequent clusterings of the eigenvectors for different k's reuse the eigenvectors. This is in contrast to successive bipartitioning (SB) which requires an eigenvector computation for every value of k. Also, in Table 1 we are reporting the run time for SB as k gradually increases. So the total run time for say $k = 5$ should be the cumulative run time for k=2 to 5.

k	SB hyper-edge cut	KP hyper-edge cut	Imprv (%)	SB ratio Cut ($\times 10^{-5}$)	KP ratio-cut ($\times 10^{-5}$)	Imprv (%)	SB time(s)	KP time(s)
2	15	14	7.1	14.6413	13.4564	8.8	32.08	134.17
3	35	17	105.9	27.3390	15.8998	71.9	31.00	1.57
4	59	20	195	36.5668	25.7352	42.1	27.00	2.52
5	75	43	74.4	41.0133	25.8631	58.6	27.00	4.38
6	78	95	-17.9	37.0844	36.7783	0.8	23.86	5.08
7	88	84	4.8	40.5960	32.1328	26.3	23.37	3.91
8	129	112	15.2	44.5272	51.0548	-12.8	22.57	4.67
9	131	87	50.6	48.7913	44.9951	8.4	12.42	5.09
10	140	128	9.4	53.5623	45.0939	18.8	10.78	6.20
11	145	103	40.8	60.4062	49.8164	21.3	16.73	6.54
12	147	126	16.7	64.8198	48.3830	34.0	15.09	7.72
13	180	126	42.9	69.6391	47.6779	46.1	14.89	6.90
14	181	173	4.6	65.4507	50.9832	28.4	9.90	7.39
15	181	139	30.2	76.0641	55.6523	36.7	9.16	9.37
16	183	140	30.7	81.0057	56.3116	43.9	9.84	9.56
17	195	153	27.5	85.9584	53.3230	61.2	10.87	9.83
18	201	154	30.5	90.9971	54.4199	67.2	9.14	11.42
19	204	139	46.8	93.1573	64.9672	43.4	9.46	12.72
20	216	191	13.1	94.8638	59.0238	60.7	8.95	12.41
21	218	189	15.3	99.1461	58.4123	69.7	11.07	12.58
22	220	191	15.2	96.1021	65.9409	45.7	9.95	13.80
23	222	200	11.0	93.2668	70.8048	31.7	9.08	14.59
24	224	220	1.8	91.5474	68.6913	33.3	9.07	15.15
25	230	204	12.7	93.1405	77.4282	20.3	10.09	16.29
Avg.			33			36		
Total							373	334

Table 1: Comparison of the hyperedges-cut and ratio-cut *SCOST* of SB and KP on graph `p1`

The effectiveness of both partitioning algorithms can be demonstrated visually. Figs. 5 and 6 plot the embedding of the vertices implied by eigen-

Graph	Hcut Improv (%)	Hratio Improv (%)
p1	33	36
p2_ga	0	0
t2	29	16
t3	-12	11
t4	-8	29
t5	-22	-4
t6	21	-2
AVERAGE	5.7	12.3

Table 2: A summary of hyperedges-cut and ratio-cut improvements of k-way partitioning (KP) over successive bipartitioning (SB)

vectors v_2' and v_3' of the graph of p1. Each dot in the diagram corresponds to a vertex in the graph. This wishbone-like plot in Fig. 5 has three distinctive components. The three partitions are identified by our KP clustering algorithm and are marked as "$\diamond$", "+", and "$\square$" respectively. Fig. 6 shows the 3-way partitioning of the same graph by SB. For this benchmark, it is evident that SB produces inferior partitions to those produced by KP due to a lack of global perspective. SB dissected the "wishbone" into three pieces. However, SB does produce better results on graphs that are less structured (e.g. t5), because SB's EIG1 heuristic can afford to evaluate many different possible partitions.

6 Conclusion

We have developed a general theory of spectral ratio-cut k-way partitioning. Our theory provides the backbone for efficient heuristics to find k-way partitions of a graph based on the eigenvectors of the Laplacian of the graph. We have presented some experimental results to support our development of the theory and heuristics. Our method compares favorably to the results obtained by recursive spectral 2-way partitioning.

Acknowledgement

Pak K. Chan is supported in part by the National Science Foundation Grant MIP-9196276, Martine Schlag is supported in part by the National Science Foundation Presidential Young Investigator Grant MIP-8896276. The authors also acknowledge the support of Xilinx, Inc.

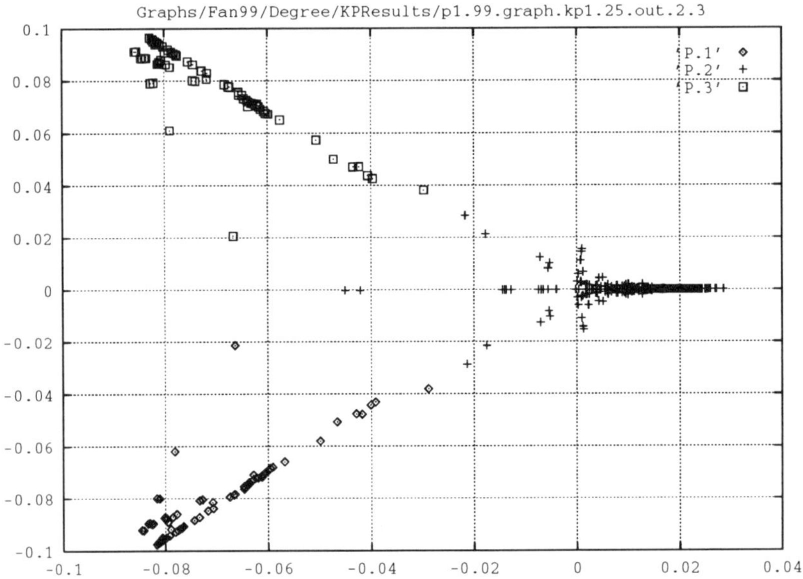

Figure 5: Embedding of nodes given by v_2' v_3' of benchmark **p1** with "◇", "+", and "□" indicating the three partitions obtained by K-way partitioning method KP

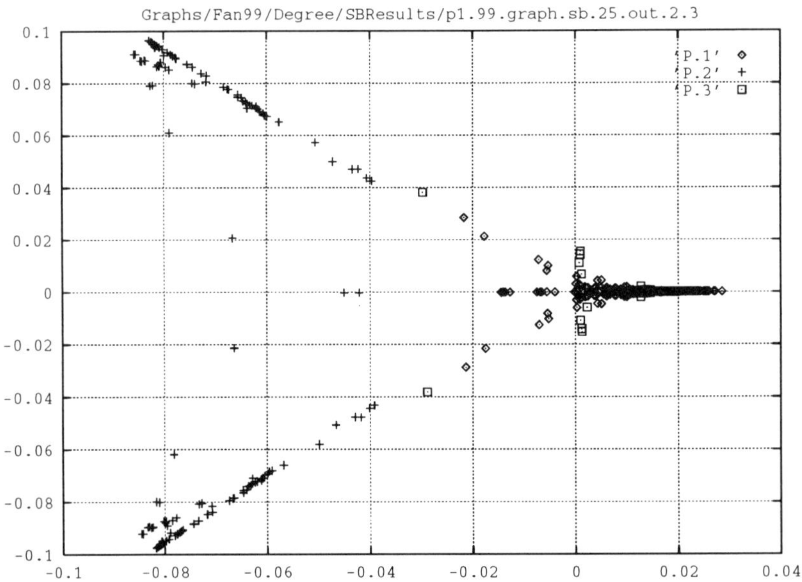

Figure 6: Embedding of nodes given by v_2' v_3' of benchmark **p1** with "◇", "+", and "□" indicating the three partitions obtained by successive bipartitioning method SB

References

[1] E. R. Barnes. An algorithm for partitioning the nodes of a graph. *SIAM Journal on Algorithm and Discrete Method*, 3(4):541–550, Dec. 1982.

[2] E. R. Barnes. Partitioning the nodes of a graph. In *Proceedings of Graph Theory with Applications to Algorithms and Computer Science*, pages 57–72, Wiley, 1985.

[3] R. Boppana. Eigenvalues and graph bisection: An average-case analysis. In *Proc. IEEE Symp. on Foundations of Computer Science*, pages 280–285, 1987.

[4] P. K. Chan, M. Schlag, and J. Zien. Spectral K-way ratio-cut partitioning. Technical Report UCSC-CRL-92-19, University of California, Santa Cruz, May 1992.

[5] C. K. Cheng and T. C. Hu. The optimal partitioning of networks. Technical Report CS89-146, University of California, San Diego, March 1989.

[6] C.-K. Cheng and Y.-C. A. Wei. An improved two-way partitioning algorithm with stable performance. *IEEE Transactions on Computer-Aided Design of Integrated Circuits and Systems*, 10(12):1502–1511, Dec. 1991.

[7] J. Cong, L. Hagen, and A. Kahng. Net partitions yield better module partitions. In *ACM IEEE* 29th *Design Automation Conference Proceedings*, pages 47–52, Anaheim, California, June 1992.

[8] D. M. Cvetkovic, M. Doob, and H. Sachs. *Spectra of Graphs: Theory and Application.* Academic Press,Inc., New York, New York, 1979.

[9] W. Donath and A. Hoffman. Lower bounds for the partitioning of graphs. *IBM Journal of Research and Development*, pages 420–425, 1973.

[10] C. M. Fiduccia and R. Mattheyses. A linear-time heuristic for improving network partitions. In *ACM IEEE* 19th *Design Automation Conference Proceedings*, pages 175–181, Las Vegas, Nevada, June 1982.

[11] M. Fiedler. *Special matrices and their applications in numerical mathematics.* Martinus Nijhoff Publishers, 1986.

[12] J. Frankle and R. M. Karp. Circuit placements and cost bounds by eigenvector decomposition. In *IEEE International Conference on Computer-Aided Design ICCAD-86*, pages 414–417, Santa Clara, California, November 1986.

[13] G. H. Golub and C. F. V. Loan. *Matrix Computations: 2nd Edition.* The Johns Hopkins University Press, Baltimore, Maryland, 1989.

[14] S. W. Hadley, B. L. Mark, and A. Vanelli. An Efficient Eigenvector Approach for Finding Netlist Partitions. *IEEE Transactions on Computer-Aided Design of Integrated Circuits and Systems*, CAD-11(7):885–892, July 1992.

[15] L. Hagen and A. Kahng. New spectral methods for ratio cut partitioning and clustering. *IEEE Transactions on Computer-Aided Design of Integrated Circuits and Systems*, Sept. 1992.

[16] K. M. Hall. An r-dimensional quadratic placement algorithm. *Management Science*, 17(3):219–229, Nov. 1970.

[17] A. Kahng. Fast hypergraph partition. In *ACM IEEE* 26th *Design Automation Conference Proceedings*, pages 762–766, June 1989.

[18] B. W. Kernighan and S. Lin. An Efficient Heuristic Procedure for Partitioning Graphs. *The Bell System Technical Journal*, pages 291–307, February 1970.

[19] S. Kirkpatrick, C. D. Gelatt, and M. P. Vecchi. Optimization by Simulated Annealing. *Science*, 13(220):671–680, May 1983.

[20] K. Kozminski. *Layout Synthesis Benchmarks*. Microelectronics Center of North Carolina, MCNC P.O. Box 12889, Research Triangle Park, NC 27709, 1992.

[21] B. Krishnamurthy. An Improved Min-Cut Algorithm For Partitioning VLSI Networks. *IEEE Transactions on Computers*, C-33(5):438–446, May 1984.

[22] T. Leighton and S. Rao. An approximate max-flow min-cut theorem for uniform multicommodity flow problems with applications to approximate algorithms. *FOCS 29*, 1988.

[23] T. Lengauer. *Combinatorial Algorithms for Integrated Circuit Layout*. Wiley-Teubner, 1990.

[24] B. Mohar. The Laplacian spectrum of graphs. In *Proceedings of the 6th Quadrennial International Conference on the Theory and Applications of Graphs: Graph Theory, Combinatorics and Applications, Vol. 2*, pages 871–898, Wiley, 1988.

[25] B. N. Parlett and D. S. Scott. The Lanczos algorithm with selective orthogonalization. *Mathematics and Computations*, 33(11):217–238, 1979.

[26] A. Pothen, H. D. Simon, and K.-P. Liou. Partitioning sparse matrices with eigenvectors of graphs. *SIAM Journal of Matrix Analysis and Applications*, 11(3):430–452, 1990.

[27] Quickturn Systems Inc. 325 East Middlefield Road, Mountain View, CA 94043, 1991.

[28] L. A. Sanchis. Multiple-Way Network Partitioning. *IEEE Transactions on Computers*, C-38(1):62–81, Jan. 1989.

[29] C. Sechen. *Placement and Global Routing of Integrated Circuits Using Simulated Annealing*. PhD thesis, Univ. of California, Berkeley, 1986.

[30] R. Tsay and E. Kuh. A unified approach to partitioning and placement. In *Proc. Princeton Conf. on Inf. and Comp.*, 1986.

[31] Y.-C. Wei and C.-K. Cheng. Multiple-level partitioning: An application to the very large-scale hardware simulator. *IEEE Journal on Solid-State Circuits*, 26(5):706–716, May 1991.

[32] J. H. Wilkinson. *The Algebraic Eigenvalue Problem*. Oxford Science Publications, 1965.

[33] C.-W. Yeh, C.-K. Cheng, and T.-T. Y. Lin. A probabilistic multicommodity-flow solution to circuit clustering problems. In *IEEE International Conference on Computer-Aided Design ICCAD-92*, Santa Clara, CA, November 1992.

[34] J. Y. Zien. Spectral k-way ratio cut partitioning. Master's thesis, University of California, Santa Cruz, In preparation, Jan, 1993.

High Speed 64-b CMOS Datapath

John Wawrzynek
Bertrand Irissou
EECS, Computer Science Division
University of California, Berkeley
Berkeley, CA 94720

Abstract

This paper reports on our investigations into the performance limits of CMOS datapaths. We have used a combination of single phase clocking, reduced voltage swing logic, moderate pipelining, and custom layout to achieve dramatic speed improvements over conventional design techniques. We have also used a novel fast adder structure and register file. To demonstrate the feasibility and effectiveness of these techniques and circuits, we have designed a test chip including a 64-bit integer datapath and a PLA-based finite state machine for testing. The chip layout was generated using MOSIS design rules and fabricated in the HP CMOS34 1.2-μm process. It has been tested and is fully functional at 180MHz.

1 High-speed circuit techniques

Industry circuit designers commonly use complementary CMOS gates and multi-phase clocking with level-sensitive latches or single phase clocking with edge-triggered flip-flops. These techniques are popular for their robustness to noise and process variations. We have found that abandoning these traditional techniques in favor of a more aggressive design style can lead to significant performance improvements.

Similar performance improvements have recently been reported independently by the DEC Alpha team[4, 3]. Their processor is a 64-bit RISC-style microprocessor that operates up to 200MHz with a 3.3 volt power supply. It was fabricated in 0.75-μm n-well CMOS with three layers of metalization. An accurate performance comparison between our chip and the the Alpha chip is not possible because of differences in the fabrication processes.

Factors under the control of circuit level designers that limit the speed of circuits include overhead due to clock distribution and latches, gate delay, capacitive loading due to busses, and the parasitics associated with transistor drains. The techniques we have used to overcome these limitations and achieve high-speed operation fall into three categories. First, we have systematically employed reduced voltage swing circuitry through the use of precharged logic. At the circuit level, reduced voltage swings help to reduce gate delay and decrease the capacitive delays of long lines. The second technique we use is single phase clocking. The technique relies on both the high and low phases of the clock to achieve the same effect of two-phase clocking. Thirdly, parasitics that limit high-speed operation are reduced through the careful custom layout of circuitry. Large folded transistors are used to counteract the effects of gate resistance and parasitic drain capacitance. Additional care has also been taken to minimize noise due to coupling of the fast switching nodes.

1.1 Precharged Logic—Reduced Voltage Swings

The first technique that we employed is reduced-voltage-swing dynamic logic for the speed-critical nodes. Precharged logic takes advantage of the fact that precharged gates switch closer to the rails than complementary gates. The switching threshold of a normal CMOS inverter is approximately $V_{dd}/2$. This threshold may be increased or decreased by making the pullup or the pulldown transistor wider. However, because the change inverter threshold voltage is proportional to the square root of the ratio of the strengths of the pullup to pulldown, in practice only a limited change is possible. On the other hand, an n-type precharged circuit (pulldown network with clocked precharge to V_{dd}) has a switching threshold much closer to ground, near the transistor threshold for the n-type transistor. In the complementary CMOS case, an input signal that is rising from ground to V_{dd} does not cause the gate to switch until the voltage makes it to $V_{dd}/2$. However, for the precharged case, the output switches at the n-type transistor threshold voltage (around 1.0v for our process). The resulting noise margin is lower than in conventional designs, but problems due to noise can be kept to a minimum by careful layout.

A good example of how this circuit family improves performance is the carry lookahead structure of our adder. We use alternating n-type and p-type NOR gates. During precharge, the n-type and p-type gates precharge high and low, respectively. During evaluation, the n-type gate begins to switch as soon as one of its inputs is above an n-type transistor threshold voltage. In turn, the p-type gate switches as soon as one of its inputs from the n-type gate drops below a p-type transistor threshold voltage. For the technology that we are using for this project, the effective gate delay through this type of circuit is approximately 200ps, less than half of the gate delay of standard CMOS circuits.

We also use precharged logic to speed up the recovery of signals from highly capacitive busses and other nodes—as one might use a sense amplifier. For a bus receiver we use a simple precharged inverter. In the case of a bus that precharges high, we use an inverter that precharges low. Both the bus and the receiver precharge on the same phase of the clock. On the next phase, the bus driver either begins to pull the bus low or allows it to float high. If the bus is pulled low, the p-type transistor of the precharged inverter detects the voltage as it falls lower than the p-type transistor threshold voltage, allowing the inverter to switch. In this situation we make the p-type pullup very wide. The reason is that circuit operation relies on the fact that as soon as the bus voltage falls below the threshold of the p-type transistor it starts conducting. However, the conductance is normally low in that regime; a wider pullup is more effective. Because the bus is a already a highly capacitive node, the extra gate capacitance added to the bus has little effect.

1.2 Single Phase Clocking

The other technique that we use is a form of single phase clocking. The technique was first introduced as "true single phase clocking" (TSPC) clocking [6, 1]. It was demonstrated for very high-speed bit-serial processing and other simple circuits. A single clock is distributed throughout the system, and never inverted. Very simple latches result in fast circuit operation, less latch setup and delay times, and less clock load. Single clock signal distribution also simplifies the task of clock signal generation and distribution.

In the TSPC clocking methodology there are two latch types: a p-latch which is transparent when the clock is low, and latched when the clock is high; and an n-latch which is transparent when the clock is high, and latched when the clock is low. Both types are shown in Figure 1 along precharged gates. As shown, the latches are used to follow two different types of precharged combinational logic blocks. The p-block precharges low when the clock is high, and evaluates when the clock goes low; the n-block precharges high when the clock is low, and evaluates when the clock is high. Each block is followed by a latch of the same type; a p-block is followed by a p-latch and an n-block by an n-latch. If the block is precharging the latch is locked, preventing the precharged value from propagating through to the next stage. The block evaluates when the latch is transparent. Systems built from blocks are arranged in pipelines of alternating p- and n-stages. When the n-blocks are precharging the p-blocks are evaluating and vice-versa. This approach imposes a constraint that cycles must be made from an even number of blocks.

The use of TSPC allows systems with no invertered clocks. While it is possible to build modules in the TSPC methodology using no inverted clocks, in practice it is very difficult to build efficient systems this way. Consequently we have relaxed this no clock inversion rule to allow an inverted clock on precharge transistors and power and ground switches, but not on the latches. This modification retains the nice properties of TSPC and its resistance to skew while permitting simplified circuits. It is simple to show that this modification is safe; the worst potential problem with this modification is extra power consumption.

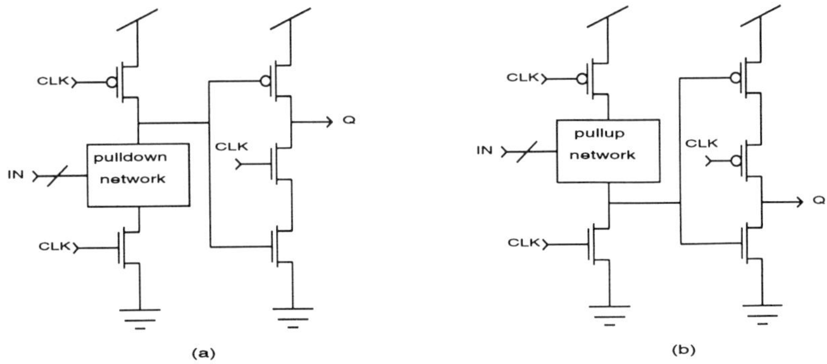

Figure 1: Precharged gates with TSPC n-type latch (a) and p-type latch (b)

2 Datapath Design

We strongly believe that system design and fabrication are imperative to understand the issues involved in successfully fabricating high speed circuits. To demonstrate that the techniques presented are feasible for CPU-sized chip design, we integrated some basic components of microprocessors into a datapath test chip. Clock distribution, power distribution, noise, input and output communication and testing are some of the issues that needed to be solved to make a functional chip.

The experimental 64-bit integer datapath consists of two of the basic building blocks of the integer core of a microprocessor: a 64-bit triple-ported register file and a 64-bit integer adder. Of course, a complete datapath is more complex that this test datapath, but we believe that these blocks are sufficient to demonstrate that our proposed techniques are viable.

Speeding up the carry chain of the 64-bit adder is a challenge for a high speed design. The register file is also difficult to design for high-speed operation because of the highly capacitive bit-lines. The width of the datapath was motivated by the fact that the popular 32-bit architectures are likely to be supplanted by 64-bit architectures. During the design phase we used several circuit-level simulators to help verify the delays in our circuits. The delays listed in this section refer to delays in simulated circuits. A later section in the paper reports on the actual running speed of the fabricated chip.

The basic structure of the test chip is shown in Figure 2. The datapath consists of the register file and the adder. Two buses are used for internal communication. The write-back bus, **wbus**, is used to write a result from the adder or the register file back into the register file. A bypass bus, **bbus**, bypasses the register file and forwards results directly into the adder.

As our clocking discipline dictates, each functional block is broken into p and n stages. The timing diagram of Figure 2 shows the operation of the pipeline. After the register addresses are decoded in the low phase, the

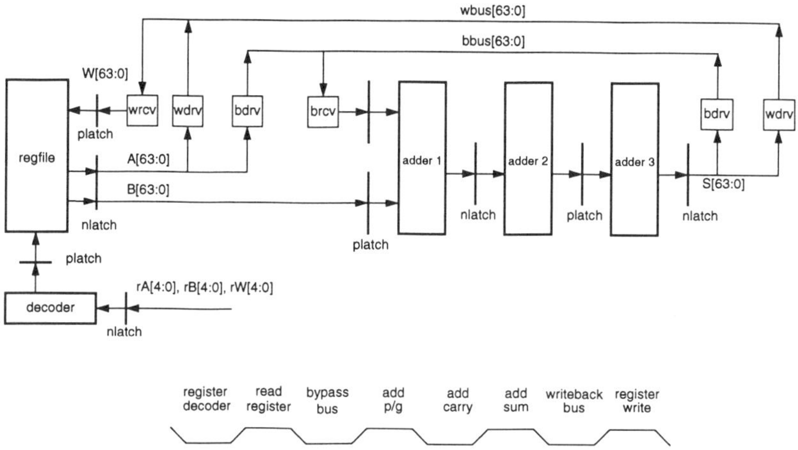

Figure 2: Fast Datapath Block Diagram

register file RAM array is read during the following high phase. The data read is latched on the falling edge of the clock. Then, the low phase is used to drive data on the bypass bus to the adder. The adder then takes three phases to evaluate, but a new addition can be started every cycle. Finally, the result from the adder is driven on the writeback bus during the low phase and the register file is written during the following high phase. All blocks precharge on the phase immediately before they evaluate.

2.1 Register File

The operation of the register file is broken down over two half-cycles. First, the low phase of the clock is used to decode the address and drive one of the 32 row-select lines. The inputs to the row select lines are then latched on the rising edge of the clock. These inputs are stable during the high phase.

For a read operation, the high phase is used by the bit-cell to drive one of the precharged bit-lines; the data is sensed by a cascaded pair of precharged inverters and latched on the falling edge of the clock. For a write operation, the data to be written must be ready by the beginning of the high phase. The write bit-lines are driven differentially during the high phase of the clock to write the cell.

The register file decoder is a p-block. It is basically a NAND decoder. The partial schematic of Figure 3 shows one branch of the decoder tree for selecting register 0. The rest of the tree can be easily completed by symmetry.

On the phase before evaluation, a high phase, the internal address lines in the decoder are grounded, causing the internal nodes of the decoder to be precharged high. Precharging the internal nodes prevents charge sharing

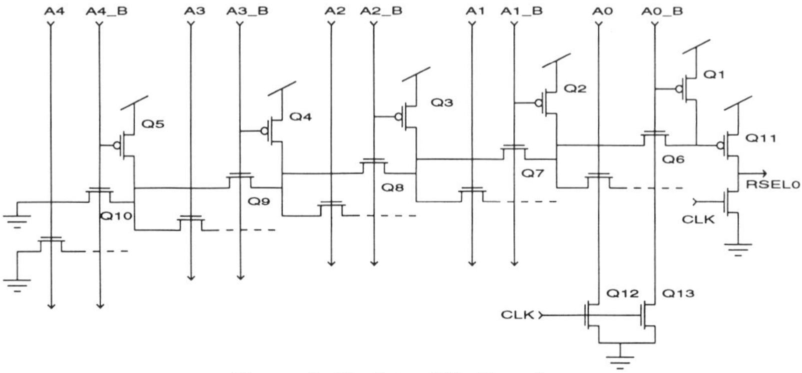

Figure 3: Register File Decoder

in the tree during evaluation. The precharge also turns off all the row select lines in the RAM array to avoid writing the bit-cells. During the low phase of the clock, the internal address lines are driven differentially, turning on only one path to ground in the decoder tree. The speed of the decoder resides in the exponentially increasing pulldown n-type transistor chain constituted by Q6–10. Each transistor is sized twice as large as the previous one in the chain. Also, the large p-type transistor Q11 at the end of the chain acts as a sensing device that turns on as soon as the output of the decoder is a threshold below V_{dd}. The layout of the decoder is very regular. Each decoded bit is made of 5 $\frac{16}{2}\lambda$ ($\frac{9.6}{1}\mu$m) n-type pass transistors and 5 $\frac{16}{2}\lambda$ p-type pullup transistors. The larger pass transistors and pullups are constructed by connecting in parallel 2, 4, 8 or 16 $\frac{16}{2}\lambda$ n-type or p-type transistors at the different stages of the chain. In simulation, the decoder evaluates in 1.7 ns from the time the addresses are asserted.

The array part of the register file is an n-block. Figure 4 shows the circuitry of one bit-cell and the sense circuitry for one port. The basic RAM bit cell is made of the static cross coupled inverters Q1–4. Two pass transistors, Q5 and Q6, provide two single ended read ports. Two additional pass transistors (not shown) are used to differentially write the cell. The bit-cell was carefully sized so that its contents are not altered on a single-sided read. The sense circuitry is precharged by Q7 and Q8 during the low phase of the clock. During the high phase, the bit-lines associated with the select row lines are pulled down or remain high depending on the content of the bit cell. The bit-line is then sensed by Q9. The data is ready on the latch 1.9 ns after the rising edge of the clock.

2.2 Adder Architecture

The adder is pipelined over three phases, each stage separated by alternating p or n latches. The latency through the adder is one and a half cycles. The low-level circuits for the adder are fairly straightforward and consists of a

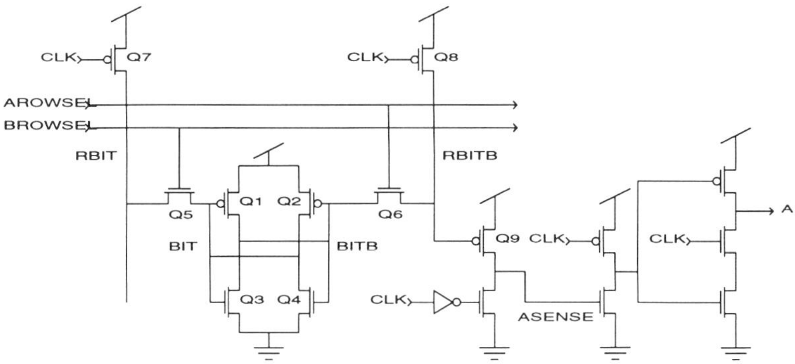

Figure 4: Register File Read Circuitry

Kogge-Stone binary lookahead tree [5]. However, a unique 2 to 1 reduction of the carry chain results in a significant reduction in area without affecting the speed.

Figure 5 shows the block diagram of the adder. The adder comprises three blocks of alternating n, p, and n stages. The first block of the adder, up to the first n-latch, has only three gate delays (in addition to the latch). We create the generate and the propagate signals. Then, the propagate and generate signals of two consecutive bits are combined in the *d-transform* block. The net result is to cut the length of the carry chain in half. Also, the first stage of the Kogge-Stone lookahead binary tree completes within this block. The second block is the last four stages of the binary lookahead tree, where each stage consists of a pair of propagate/generate blocks. In the last block of the adder, the real carry is reconstituted and the final sum calculated.

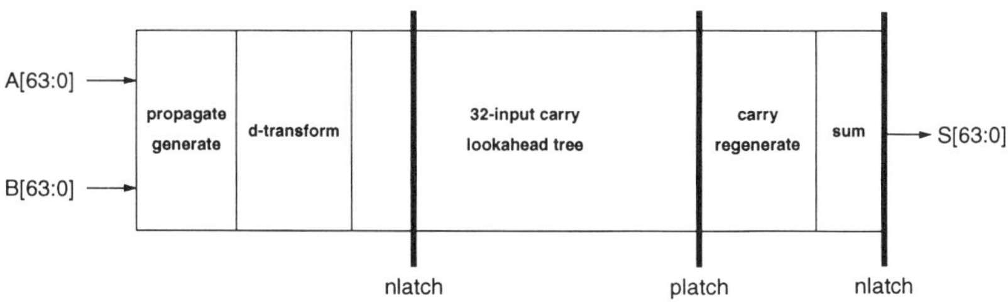

Figure 5: Adder Block Diagram

The 2 to 1 reduction in the carry chain was invented by A. Despain and extended by P. de Dood [2]. The reduction starts by rewriting the carry term of a Manchester carry chain of equation:

$$C_{2n} = G_{2n} + (P_{2n}.C_{(2n-1)}) \quad (1)$$
$$= P_{2n} \cdot (G_{2n} + C_{(2n-1)}). \quad (2)$$

By doing so on every even term, the carry chain can be modified to have adjacent OR function and adjacent AND function rather than alternating AND and OR. We can now combine these two consecutive AND or OR into one gate by creating the P'_n and G'_n terms:

$$P'_n = P_{(2n+1)} \cdot P_{2n} \quad (3)$$
$$G'_n = G_{2n} + G_{(2n-1)}. \quad (4)$$

The modified carry chain is shown in Figure 6; it is half the size of the original chain. Notice that the carry chain term are now C'_n and not C_n. The carry terms can be easily reconstructed once the carry chain has evaluated as:

$$C_{2n+1} = G_{(2n+1)} + C'_n \quad (5)$$
$$C_{2n} = P_{2n} \cdot (G'_n + C'_{(n-1)}). \quad (6)$$

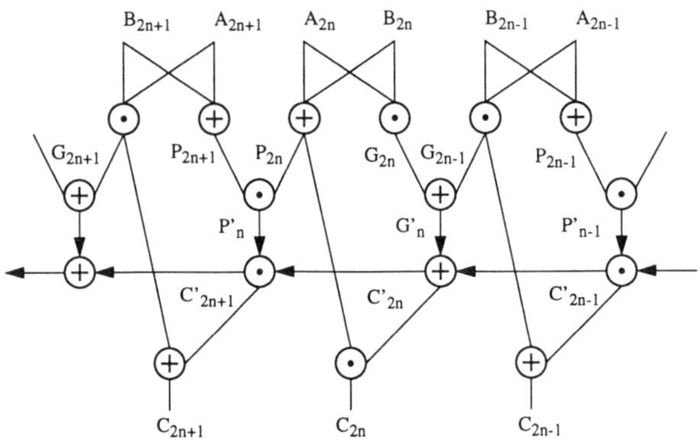

Figure 6: D-Transform

The result of applying this transformation is that, with the same gate delay, a 32-bit carry lookahead tree can be used instead of the much larger 64-bit tree.

3 Chip Layout and Implementation

Figure 8 shows a die photo of the datapath. The die is 6mm on a side. In the center the adder is $5100 \times 4000\lambda^2$ ($3060 \times 2400\mu m^2$). Pitch-matched to its left, the register file measures $5100 \times 2000\lambda^2$ ($3060 \times 1200\mu m^2$) for the RAM array and the sense and write circuitry. The decoders are to the top and bottom of the register file. The bus logic and shift registers are located to the right of the adder. To the top, a PLA state machine is used to control the operation of the datapath.

There is a triple ring of power, ground and clock circling the entire chip. It is distributed in the datapath horizontally in first layer metal interconnect (metal-1) and then down to the particular cells in metal-2. Metal-2 is used to run power lines over the tops of the cells; this organization permits us to keep the wires as wide as possible, and helps to minimize resistive effects. This is important in power and ground distribution (to prevent resistive voltage drops) and in the clock wires (to minimize clock skew). The other advantage to running power and ground perpendicular to the dataflow is that it helps in minimizing the effects due to ground noise in one stage effecting the next. Noise coupling between stages was of particular concern to us because we have dynamic nodes, surrounded by wide transistors with high currents.

Power supply inductance issues were a major challenge in the design of the test chip; our goal was to keep the bounce on the supply rail to no more than 500mV. After our first fabricated chip failed because of bounce problems, we added on chip bypass capacitors. By adding about 3nF of gate-oxide capacitance between power and ground, we were able to bring down the instantaneous clock switching current from 3.8A to 1.8A, limiting the bounce on the internal power and ground rails. Because the 84-pin PGA package used has no dedicated power or ground plane and has over 10nH of inductance per pin, all but 12 pins are used to supply power.

The chip layout was generated using MAGIC. Each block was simulated in its entirety using CaZM, which has allowed us to simulate up to 20,000 transistors, at the circuit level. We used IRSIM at the chip level to verify the function of the chip. Finally, we used HSPICE to confirm the timing of some of the critical paths of the chip.

The design process was complicated by the lack of adequate design tools. We were forced to rely on manual calculation and overly conservative design in several areas. Minimizing resistive effects in power and clock rails was a particular concern to us and, where possible, we over-sized these rails. Another layout issue of concern was in node-to-node capacitive coupling. Our layout tools did not accurately extract inter-node capacitance nor did our timing simulators efficiently estimate induced voltages. A tool that could flag trouble spots in a layout based on extracted coupling capacitance and estimated rise and fall times would be extremely useful.

The clock distribution is unique. We distribute the clock driver around the pad ring; close to the power and ground leads that supply the power to the driver and to the on-chip bypass capacitors. There is a U-shaped clock node that surrounds the core. The clock signal is brought into the core from both sides. This configuration is an attempt to minimize clock skew due to physical distance. The worst case delay is to the middle of one piece of the datapath. This represents an insignificant transmission line delay of 35ps. The more difficult problem with clock distribution is RC delays and mismatched buffer delays. We avoid mismatched buffer delays by having a single clock buffer. To minimize RC delays on the wires, we kept the clock wires very wide, throughout the clock distribution. The goal was to have one equipotential clock node. By keeping the wires wide we minimized the resistive effects, attempting to keep the clock node as one large capacitor. Of course, by making the clock node wire wide, it adds area and capacitance to the node. However, the clock node is inherently high capacitive, so adding more capacitance is not very significant. The total clock node capacitance in our case is 0.38nF.

In simulation the chip dissipates an estimated 9 Watts at 250MHz and power supply of 5V. Over a third of the 9 W is used by the clock driver (3.6 W). In comparison, the adder consumes 2.9 W, the register file, 1.8 W, and the test circuitry, 0.6 W.

4 Testing

Testing at high speed is a difficult task. We opted for a test strategy where the internal datapath is continuously clocked at high speed and controlled by an internal PLA. The *test id* and *reset* signals are applied from the pins to start the controller. When the program completes, a *done* signal is asserted. Data is shifted in and out at 15MHz via two serial shift registers, buffer_in and buffer_out. Figure 7 shows the complete chip block diagram.

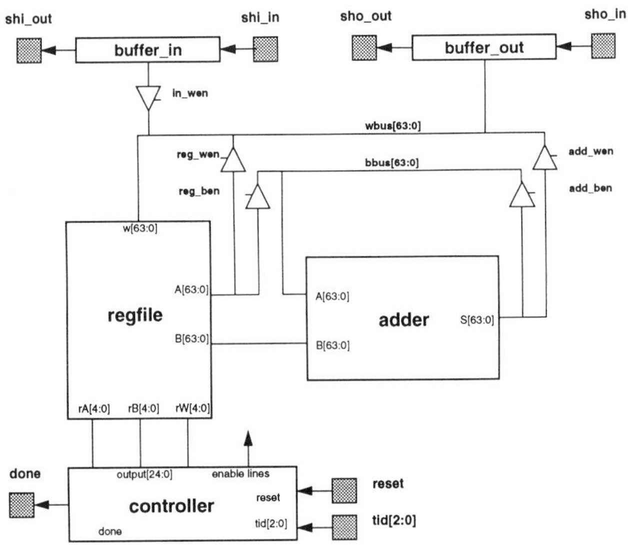

Figure 7: Fast Datapath Test Setup

5 Results

The design began as a class project in the Spring 1991. The layout was completed using MOSIS design rules and submitted for fabrication in the HP CMOS34 1.2-μm (1.0-μm gate length) in early December 1991. The first silicon was received from MOSIS in mid-January. Problems with this chip were traced to excessive V_{dd} and ground bounce. We added on-chip bypass capacitors and refabricated. The second chip returned from fabrication mid-May 1992.

The second fabrication of this design was successfully tested at 180MHz at 39°C and 5.0V power supply. A HP 8082A high speed signal generator was used to generate externally the high speed clock. A Tektronix DAS 9100 was used to drive the test id and reset signals, and to control the input and output shifters. The maximum operating frequency of the chip is limited by the adder. The register file works up to 210MHz and the PLA-based controller functions up to 230MHz. The power consumption is 1.4A at 180MHz.

Acknowledgments

We would like to thank all of the members of Spring 1991 CS254 class, in particular Krste Asanović, Brian Kingsbury, and Stuart Kleinfelder. This work was funded by the National Science Foundation through the PYI award, MIP-8958568.

References

[1] Morteza Afghahi and Christer Svensson. A Unified Single-Phase Clocking Scheme for VLSI Systems. *IEEE Journal of Solid-State Circuits*, 25(1):225–233, February 1990.

[2] Alvin Despain and Paul de Dood. Unpublished Notes on Fast Arithmetic, UC Berkeley. January 1989.

[3] Daniel Dobberpuhl et al. A 200-MHz 64-b Dual-Issue CMOS Microprocessor. *IEEE Journal of Solid-State Circuits*, 27(11):1555–1565, November 1992.

[4] Daniel Dobberpuhl et al. A 200mhz 64b Dual-Issue CMOS Microprocessor. In *IEEE International Solid-State Circuits Conference*, pages 106–107, February 1992.

[5] Tackdon Han, David A. Carlson, and Steven P. Levitan. Fast Area Efficient VLSI Adders. In *IEEE International Conference on Computer Design*, pages 418–422, October 1987.

[6] Jiren Yuan and Christer Svensson. High-Speed CMOS Circuit Techniques. *IEEE Journal of Solid-State Circuits*, 24(1):62–70, February 1989.

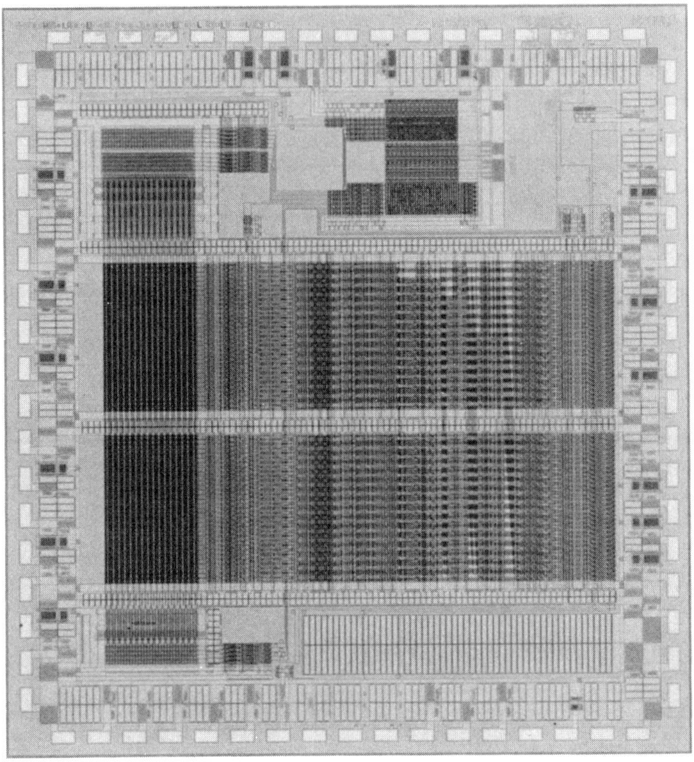

Figure 8: Test datapath chip die photograph.

A 32kb GaAs SRAM With Electronically Programmable Redundancy

Ajay Chandna and Richard B. Brown
Department of Electrical Engineering and Computer Science
University of Michigan
Ann Arbor, Michigan 48109-2122
ajayc@engin.umich.edu, brown@engin.umich.edu

Abstract

This paper describes novel circuit techniques that have been employed in a 1Kx32-bit SRAM designed in Vitesse's 0.6μm H-GaAs III E/D MESFET technology. The memory has a projected access time of 2.2ns with a nominal power dissipation of 5W.

Subthreshold leakage currents in MESFETs are orders of magnitude larger than those found in MOSFETs. The impact of these currents on circuit area and power are discussed. Based on this SRAM design, the trade-offs presented by subthreshold currents on system-level integration are examined.

A fuse circuit has been developed for this SRAM making it the first GaAs memory to employ electronically programmable redundancy. The fuse circuit is presented with SEM photograph of a programmed device. Transfer curves during programming and I-V curves of the device before and after programming are also shown.

A high-speed sense amplifier has been designed for this SRAM with a sensing speed that is independent of bit-line capacitance. Sense-amps tend to be particularly sensitive to the large device threshold variations of GaAs MESFETs. Simulation results showing the effects of these process variations on sense-amp speed and reliability are presented. A circuit design which minimizes the effects of process variations on this sense amplifier is presented.

1 Introduction

The integration level of gallium arsenide technology is now high enough to support the fabrication of a RISC CPU with some primary level cache on a single processor chip. The goal of this effort was to develop a static RAM to satisfy the needs of a GaAs microprocessor system being designed at the University of Michigan. The system [6] consisting of a CPU, a floating-point accelerator, a memory management unit, and instruction and data cache will be packaged on a multichip module (MCM) to minimize the impact of chip crossings on system speed. The initial clock frequency goal was 250MHz.

Based on preliminary architectural considerations including memory access time and page-size constraints, 4KW was determined as the optimal size of the instruction and data cache [11]. The integration levels of GaAs require that the cache be partitioned into 4 1Kx32 chips. A 4ns clock cycle time with a 1ns MCM transit delay penalty requires that the SRAM have an access time of less than 3ns. Since the SRAM will be embedded on the CPU for a primary-level instruction cache and since cache chips will be used on the MCM, the power dissipation of the system will be dominated by the SRAMs and should therefore be minimized.

The SRAM has been designed in Vitesse Semiconductor's H-GaAs III E/D MESFET technology. This process has a minimum gate length of 0.6μm and utilizes four levels of metallization. The top level of metal is used as a ground plane to minimize ground bounce across a chip. Memory cells in our design use a standard 6-T design with depletion transistor loads and enhancement transistor pass gates and pull-downs.

This paper describes novel circuits that have been developed for this SRAM to achieve high yield and high speed. The large subthreshold leakage currents of MESFETs present challenges to a circuit designer. The following section describes the impact of these currents on system power and area. A novel circuit is also presented which attempts to minimize these effects. Section 3 presents a fuse circuit which has been developed to allow electronically programmable redundancy. The sense amplifier for this RAM, whose sensing speed is independent of bit-line capacitance is discussed in section 4. Concluding remarks are made in section 5.

2 Impact of Leakage Currents on Memory Design

Subthreshold leakage currents in MESFETs and MODFETs are orders of magnitude larger than those in MOSFETs. The leakage currents consist of both gate-drain current caused by thermionic emission of carriers over the Schottky barrier [13] and drain-source conduction through the substrate [2]. Measurements of drain current as a function of gate-source voltage for a 0.8μm long 20μm wide enhancement MESFET with V_{DS}=0.6V are shown in Fig. 1. Leakage currents from the pass transistors of unselected memory cells connected to a bit-line (~20μA/bit at Vgs = 0V) can corrupt the current of a selected memory cell[1] (~400μA at Vgs=0.6V). Fig. 1 also shows that the leakage currents become a larger fraction of the active current at higher temperatures. In order to minimize the effect of leakage currents on read operation, it is clear that the access transistors of unselected rows must be reverse biased by at least 0.2V. This is achieved by applying a negative bias on the gate of the pass transistor with respect to the source (in

1. The actual pass transistors used were 2μm wide, one-tenth the size of the device measured in Fig.1.

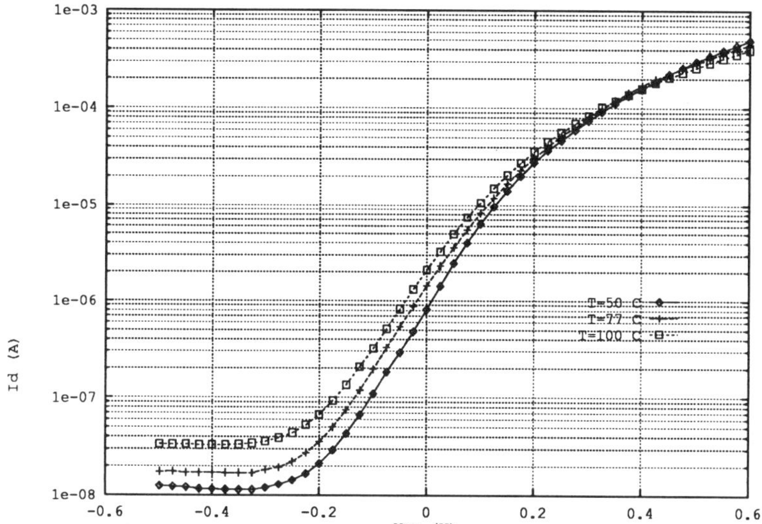

FIG. 1. Drain current vs. gate-source voltage for a 20μm wide 0.8μm long enhancement MESFET at different temperatures.

the memory cell). This can be done either by introducing a separate negative power supply to lower the gate voltage with respect to the cell ground, or by raising the cell ground with respect to the gate voltage. H. Makino et al. [5] achieved this reverse biasing by using the ground-shifting technique shown in Fig. 2. When the row is selected, the ground-shifting transistor is turned on, lowering cell ground to system ground. When the row is unselected, the ground shifting transistor is turned off, raising the cell ground to one diode drop above system ground. This reverse biases the access transistor by at least 0.6V.

While this solution addresses the problem of the leakage currents, it does so at the expense of access time. For a 2mm long word line, the extra time needed for

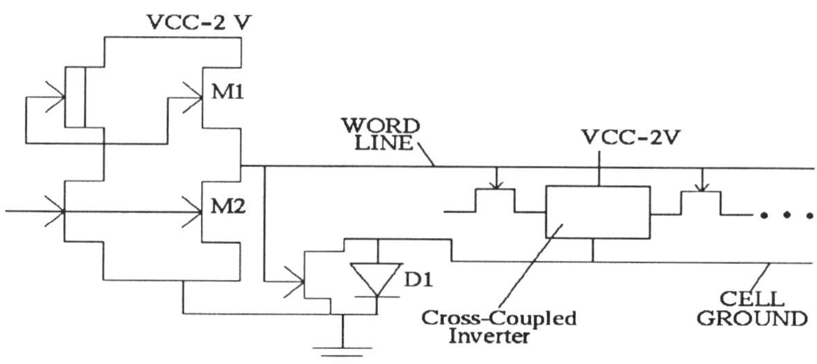

FIG. 2. Diode based ground shifting technique for minimizing leakage currents. [5]

the unselected rows to be turned off is ~500ps. If active circuitry were used to charge and discharge the cell ground line at the same speed as the word line, the power dissipation of the chip would increase. In our SRAM, the word-line driver power already accounts for 40% of the total chip power. This method of ground shifting would increase the chip power dissipation by 2W.

A more elegant solution presented in Fig. 3 has been incorporated into this chip. A single resistor is placed between cell-ground and system ground. This resistor is implemented using n-channel active material with a sheet resistance of less than 300 Ω per square. The resistor size is chosen so that there is at least a 0.3V drop across it when the row's cell current is flowing through it. When the row is selected, M1 is turned on and M2 is turned off. The word line is clamped by D1 at one diode drop above cell ground, thereby correctly biasing the access transistors. When the row is deselected, M2 is turned on and M1 is turned off which brings the word line down to 0V. The cell-ground remains at 0.3V[1] causing the access transistors to be reverse biased by at least 0.3V. This amount of reverse bias achieves the maximum suppression of leakage currents.

While the resistor-based solution achieves minimization of leakage currents at lower power and higher speed than the diode-based ground-shifting scheme, it is still ~150ps slower than a method that does not compensate for any leakage currents. The extra delay is because the word line must now swing from 0V to 950mV rather than from 0V to 650mV.

For reliable operation, the active current of a selected transistor should be at least an order of magnitude larger than the total leakage currents on a bit-line. In our SRAM, this constraint limits to 64 the number of bits that can be safely connected to a column, requiring that a significant portion of the total RAM area be devoted to sense amp/write circuitry. If the off-state leakage current could be reduced, the number of bits per column could be increased, and the circuitry devoted to a column could be amortized over more bits. Table 1 shows the impact of reducing leakage currents on SRAM area for a 1Kx8-bit quadrant of the

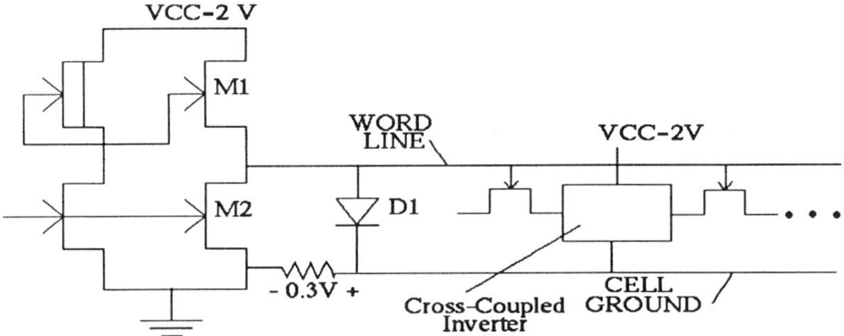

FIG. 3. Resistor based ground shifting technique for minimizing leakage currents.

1. Notice that diode D1 maintains isolation between the word line and cell-ground while reverse biased.

SRAM. The total area consists of cell area, word line drivers, sense amplifiers, write circuitry, and decode circuitry. As the number of bits per column is increased by a factor of N, the sense amp/write circuitry area decreases by N, while the word line driver area increases by N. The area occupied by the cells, predecode circuitry and drivers for both the write circuitry and sense amps remain constant. The table shows the percentage of total area occupied by the cells themselves as the number of bits/column is increased. As can be seen, at 32 bits/column only 70% of the total chip area is consumed by the memory cells. A reduction in leakage current by only 1 order of magnitude would allow 512 bits/column, reducing chip size by 25% and increasing the share of area occupied by the memory cells to 93.8%.

Table 1: Effect of reducing leakage currents on area of 1Kx8 SRAM.

Number of Bits/Column	32	64	128	256	512
Normalized SRAM Area	1.00	0.87	0.80	0.77	0.75
Cell Area Percentage of Total Area	70.6	81.6	84	92.1	93.8

A second order effect of leakage currents is an increase in cell power. In any technology, the pullup of a static RAM cell should provide just enough current to offset the leakage current of the pulldown devices. (Leakage currents, therefore, also set the lower limit for cell power.) In conventional GaAs DCFL processes, long minimum-width depletion transistors are used to keep this current small. The characteristics of these devices present an area / power trade-off. In our SRAM, the highest impedance standard-threshold depletion transistor that fits in a 400

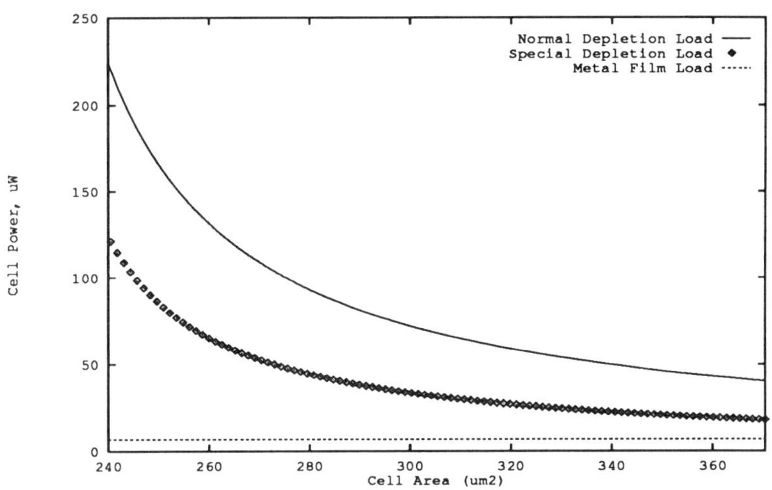

FIG. 4. RAM cell power (μW) vs. cell size (μm²) for three load devices.

μm^2 cell provides much more current than is needed to offset the leakage currents. As the area of the cell is decreased, the pullup length must be decreased, increasing the power. Fig. 4 shows the effect of varying the pullup length (cell size) on power dissipation. This plot includes curves for Vitesse's digital process pullup transistor, a special higher-threshold depletion transistor, and a Ti metal-film resistive load with sheet resistance of 100 kΩ per square [16]. As can be seen, such a resistive load is invaluable for embedding memory on-chip.

3 Electronically Programmable Redundancy

Previous GaAs MESFET SRAM designs have been plagued by low yield. While wafer lot yields of 30% have been reported [3] for 4K-bit SRAMs, the largest yield reported by manufacturers of 16K-bit SRAMs has been 10% [7], [15]. GaAs memories of 32K-bit and larger sizes definitely need to incorporate redundant rows or columns to improve yield. Until now, the GaAs MESFET implementations of redundancy have relied on an additional laser processing step to open links associated with defective rows/ columns.

We have developed the first electronically programmable fuse in GaAs. This circuitry allows the replacement of defective rows/columns to be done electronically at test time rather than requiring additional processing steps. A spare row is added to each quadrant of the SRAM. The SRAM is tested to determine whether any of the rows contain defective bits. If bad rows are found in the memory, a "bad rows exist" programming element is fused. The address of the defective row is then fused into a bank of 6 programming elements. If the decode circuitry senses that bad rows exist and that the current row address presented to the SRAM matches the fused address, the bad row is switched out and a spare row is switched in.

The programming element, a pair of inverters as shown in Fig. 5, has been fabricated and verified. The input to the first inverter is normally left unconnected which is equivalent to forcing the input low. This causes the output of the first inverter to be high, storing a logic low on the output of the second inverter. To program a logic high on the output of the second inverter, a large current is forced into the gate of M1. The gate-source and gate-drain diodes of M1 both conduct heavily eventually causing the device to short from drain to source. This short circuit persists after the input is disconnected. Transistor M1, which was off when its input was disconnected is now permanently on. The output of the first stage is thus pulled low, storing a logic high on the output of the second inverter.

The transfer curve of this programming procedure is shown in Fig. 6. This figure shows the output of the first inverter as its input voltage is swept from 0 to 5V. When the input is disconnected (Vin=0), its output is clamped by the second inverter to 0.7V (1). Below Vin=0.65V, only the gate-source diode of M1 conducts the gate current. Above Vin=0.65V (2), the gate-drain diode of M1 starts conducting forcing current into the gate of M3. Eventually at (3), transistor M1 undergoes a permanent transformation after which the input no longer has any

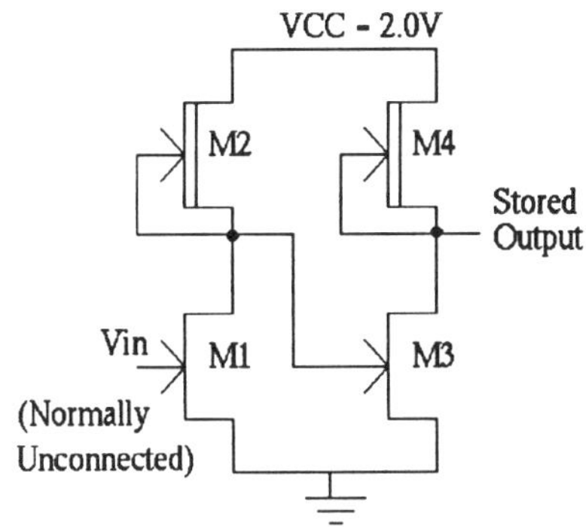

FIG. 5. Programming Element.

effect on the output. The output of inverter M1-M2 remains at 0.15V causing a one to be stored on the output of theM3-M4 inverter.

During the programming, we believe that the following sequence of events occur in M1. First, the current is slowly increased. This causes electromigration of the source/drain ohmic metal into the channel of the MESFET. Since the gate-metal can withstand higher current densities than the ohmic metal, it is capable of sustaining the large programming currents. As migration of ohmic metal into the channel continues, the effective channel length then decreases to that just below the gate of the MESFET, thus reducing the channel resistance. Finally the gate metal itself is opened allowing no further gate current to flow. The current densities at which permanent changes were observed agree with the specified limits for both the ohmic metal and gate metal.

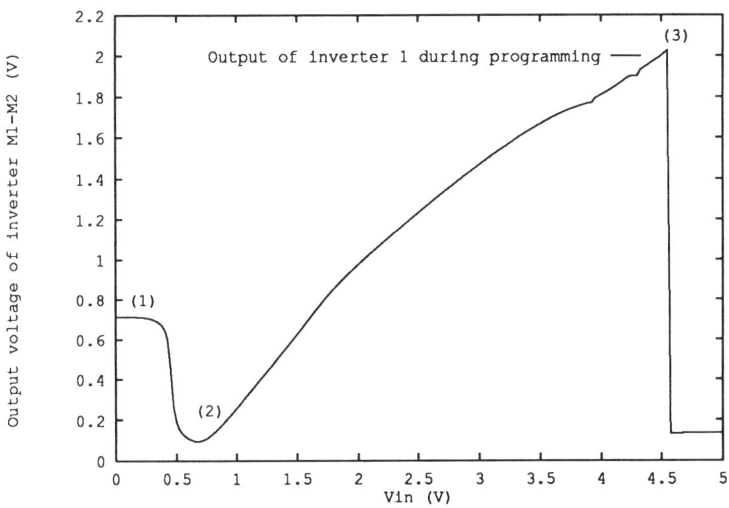

FIG. 6. Transfer characteristic of inverter M1-M2 during programming.

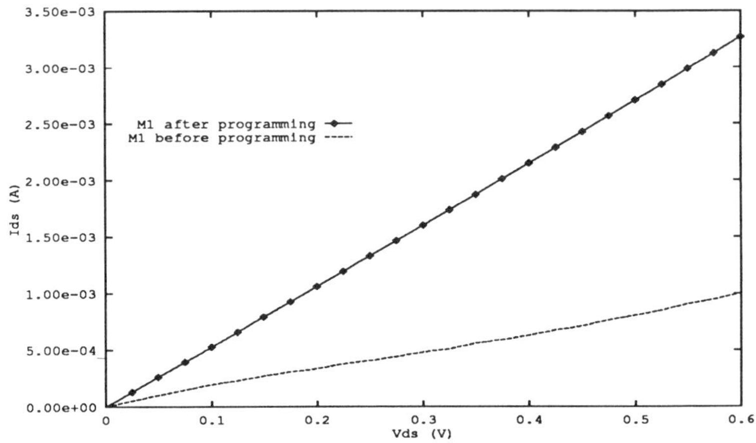

FIG. 7. I-V Characteristics of transistor M1 before and after programming.

Fig. 7 shows the drain characteristics of M1 before and after programming with the gate disconnected. Before programming, the output resistance of M1 is 600Ω. After programming, the output resistance of M1 was measured to be 185Ω. This number compares favorably with what we would be expected if the device were made up simply of n-channel active area under the gate of the transistor.

The programming procedure was optimized by varying the slew rate of the programming current over a wide range. With too high a slew rate, the input inverter showed catastrophic destruction as shown in the SEM photo of Fig. 8.

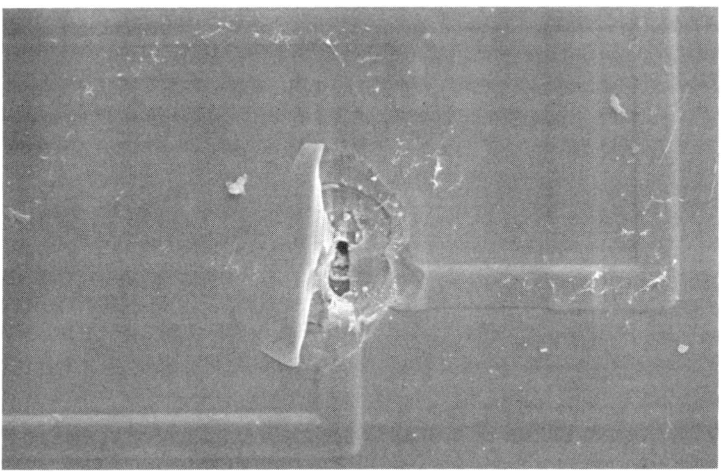

FIG. 8. Programming element transistor M1 which has undergone catastrophic,unpredictable failure.

The excessive current densities are shown to have ruptured the gate metal as well as melting away the underlying transistor. Due to the heavy gate-drain conduction, such high slew rates also damaged the second inverter. When programmed with a safe slew rate, the input inverter displayed no signs of the catastrophic behavior shown in Fig. 8. A programming slew rate of 500mV/s was experimentally found to provide 100% reproducible behavior over 15 chips. The I-V curves and transfer curves shown above were taken with this slew rate.

4 Clamped Bit-Line Sense Amplifier

Various current-mode sense amplifiers have been used recently in CMOS/BiCMOS RAMs. The advantage of these schemes over conventional voltage-based sense amplifiers is that their sensing speed is independent of bit-line capacitance. Most of these, e.g. [14], however, require complex circuit topologies that cannot be accomplished using GaAs MESFETs. The barriers to implementing these sense amps in GaAs are large source/drain resistances and back-gating, both of which preclude stacking transistors. The Schottky diodes on the gate inputs also make the design of current mirrors difficult.

We have designed a sense amplifier the operating principle of which is similar to a CMOS design [1]. This clamped bit-line sense amplifier (CBLSA) is shown in Fig. 9.

Transistors M1 through M4 form cross-coupled inverters. The bit-lines are connected to the sources of M3 and M4. Diodes D1 and D2 are used to clamp the

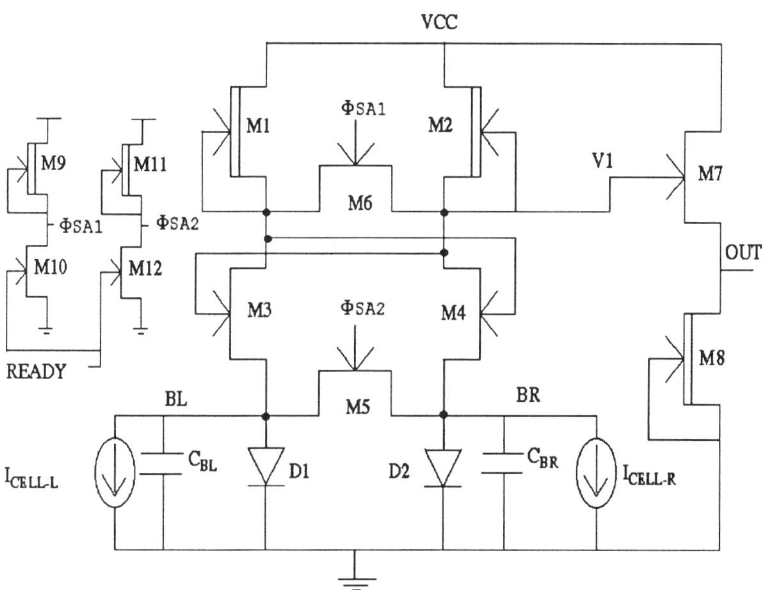

FIG. 9. Clamped bit-line sense amplifier schematic.

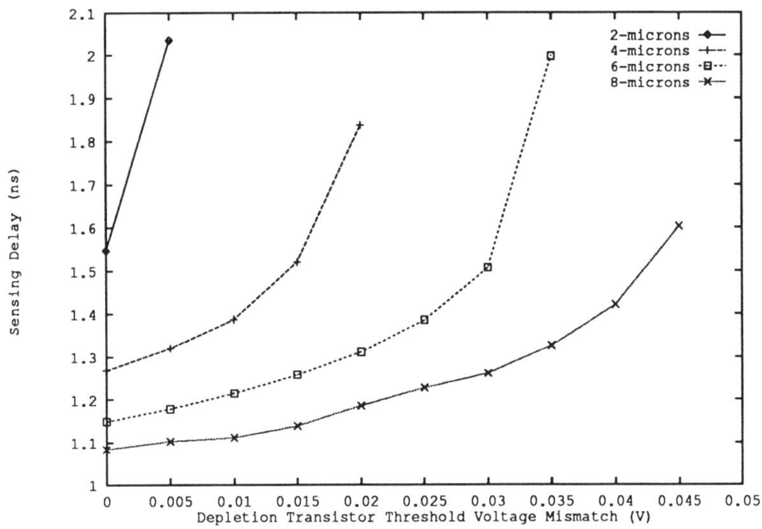

FIG. 10. Delay from word-line to sense-amp output for different sized memory cell access transistors.

bit-lines at approximately 0.65V. C_{BL} and C_{BR} represent the bit-line capacitances. $I_{CELL\text{-}L}$ and $I_{CELL\text{-}R}$ represent the active cell currents of the selected transistors.

Transistor M6 is used initially to bring the latch formed by M1 through M4 into a metastable state. M5 is used to clamp the two bit-lines at the same voltage, and to provide a path for the current flowing from the memory cell pass transistors. Once the word-line has been asserted long enough for the selected cell to establish a current flow through M5, the READY signal is asserted. This deasserts Φ_{SA1} and Φ_{SA2}. Depending on the sense of the cell current, the sense amplifier is tipped into one of its two stable states.

Since the highly capacitive bit-lines are connected to a low-impedance point (diodes D1 and D2), and since the outputs of the latch are not connected to these bit-lines, the switching speed of this sense-amp is dependent only on the load capacitance seen by the latch. The output from the sense amplifier is taken single-endedly off of V1. This is passed through a source follower stage (M7,M8) which provides both high-gain and level shifting for the output for successive DCFL stages.

The biasing circuits for Φ_{SA1} and Φ_{SA2} perform dynamic biasing to adjust with fluctuations in VCC down to 1.5V. These circuits also adjust to fluctuations in the bit-line voltages BL and BR so that data can be written through the sense amplifiers if needed. The most critical aspect of the sense-amplifier design is its performance in the presence of mismatches between M1 and M2. Fig. 10 shows how the access time varies with mismatches in the depletion transistor threshold voltages.

The weaker the memory cell pass transistor is, the more sensitive is the sense-amplifier to threshold voltage mismatches. For each of these cell sizes, the sense

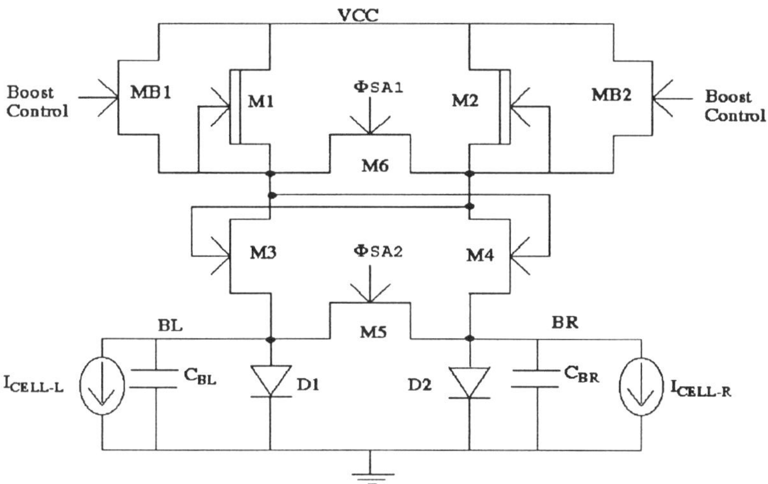

FIG. 11. Clamped bit-line sense amplifier with boosting circuitry.

amplifier fails above a certain threshold voltage mismatch. For instance, a 4μm access transistor can not withstand more than 20mV mismatch in threshold voltage and still function correctly. Based on measured data, adjacent depletion devices have a σ_{VT} of less than 30mV. As a result, memory cells with access transistors larger than 6μm are likely to produce functional sense amplifiers. The sensitivity of functionality and access time to threshold voltage deviations can be reduced by increasing the bias applied to the pass transistors, by weakening the sense-amp pullup devices, or by laying out the devices so that they have the same center of mass. This last method ensures that no matter which way the process gradients go, both pullup transistors will be affected in the same manner. By making the pullup devices M1 and M2 weaker, a given cell current will have a larger influence on the sense amp. This leads to a sense amplifier that is more tolerant of process variations. Since the size of these transistors also determines the delay of the sensing stage, this greater tolerance is achieved at the expense of speed.

A solution to this problem which attains the process variation immunity associated with a weak pullup device and the speed associated with a stronger pullup device is shown in Fig. 11. The transistors M1 and M2 are weak pullup devices. Two booster transistors, MB1 and MB2 have been added to the sense amplifier. These transistors are initially turned off when sensing begins. Two gate delays after sensing begins, MB1 and MB2 are turned on very hard to force the output to the desired state. Simulations showing the impact of these transistors on the sensing delay are shown in Fig. 12.

The first three curves show the effect of varying the strength of the pullup transistors in the original sense amplifier. This data is for a memory cells with pass transistor of width 4μm with sense-amp MESFETs M1 and M2 of size (WxL) 2μmx4μm, 2μmx6μm, and 2μmx8μm. While the strongest pullup device (2μmx4μm) shows the best access times, it fails with only a 45mV threshold voltage mismatch. The 2μmx8μm pullup sense amp is shown to be reliable well

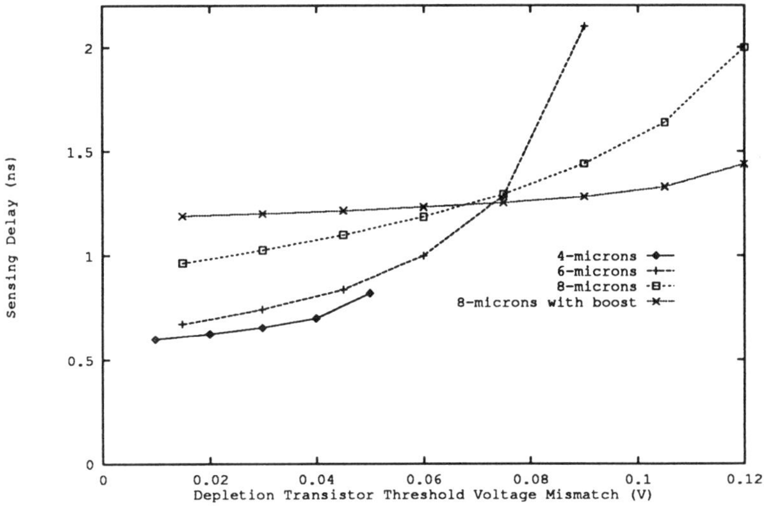

FIG. 12. Sensing delay for different length sense-amp pullup devices for a 4μm wide memory cell pass transistor.

beyond the range of expected transistor mismatch. It does, however, have a worst-case sensing delay of 2ns. The fourth curve shows the performance of the boosting transistor solution. It exhibits high tolerance to threshold voltage mismatch with a very small dependence of sensing speed on threshold voltage mismatch.

5 Conclusions

We have developed a 1Kx32-bit SRAM for use in a high performance microprocessor system. The memory achieves an access time of 2.2ns at a power dissipation of 5W. Ground shifting circuitry has been developed to minimize the impact of leakage currents on chip area and speed. We have shown the impact of subthreshold leakage currents on both the power and area of a GaAs SRAM. A fuse element has been developed to give this memory electronically programmable redundancy. The procedure for programming this element has been optimized for reproducibility. A high speed sense amplifier has also been developed. Simulation results have been presented showing the impact of process variations on the speed and reliability of this sense amplifier. A circuit which reduces the impact of these process variations on performance has also been presented.

Acknowledgment

This research has been supported in part by the U.S. Defense Advanced Research Projects Agency under DARPA/ARO Contract No. DAAL0390-C-

0028, by the U.S. Army Research Office under Contract No. DAAL03-87-K-0007, and by an NSERC 1967 Science and Engineering Scholarship.

References

[1] T.N. Blalock and R.C. Jaeger, "A High-Speed Clamped Bit-Line Current-Mode Sense Amplifier," IEEE Journal of Solid-State Circuits, Vol. 26. No. 4. April, 1991. pp. 542-548.

[2] T. Enoki, et al., "Characteristics including velocity overshoot for 0.1-um-gate-length GaAs SAINT MESFET's," IEEE Trans. Electron Devices, vol. 37, no. 4, pp. 935-941, 1990

[3] R.S. Hinds et al., "A 20K GaAs Array with 10K of Embedded SRAM", IEEE Journal of Solid-State Circuits, Vol. 26, No. 3, March 1991.

[4] T. Iizuka, "Embedded Memory: A Key to High Performance System VLSIs", 1990 Symposium on VLSI Circuits,pp. 1-4.

[5] H. Makino, et al., "A 7ns/850mW 4Kb SRAM Fully Operative at 75°C," IEEE GaAs Symposium Digest of Technical Papers Nov. 1988.

[6] T.N. Mudge, R.B. Brown, et al., "The Design of a Micro-Supercomputer," IEEE Computer Magazine, January 1991, pp. 57-64.

[7] H. Nakano et al., "A High-Speed GaAs 16Kb SRAM of 4.4ns/2.2W Using Triple-Level Metal Interconnection", IEEE GaAs Symposium Digest of Technical Papers, Nov. 1987.

[8] H. Nambu, et al., "A 1.5ns, 64Kb ECL-CMOS SRAM", 1991 Symposium on VLSI Circuits, pp. 11-12, 1991.

[9] P.G. Neudeck, et al., "Significant Long-Term Reduction in n-Channel MESFET Subthreshold Leakage Using Ammonium-Sulfide Surface Treated Gates," IEEE Electron Device Letters, vol. 12, no. 10, pp. 553-555, October 1991.

[10] K. Nummila, et al., "Short-Channel Effects in Sub-100nm GaAs MESFETs," Electronic Letters, vol. 27, no. 17, pp. 1519-1521, August 1991.

[11] O.A. Olukotun, T.N. Mudge, R.B. Brown, "Implementing a Cache for a High-Performance GaAs Microprocessor," 18th International Symposium on Computer Architecture, Toronto, May 27-30, 1991.

[12] M. Suzuki, "A 1.2-ns HEMT 64-kb SRAM," IEEE Journal of Solid-State Circuits, p.p. 1571-1576, Nov. 1991.

[13] S.M. Sze, Physics of Semiconductor Devices, 2nd ed. New York: Wiley, 1981 pp. 245-284.

[14] K. Takayuki et al., "A Circuit Technology for Sub-10-ns ECL 4-Mb BiCMOS DRAM's," IEEE Journal of Solid State Circuits, Vol. 26, No. 11, November, 1991. pp. 1530-1537.

[15] C.T. Tsen et al., "A Manufacturable Low-Power 16K-bit GaAs SRAM", IEEE GaAs Symposium Digest of Technical Papers, Nov. 1987.

[16] M.R. Wilson, Cray Computer Corporation, personal communications.

A Scalable Systolic Multiprocessor System for Analysis of Biological Sequences

Raj K. Singh[*] [†] Stephen G. Tell[*] C. Thomas White[§]
Doug Hoffman[*] Vernon L. Chi[*] Bruce W. Erickson[§¥]

Department of Computer Science[*]
Program in Molecular Biology and Biotechnology[§]
Department of Chemistry[¥]

University of North Carolina at Chapel Hill
Chapel Hill, NC 27599

Abstract

The design and implementation of an application-specific, fault-tolerant, and scalable multiprocessor system called BioSCAN (**Bio**logical **S**equence **C**omparative **A**nalysis **N**ode) are described. Discussed are system partitioning and integration, functional decomposition between hardware and software, the algorithm and its implementation in VLSI, the early results of using the system, and comparison with other hardware and software solutions for biological sequence analysis.

1. BIOLOGICAL SEQUENCE COMPARISON

In the last decade, the field of molecular biology has seen a virtual explosion in the amount of data collected on a variety of living organisms. Much of this information identifies the primary structures of specific biopolymers of three classes: DNA (deoxyribonucleic acid), RNA (ribonucleic acid), and protein. Each class has its own distinct set of monomers. Members of the class are linear chains of monomers. The primary structure of a specific biopolymer is determined by the order of its monomers. Primary structural information does not directly represent the complex three-dimensional structures into which these molecules fold.

[†]Correspondence: CB#3175, UNC-Chapel Hill, NC 27599-3175.
E-mail: singh@cs.unc.edu

The primary structure of these biopolymers is uniquely and concisely represented as a *sequence*, an ordered string of letters from an alphabet. Each letter stands for a distinct monomer appearing in the biopolymer chain. The four distinct monomers of DNA sequences are represented by the letters **A**, **C**, **G**, and **T**. DNA sequences are typically a few thousand letters long, although the largest known DNA sequence contains over two hundred thousand letters. RNA sequences are represented using the letters **A**, **C**, **G**, and **U**. Protein sequences are represented by strings using twenty letters from the English alphabet, each letter denoting one of the twenty different amino acid residues in proteins. Protein sequences average a few hundred letters in length. However, the largest known protein sequence contains over several thousand letters.

Central repositories maintain databases containing nearly all known biological sequences. The GenBank database [7], with more than eighty thousand DNA and RNA sequences, contains over one hundred million letters. The SWISS-PROT database [7] contains over twenty-six thousand protein sequences with a total of eight million letters. These databases have been growing at an exponential rate with a doubling time between 18 and 24 months [3].

Analyzing relationships between sequences within a database is an important task. Sequences may be related because they share a common ancestral sequence from which both known sequences have subsequently evolved by a series of substitutions, deletions, insertions, and other rearrangements. Alternatively, small fragments of two sequences may be related because they each interact with a third molecule. Determining new relationships between known sequences is an active area of research. The computer algorithms employed for these purposes are time intensive due to the large amount of data involved. The sequence databases continue to grow at a rate similar to the rate at which general-purpose computers continue to get faster. So, we cannot expect advances in conventional computer technology to significantly reduce the time required to perform a rigorous search of a sequence database. The search algorithms are better suited to acceleration by application specific computer system hardware.

2. SYSTEM ARCHITECTURE

The BioSCAN system [12] accelerates a class of specialized search algorithms for biological sequences. Common among these algorithms is an optimized core procedure. The C program fragment of Fig. 1 represents this procedure as implemented by the BioSCAN ASIC [5], where the inner loop is performed in parallel. The system software generates the sequence A by concatenating all sequences in a database to be scanned for similarities with the query sequence B. Sequence A is padded with a special letter to force threshold-exceeding scores, S, in the pipeline to be flushed to the end of the accumulator chain. The inner loop calculation is implemented in VLSI by a small processing element (PE) that is replicated 812 times in the BioSCAN ASIC. A 16-ASIC circuit board is installed in a host computer that provides power and cooling, access to storage and network services, and general housekeeping. The hardware subsystem and the associated system software effectively speed up the inner loop calculation several orders of magnitude over what is currently possible with software alone on general-purpose computers. The results of these comparisons are presented near the end of this paper.

```
/* Inputs */
     unsigned int      La, Lb;         /* length A, B */
     char              A[La],          /* sequence A-padded */
                       B[Lb];          /* sequence B */
     signed int        T[M][N],        /* similarity table */
                       Th;             /* threshold */
/* Outputs */
     signed int        S[Lb+1];        /* scores 1..Lb */
/* Temporaries */
     int               i, j;           /* index A and B */
/* Algorithm */
for (i=0; i<Lb; ++i) S[i]=0;           /* initialize */
for (i=0; i<La; ++i) {                 /* elements of A */
   for (j=Lb-1; j>=0; --j)             /* elements of B */
      if (S[j] < 0) S[j+1] = T[A[i]][B[j]];
      else if (S[j] >= Th) S[j+1] = S[j];
      else S[j+1] = S[j] + T[A[i]][B[j]];
   if (S[Lb] >= Th) report(i);
}
```

Figure 1: BioSCAN Algorithm

The BioSCAN system compares a given query sequence with each sequence in a specified database. The system flags locations in database sequences which are sufficiently similar to the query sequence to satisfy a preprogrammed *goodness-of-fit* measure specified by the table T [1] and the threshold Th of Fig. 1. The resulting information is used in various ways depending on the target application. Generally the system helps reduce the amount of data to be processed by software by several orders of magnitude. The BioSCAN system serves as a prefilter for a broad range of existing and future applications.

Each segment, defined as a contiguous substring of any length, of the query sequence is compared with every equal-length segment of every database sequence. The algorithm, in effect, traverses a *search space* that is proportional to the product of the length of the query sequence and the total length of the scanned database. For each database sequence comparison, the search space can be visualized as the interior of the rectangle in Fig. 2. Pairs of segments (one from each sequence) deemed similar are represented as diagonally-oriented line segments of various lengths. Segment pairs represented on the same diagonal correspond to the same alignment of sequences A and B. For any particular alignment between A and B, there is a constant difference in the indices of each pair of aligned letters. This provides a convenient way to number the alignments (and corresponding diagonals). Hence, diagonal *(i-j)* in Fig. 2 contains the aligned letters *A[i]* and *B[j]*.

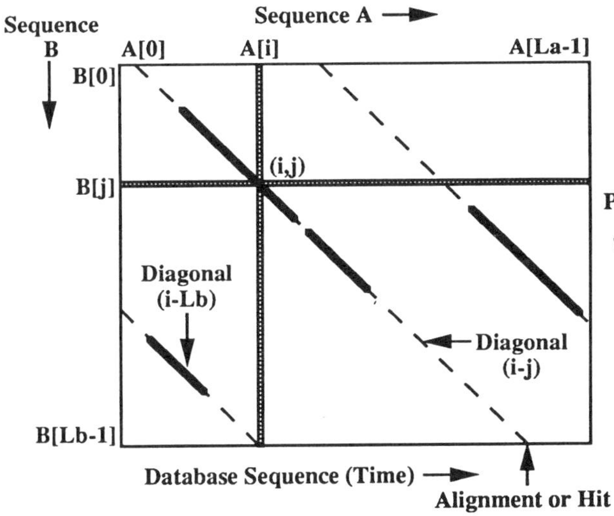

Figure 2: Diagonal Path Graph

Searching for high-scoring segments in software requires time proportional to the product of the lengths of the two sequences. Given sufficient BioSCAN PEs (one per query sequence letter), the query sequence dimension of this space (vertical axis) is processed in parallel. Hardware completes the search in time proportional to the size of the database. This may be visualized in Fig. 2 where database letter *A[i]* is completely processed in the ith cycle. The information state of the system after the *i*th cycle is represented by the vertical bar beneath *A[i]*. The *j*th PE computes partial sums on the *j*th row. The sum computed by PE_j at step i is passed to PE_{j+1} to use in step $i+1$. At the end of each cycle, one bit result is generated. This bit specifies whether the diagonal just completed (e.g. diagonal *(i-Lb)* at step *i* in Fig. 2) contains at least one sufficiently high-scoring segment pair. If necessary, the locations and scores of significant segment pairs can be easily computed from the diagonal index using software. This system partitioning has lead to a hardware architecture that is exceedingly efficient at filtering the initial search space, relegating application-dependent processing to software executed on general-purpose machines. If any more information were to be maintained and reported by the hardware, the PEs would have to be much larger, making the overall system more complex.

3. ASIC ARCHITECTURE & IMPLEMENTATION

The BioSCAN ASIC architecture features a scalable one-dimensional systolic array of identical PEs. The internal data path of the ASIC is shown in Fig. 3. The architecture does not constrain the size of the PE array. Thus, incremental improvements in layout or technology immediately translate into higher computational density. Moreover, since the ASIC pinout is independent of the array size, the system offers scalability without requiring a redesign of the circuit board. The BioSCAN architecture also allows any number of BioSCAN ASICs to reside on a circuit board, requiring only a single-wire communication path between adjacent devices.

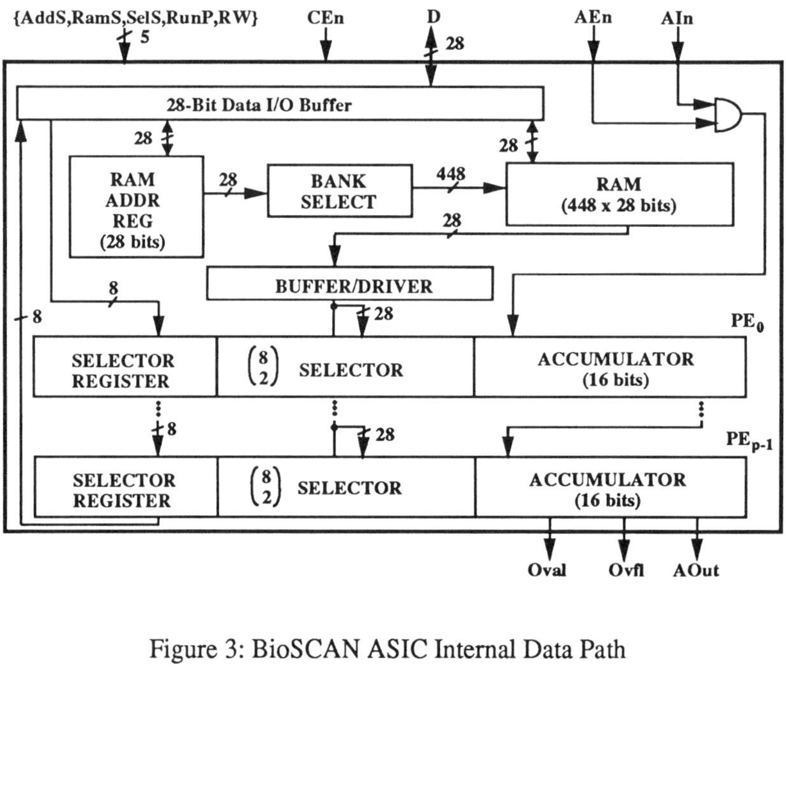

Figure 3: BioSCAN ASIC Internal Data Path

Prior to a database scan, the ASIC is programmed with a similarity table and a query sequence. Since the threshold value in each PE is hard-wired to 16,384, the similarity table entries are scaled before initialization. The similarity table is stored in an SRAM array with 28 banks, each with 16 rows and 28 columns. Logically the SRAM is a 28x28 table of 16-bit signed integers. Each database character received by the ASIC selects 1 of the 28 banks, which corresponds to an entire row of the logical table (28 16-bit integers). During each of the next 16 clock periods, 28 column bits from successive rows of the selected bank are sensed, amplified, and broadcast in parallel within the ASIC.

Data from SRAM is broadcast to all PEs of the array. Each PE, shown in Fig. 4, stores one character of the pattern sequence encoded in a 2-of-8 encoding scheme and uses it to select one of the 28 broadcast lines as its input for the duration of the database scan. The 2-of-8 encoding was chosen over binary encoding because 2-of-8 decoding logic, replicated in each PE, is much smaller than binary decoding logic. This encoding scheme limits the alphabet size to 28 (8*7/2). During the scan cycle, each PE bit-serially adds the selected value to the similarity score in its accumulator, which was received from the previous PE in the previous scan cycle. Each bit of the calculated score is immediately passed to the next PE in the array. Simple logic in each PE handles two exceptions: (1) A negative score input to a PE is zeroed before accumulation. (2) A score of 16,384 or greater is propagated through the entire array unchanged.

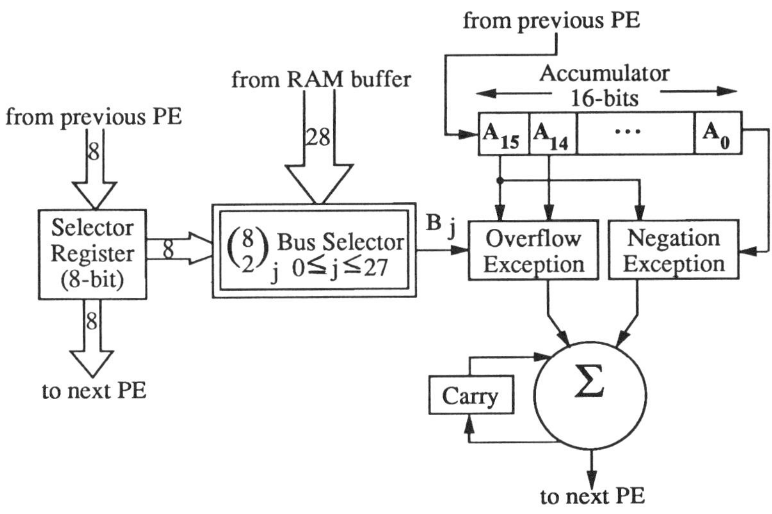

Figure 4: Block Diagram of a Processing Element

The device pad-frame was laid out with due consideration to the design, manufacturing and assembly of the circuit boards. The bit-serial accumulator output was placed on the opposite side of the IC package from the accumulator input, which helps the placement and routing of signals on the circuit board.

The device implementation offers testability and fault tolerance. The SRAM is independently readable and writable. The pattern sequences held by the PEs can be scanned out to verify their contents. The PEs can be controlled and observed individually to obtain 100% stuck-at-fault coverage. Both for device testing during fabrication and accommodating PE failures in the field, individual PEs can be shunted through by storing a special code in the selector register to activate logic not shown in Fig. 4. Several adjacent PEs can be shunted through to improve the fabrication and system yield. During tests, up to 15 PEs have been successfully shunted through without impacting the system performance. PE or chip failure can be diagnosed in the field and the defective component can be removed from the array using software.

The BioSCAN ASIC was designed using 1.2-μm scalable CMOS design rules from MOSIS and fabricated by Hewlett-Packard. Each 7.75 mm x 9 mm die contains 812 PEs. Sixty-five of the 108 devices tested with zero defects. Eight other devices had a few defective PEs that can be programmed out from the array. Thus, the overall processing yield of usable devices was 67%. Table 1 shows the design and process parameters for the chip.

Die Size	7750μm x 9050μm
Transistors	537,675 (89,736 in RAM)
Processors	812 (536 transistors/PE)
Pins/Package	84 pin PGA (42 Vdd/GND)
Clock	32 MHz @ 50 °C (ambient)
Voltage/Power	4.75-5.25V (3.0W max)
Interface	TTL-compatible
Process	1.2μm N-well CMOS
Gate length/tox	1.2μm/20nm
Poly/Metal1/Metal2	2.4/3.6/4.2μm pitch

Table 1: BioSCAN ASIC Device Summary

4. BOARD ARCHITECTURE & IMPLEMENTATION

The BioSCAN circuit board architecture was designed with the goals of modularity and reusability of subsystems, testability during manufacturing and field installation, ease of installation and configuration, and scalability and parallelism using multiple ASICs.

The major subsystems of the board, shown in Fig. 5, are the configurable BioSCAN processing array, the input/output buffers, and the host interface. The circuit board contains 16 sockets that can be fully or partially populated with BioSCAN ASICs. These ASICs are accessed from the host machine through a static memory interface of three registers. Using this interface, the on-chip SRAM is loaded with scaled similarity table and each PE is initialized with a letter from query sequence. After the initialization is completed, the compressed database is loaded in the input data FIFO on the circuit board. The ASICs may be enabled individually for testing and initialization, and simultaneously for searching with a long query sequence. In the present implementation, a query sequence can contain as many as 12,992 (812*16) letters.

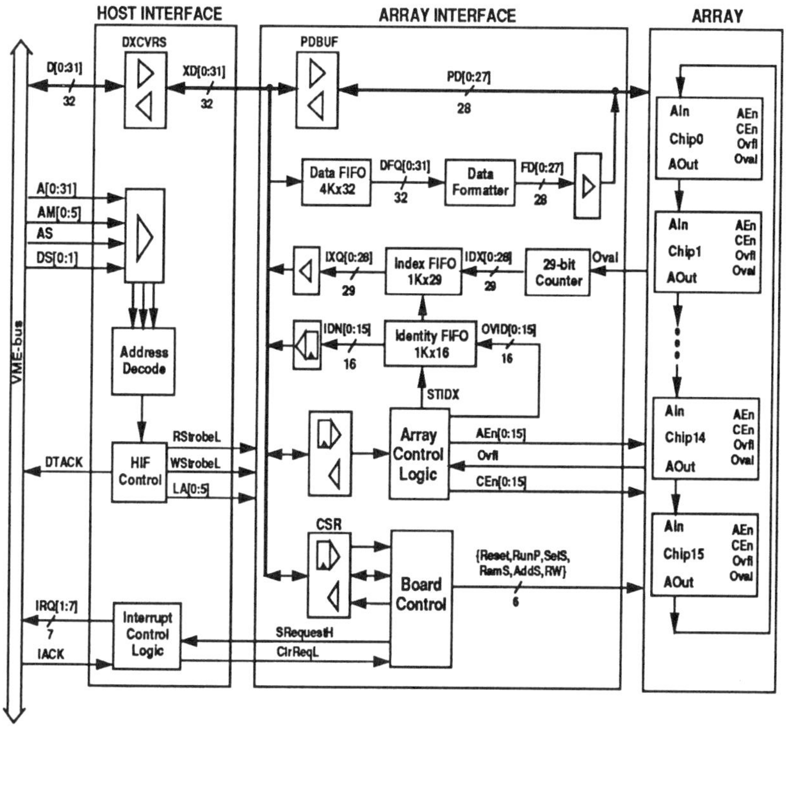

Figure 5: Block Diagram of BioSCAN Circuit Board

The ASICs are organized in a ring which can be broken at any point to form the linear systolic array. In case of complete failure of one ASIC, the ring organization allows the remaining ASICs to be used as a contiguous array of 15 ASICs. The array may be reconfigured into smaller subarrays of adjacent ASICs. This is useful when the query sequence can be contained in few chips because it allows multiple searches to be performed in parallel. The gate logic that extends or breaks the array is contained within each ASIC, allowing high-speed operation by eliminating logic on the circuit board from the data path.

The input to the BioSCAN array is a stream of fully-decoded 28-bit words comprising the database. To use the host VMEbus more efficiently and to reduce disk space requirements, databases are compressed and the board implements a simple hardware decompression scheme. The decoding logic broadcasts the data to all 16 ASICs. The sequence databases are transferred over the VMEbus and buffered in a 32-bit FIFO. Each FIFO word contains multiple encoded database characters packed into 2, 3, 4, or 5-bit fields. The character size is selected based on the size of the alphabet required by a particular database.

The output from the systolic array consists of *overflow indications* from each individual ASIC. An overflow, defined as an accumulated score exceeding the threshold, indicates a significant alignment. In order to be meaningful, the overflow needs to be related back to the location of the alignment in the database. When an overflow occurs, a word is written in the output FIFO. The output word consists of two fields: a 29-bit index field and a 16-bit identity field. The index field is the value of a counter which is incremented whenever a database word is written to the array, thus counting diagonals. The identity value consists of one bit per ASIC indicating that the corresponding ASIC registered an overflow at the current diagonal.

The relative sizes of the input and output FIFOs were selected with the expectation that the similarity table and threshold would be designed to identify less than one percent of the diagonals as having significant alignments. Smaller input FIFOs result in reduced throughput on low-performance VMEbus hosts like a SUN 4/280. On the other hand, the size of the input FIFOs is less of a factor with higher performance hosts such as a SUN SparcSERVER 690MP.

The board control subsystem supervises the input and output FIFO status and changes operating modes accordingly. Database characters are written to the array whenever the board has been enabled by software and the input and output FIFOs are not empty and not full, respectively. The host is interrupted whenever either FIFO requires servicing.

The host interface subsystem is designed as a generic VMEbus interface module that could be replaced with a similar module in a board designed for a different bus. The device driver software provides access to the BioSCAN circuit board through a set of 9 registers. Some registers access the ASICs directly for testing and initialization. Other registers are used to transfer data and operating parameters to the circuit board. The input FIFO register is replicated to form an 8K-byte page of simulated memory that may be useful for direct memory access (DMA).

The BioSCAN board achieves a high degree of test coverage through the use of a small amount of extra control logic and no additional data paths. An input subsystem diagnostic mode allows the input FIFO and decoder to be tested by reading back each decoded word as it is presented to the array. Output subsystem diagnostic modes allow overflow indications to be written to the FIFO, independent of the array. Together, these diagnostic modes allow the board logic to be tested before installing any BioSCAN ASICs.

The BioSCAN board consists of over 250 equivalent ICs including 23 PLDs. These devices are connected by over 1700 signals on a circuit board with 4 signal layers and 2 power planes. The system was extensively simulated using the Verilog-XL simulator and worked the very first time it was powered up.

5. SOFTWARE ARCHITECTURE & IMPLEMENTATION

The two major objectives of the project were to provide a unique and useful resource for computational molecular biologists and to offer a platform for future research in computer science. Accessibility and reliability of the resource is important in this regard. For wider accessibility, the BioSCAN system is designed to operate as a server node on the Internet. Any researcher with a network connection and the appropriate application software can access the BioSCAN system.

The client-server approach was chosen to most efficiently utilize the processing bandwidth of the BioSCAN system. The functional components of the software architecture, shown in Fig. 6, reflect the hierarchical and layered nature of the design. We see three distinct groups of users of the BioSCAN system: end users, application developers, and system programmers. The primary design goal for the BioSCAN software was to provide a stable and consistent interface to the BioSCAN hardware for the first two types of the users while preserving flexibility in the underlying implementation to accommodate the third.

The end users access the BioSCAN system through application programs. No specific knowledge of the hardware and software configuration is required. The application developers use the Applications Programming Interface (API) provided by the Application Subroutine Library (ASL) to interface existing sequence analysis software to the

BioSCAN system or to create new application programs based on the capabilities of the system. The ASL translates API calls into transactions to the BioSCAN server system.

Transactions in the BioSCAN server system are implemented as TCP/IP stream sessions of short duration. TCP/IP streams are used for reliable transfer of the transactions, which are often larger than the allowable packet size in connectionless datagram protocols. This structure was inspired partly by the Gopher internet information system.

The BioSCAN server system is composed of three types of processes: server processes, one scheduler process, and run processes. Server processes accept transactions from clients and pass job requests to the scheduler process. The scheduler process organizes user jobs into runs that can be accommodated by the BioSCAN hardware. One aspect of ongoing research is to devise scheduling algorithms that optimize the throughput of the server.

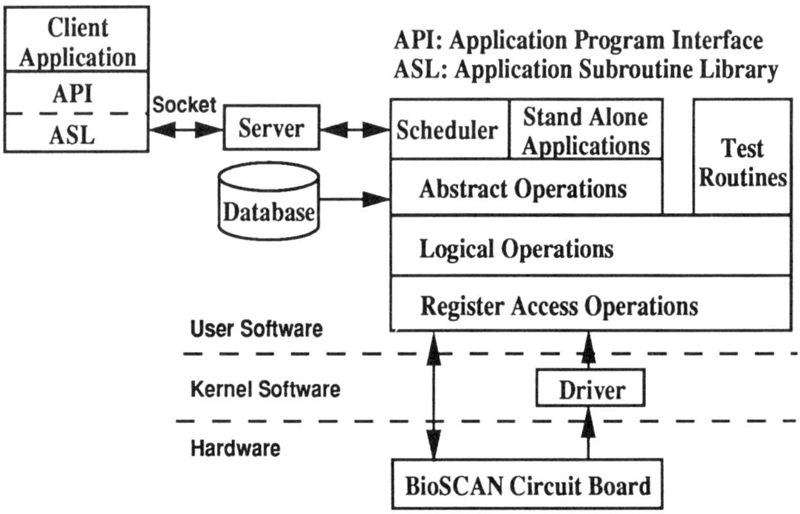

Figure 6: Software Architecture of BioSCAN System

After a run is constructed, a run process is started to handle interaction with the hardware. The run is passed to the first of several objects that form a hierarchy of abstract BioSCAN machines. This layered approach allows changes to the BioSCAN hardware and low-level software without impacting the scheduling and the interface software. It is even possible to incorporate software simulations of unbuilt hardware devices with existing server software to assess impact of the new design on the overall system function. The server-level software is being developed using C++ while the ASL is being written in C.

6. SYSTEM PERFORMANCE & COMPARISON

Many other special-purpose hardware [4,6,8,9] and software [2,11] tools for the analysis of biological sequences have been previously reported in the literature. However, due to the differences in algorithms used and the diversity of implementation technology, an objective basis for evaluating system performances is not evident. For example, software tools such as BLAST [2] generally use heuristic methods to improve the system throughput at the cost of selectivity and sensitivity. Moreover, the software performance is directly coupled with the performance of the host computer on which it is executed. The BLAST system running on a Sun-4 workstation is reported to scan the protein database with a query sequence of a few hundred letters at about half a million letters per second and to search DNA database with a query sequence of a few thousand letters at about two million letters per second [2]. Larger query sequences are analyzed more slowly. For comparison, the BioSCAN system can scan either database with a query sequence of almost 13,000 letters at nearly two million letters per second. Dynamic reconfiguration of the BioSCAN system, however, allows it to scan 16 812-letter query sequences simultaneously at 1.95 million letters per second, which is equivalent to an effective rate of about 31 million letters per second.

In comparing the BioSCAN system with other special-purpose parallel computing systems, we note that these systems use different core algorithms (variations of dynamic programming algorithm [10]) and different VLSI technology. Since the basic architecture of these systems is systolic in nature, we have devised a performance measure that is based on the number of *million cell operations per second* (**MCOPS**), where a cell is defined as a PE in a hardware implementation or the equivalent block of instructions (basic inner loop) in a software implementation.

Based on this measure, Table 2 shows the performance of several general-purpose and special-purpose systems. The BioSCAN system, operating at clock rate of 32 MHz, performs a cell operation in each PE every 16 clock cycles for a theoretical processing rate of 2 MHz. Each 812-PE chip delivers a peak theoretical performance of 1,624 MCOPS. The current BioSCAN system with 16 chips on a single board can deliver a peak performance of 25,984 MCOPS. In practice, the observed processing rate of the BioSCAN system is 1.95 million letters per second. Thus the 812 PEs on a single chip run at an effective rate of 1,583 MCOPS and the current 16-chip BioSCAN system runs at 25,334 MCOPS.

The performance data for the Sun-4/690MP system was obtained by running an optimized version of the BioSCAN algorithm listed in Fig. 1. For the software implementation of the BioSCAN algorithm, a cell corresponds to the actions needed to perform a single letter to letter comparison. The BioSCAN algorithm, modified appropriately to improve vectorization, was also used to measure the performance of the Convex C-240 system. The performance figures of P-NAC and SPLASH systems are derived from measurement results in Gokhale *et al* [6]. The data for the B-SYS and BISP systems are derived from the system specifications given in Hughey *et al* [8] and Chow *et al* [4], respectively, and denote the theoretical maximum performance.

System	Date	Technology	Performance (MCOPS)
Sun-4/690 MP	1991	1 CPU	0.56
Convex C-240	1989	1 CPU	2.67
P-NAC [9]	1987	4.0 μ NMOS	1.1
SPLASH [6]	1990	1.0 μ FPGA	50
B-SYS [8]	1991	2.0 μ CMOS	1,551
BISP [4]	1991	1.0 μ CMOS	9,600
BioSCAN [12]	1991	1.2 μ CMOS	25,334

Table 2: Performance of Special-purpose & General-purpose Systems

Based on the performance numbers of the Table 2, the BioSCAN system (with estimated cost of $7,000 per board in quantities of 50), offers a competitive price-performance advantage over general-purpose as well as other special-purpose systems.

7. SUMMARY & FUTURE WORK

We have described a fault-tolerant, scalable systolic multiprocessor system called BioSCAN for rapid similarity analysis of DNA, RNA, and protein sequences. The system consists of 16 custom ASICs, each with a systolic array of 812 processing elements implemented in 1.2-μm CMOS technology. The system is field-programmable and fault-tolerant. Under software control, faulty ASICs or faulty processing elements on a given ASIC can be eliminated from the active array. The large systolic array of 12,992 processors can be dynamically reconfigured as multiple smaller arrays of adjacent ASICs. The system scans a given database at nearly 2 million letters per second.

Future directions for the BioSCAN project work involve detailed characterization of system performance and comparison with other software and hardware systems. We plan to study architectural enhancements for scalability and improved throughput, database caching, performance tradeoffs between interrupt and polling driven service, buffer management strategies, and design of a marketable production system. We hope also to explore implementation of other algorithms from the general class of pattern matching applications on this architecture.

ACKNOWLEDGMENTS

This research was supported in part by NSF grant MIP-9024585. The authors would also like to thank Cadence Design Systems, Inc. for providing the Verilog-XL simulator.

REFERENCES

[1] S. F. Altschul, "Amino Acid Substitution Matrices from an Information Theoretic Perspective," J. Mol. Biol. 219, pp. 555-565, 1991.

[2] S. F. Altschul, W. Gish, W. Miller, E. W. Myers, & D. J. Lipman, "Basic Local Alignment Search Tool," J. Mol. Biol., 215, pp. 403-410, 1990.

[3] C. Burks, "How much sequence data the data banks will be processing in the near future," Biomolecular Data, R. R. Colwell, ed., pp. 17-26, Oxford University Press, NY, 1989.

[4] T. E. Chow et al., "A Systolic Array Processor for Biological Information Signal Processing", Procs. 1991 International Conference on Supercomputing, June 17-21, Cologne, Germany, pp. 216-223, 1991.

[5] W. D. Dettloff, R. K. Singh, C. T. White, & B. W. Erickson, "A 50 MHz 1.5M Transistor ASIC for Biosequence Analysis," ISSCC Digest of Technical Papers, pp. 40, Feb. 1991.

[6] M. Gokhale et al., "SPLASH: A Reconfigurable Linear Logic Array", Proc. 1990 ICPP, pp. I-526, 1990.

[7] M. Gribskov &J. Devereux, eds., "Sequence Analysis Primer", Stockton Press, 1991.

[8] R. P. Hughey & D. P. Lopresti, "B-SYS: A 470-Processor Programmable Systolic Array", Procs. 1991 ICPP, pp. I-580, 1991.

[9] D.P. Lopresti, "P-NAC: A Systolic Array for Comparing Nucleic Acid Sequences," Computer, pp. 98-99, 1987.

[10] S. B. Needleman & C. D. Wunsch, "A General Method Applicable to the Search for Similarities in the Amino Acid Sequence of Two Proteins", J. Mol. Bio., 48, pp. 443-453, 1970.

[11] W. R. Pearson & D. J. Lipman, "Improved tools for biological sequence comparison," Proc. Natl. Acad. Sci. USA 85, pp. 2444-2448, 1988.

[12] C. T. White et al., "BioSCAN: A VLSI-Based System for Biosequence Analysis", Procs. 1991 ICCD: VLSI in computers & Processors, pp. 504-509, 1991.

A Field-Programmable Gate Array for Systolic Computing

Frederick Furtek
Concurrent Logic, Inc.
1092 East Arques Avenue
Sunnyvale, CA 94086-4601, USA
fred@clogic.com

Abstract

There is a growing awareness that reconfigurable logic, in the form of SRAM-based field-programmable gate arrays (FPGA's), is an ideal vehicle for implementing a wide range of compute-intensive algorithms. The CLi6000 series of SRAM FPGA's from Concurrent Logic is especially well suited to the special needs of this area. We describe the architecture of the CLi6000 series of SRAM-based FPGA's, emphasizing those features that support efficient implementation of pipelined arithmetic circuits. These features are illustrated through a massively parallel, highly pipelined algorithm for motion estimation, an especially compute-intensive algorithm used in digital video compression.

1 Introduction

Recent work [1,2,3] has shown that a handful of SRAM-based field-programmable gate arrays (FPGA's) wired together can achieve extraordinary levels of performance, often outperforming supercomputers at a tiny fraction of the cost. And, unlike ASIC solutions, these systems are general purpose and reprogrammable—they can be reconfigured in milliseconds to perform a completely new task.

The CLi6000 series of SRAM FPGA's from Concurrent Logic has evolved through several generations and is the product of two joint-development efforts, with Apple Computer [4] and National Semiconductor [5]. The original motivation for the technology, and still a key application area, is the acceleration of compute-intensive algorithms by exploiting the parallelism inherent in hardware. The capabilities of the technology are illustrated through a massively parallel algorithm for performing motion estimation, an especially compute-intensive algorithm used in digital video compression. The algorithm—which determines the best estimate of how blocks of pixels move from frame to frame—requires about 4,000 MIPS to perform in real time on a standard video signal.

The algorithm is implemented as a systolic array of 256 processing elements, each achieving 100% efficiency—with no wasted clock cycles—through pipelining and a clever synchronization of input pixel streams. The CLi6000's high register count, NAND/XOR cell states, multiplexer states, fine-grained architecture, array symmetry and high cell connectivity all work together in efficiently implementing the heavily pipelined, bit-serial arithmetic operations used in each processing element.

2 CLi6000 Architecture

At the heart of the CLi6000 architecture[1] is a symmetrical array of identical cells (Figure 1). Routing within the array is either by bus or by direct connection between neighboring cells (Figure 2). Direct connections are most useful in tight spaces and over short distances where maximum flexibility is required. Busses support fast, efficient communication over longer distances.

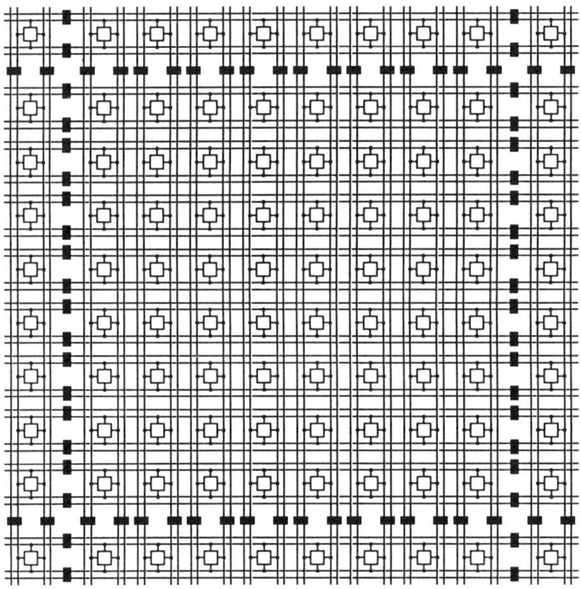

Figure 1 CLi6000 Array (10X10 Section)

There are two kinds of busses: local and express. Local busses are the link between the array of cells and the bussing network. Express busses are not connected directly to cells and hence provide higher speeds. Connective units, called repeaters, spaced every eight cells, divide each bus—both local and express—into segments spanning eight cells. Repeaters can: isolate bus segments, connect local-bus segments, connect express-bus segments and provide local/express transfers.

1. See [6] for a more complete description of the CLi6000 architecture.

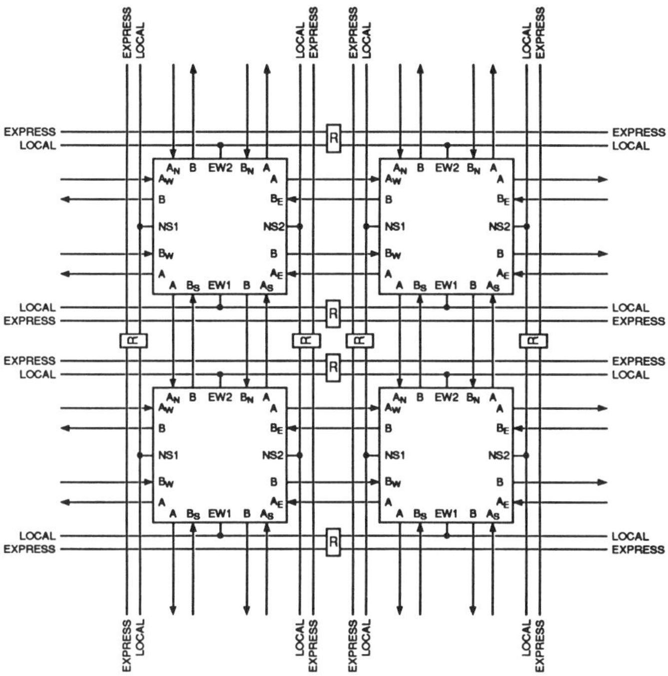

Figure 2 Cell Connections

The cell (Figure 3) is simple and small and yet provides many important and commonly used logic functions, including those most useful in pipelining and arithmetic: a D-type register, a NAND/XOR pair and a two-input multiplexer. The register is essential in pipelining, which is the most effective way to achieve high system throughput. The high register count of the CLi6000 architecture (there are 3,136 in the CLi6005) makes pipelining simple and efficient. The NAND and XOR, available simultaneously in a single cell, provide efficient implementation of full adders and other arithmetic functions. The multiplexer, available by itself or feeding into a register, finds many uses in a broad range of circuits.

Because its four sides are functionally identical, each cell is completely symmetrical. Cell symmetry—together with array symmetry, high cell connectivity and a fine-grained architecture—make the design of hard macros (circuits with fixed layouts) especially easy. More specifically, it facilitates the creation of bit-slice macros, hard macros in which the inputs/outputs on one side are aligned with the outputs/inputs on the opposite side. A larger circuit is constructed by simply abutting these bit-slice macros. The remainder of this paper describes just such an approach in implementing a massively parallel algorithm.

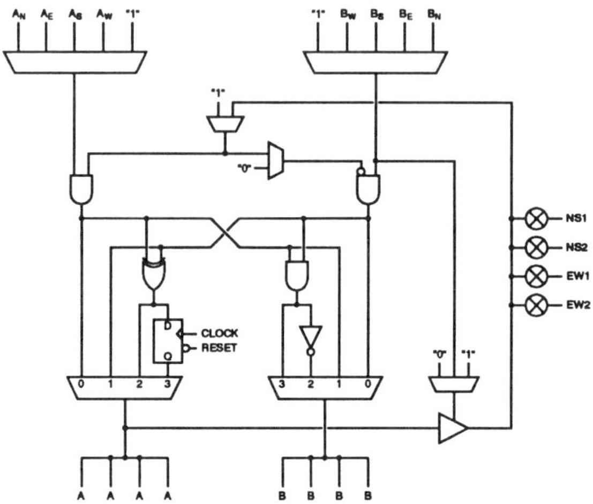

Figure 3 Cell Structure

3 Motion Estimation

Motion estimation is concerned with finding the best estimate of how blocks of pixels move from frame to frame in a video signal. For a block of pixels in the current frame (the *current block*), the object is to find the best match among *search blocks* the same size within a *search window* of the preceding frame. The best match is determined by finding the minimal *error* (*distortion*) between the current block and the search blocks. For a block C in the current frame and a block P in the preceding frame:

$$\text{Error}(\Delta x,\Delta y) = \sum_{x=0}^{X-1} \sum_{y=0}^{Y-1} |C(x,y) - P(x+\Delta x, y+\Delta y)| \qquad \text{Eq. 1}$$

where Δx and Δy are the horizontal and vertical displacements, respectively, of the search block relative to the current block, and X and Y are the horizontal and vertical dimensions, respectively, of the blocks being compared. The motion-estimation algorithm returns both the displacement vector $\langle \Delta x,\Delta y \rangle$ and Error($\Delta x,\Delta y$) of the best match. This calculation is repeated for each block in the current frame, which is assumed to be tiled with blocks—except for an indentation around the edges to allow for the displacement of search blocks.

We assume a standard video signal:

Frame rate:	30 frames / sec
Frame size:	360 pixels X 288 pixels

Pixel depth:	8 bits
Block size:	16 pixels X 16 pixels
Max hor. displ:	–8,+7
Max vert. displ:	–8,+7

The frame rate and frame size lead to a pixel rate of 3.11 MHz, which, in the bit-serial algorithm described here, means a bit rate and clock frequency of 24.88MHz.

The algorithm assumes four bit-serial inputs:

- Preceding Frame: Upper Band
- Preceding Frame: Lower Band
- Current Frame
- Start of Block

The ordering of pixels in the first three inputs is by horizontal *bands*. Each band is 16 pixels high, and the pixels within a band enter serially by column, from top to bottom (Figure 4). The two bands in the preceding frame straddle the band in the current frame. The Upper Band is offset from the current band by –8 pixels, while the Lower Band is offset by +8 pixels. The scanning of all three bands starts at the same time.

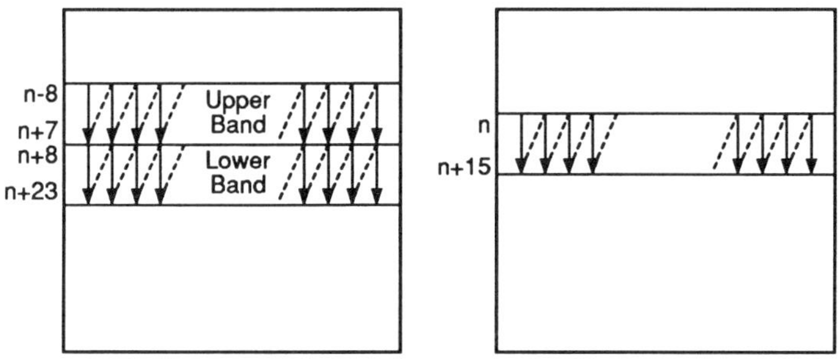

(a) Preceding Frame (b) Current Frame

Figure 4 Input Ordering

Start of Block (SOB) signals the start of a new 16pixelX16pixel block in the current frame. SOB is assumed to be active high, a high value coinciding with the first (8-bit) pixel of the new block.. The first 16X16 block in a band of the current frame is indented by 8 columns—to permit search blocks to be displaced by –8 pixels (Figure 5). Thereafter, blocks arrive every 16 columns. (In the algorithm described below, blocks need not abut, but they must not overlap.) Figure 6 illustrates the *search window*, containing 256 search blocks, corresponding to the current block. (The search window does not reach the bottom of the Lower Band because search blocks are displaced a maximum of +7 pixels.)

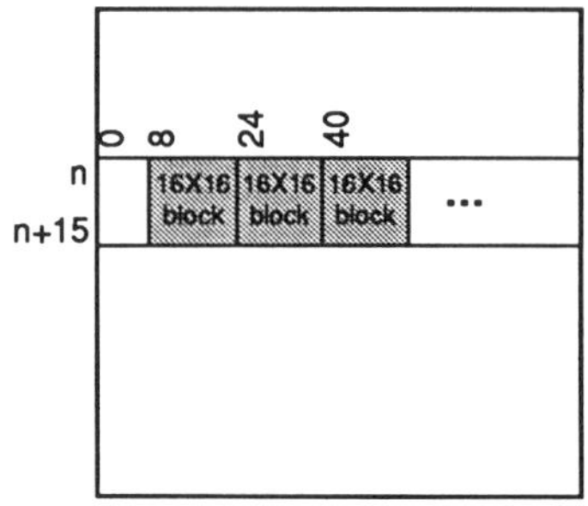

Figure 5 16X16 Blocks in the Current Frame

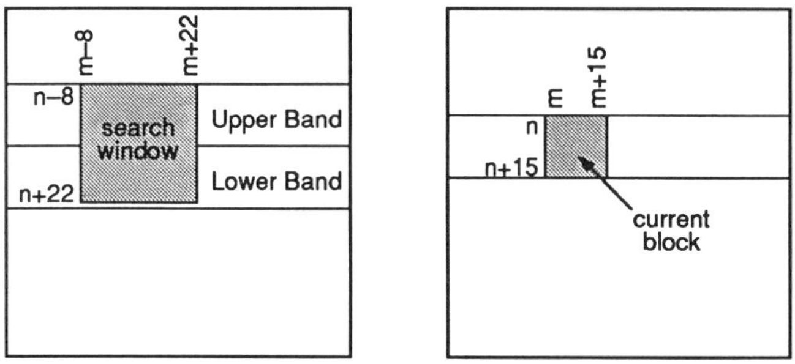

(a) Preceding Frame (b) Current Frame

Figure 6 Current Block & Corresponding Search Window

4 Algorithm Overview

The algorithm is bit serial and massively parallel. Figure 7 is a block diagram showing the heart of the algorithm: a linear array of 256 processing elements (PE's). The array, as well as each PE, has nine inputs:

- Current_Frame (CF)
- Select_Upper_Band (SUB)
- Select_Lower_Band (SLB)
- Upper_Band (UB)
- Lower Band (LB)
- Start_Of_Block (SOB)
- End_Of_Block (EOB)
- Error_Low_Order_Byte (ELOB)
- Error_High_Order_Byte (EHOB)

The Current-Frame pixel stream is supplied directly to the CF input of the linear array. The Upper-Band pixel stream, however, is delayed by 1024 clock

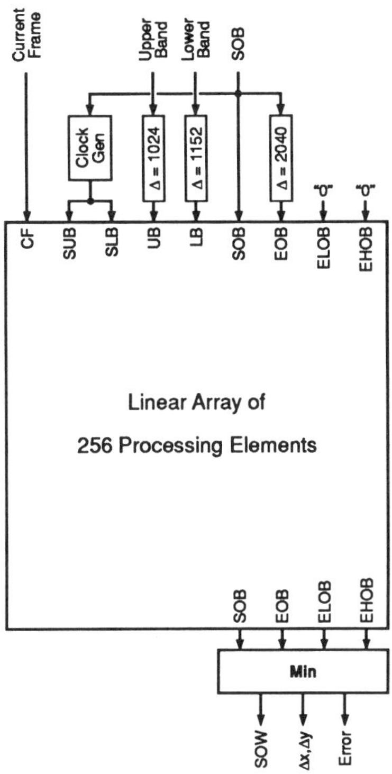

Figure 7 Block Diagram of Motion-Estimation Algorithm

cycles before being supplied to the UB array input. The delay retards the Upper-Band pixels by 8 columns relative to the current frame. (8 columns X 16 pixels/column X 8 cycles/pixel = 1024 cycles) The Lower-Band input is delayed by 1152 clock cycles, which retards the Lower-Band pixels by 9 columns relative to the current frame. (9 columns X 16 pixels/column X 8 cycles/pixel = 1152 cycles) Clock Gen provides an 8-cycle high value every 128 clock cycles to both the SUB and SLB inputs of the array. Start of Block synchronizes Clock Gen so that this high value coincides with the first pixel of each column on the CF, UB, and LB inputs. Start of Block is also fed directly to the SOB input and is delayed by 2040 clock cycles to provide the EOB input. The values on the ERL and ERH inputs are immaterial since they are replaced by PE-supplied values. ERL and ERH are arbitrarily set to "0".

The array outputs SOB and EOB are each delayed by 2304 (9X256) clock cycles from the corresponding inputs. The 256 errors associated with the 256 search blocks appear serially on the ELOB and EHOB outputs. **Min** looks for the minimal error and supplies that error along with the corresponding displacement vector.

5 Processing Elements

Each processing element has nine outputs to match its nine inputs. The array of 256 PE's is formed by connecting the outputs of one PE to the matching inputs of the next PE as shown in Figure 8. The computational demands of motion estimation are handled through massive parallelism: each PE computes the error between the current block and one of the 256 search blocks. Table 1 shows the correspondence between PE's and displacement vectors.

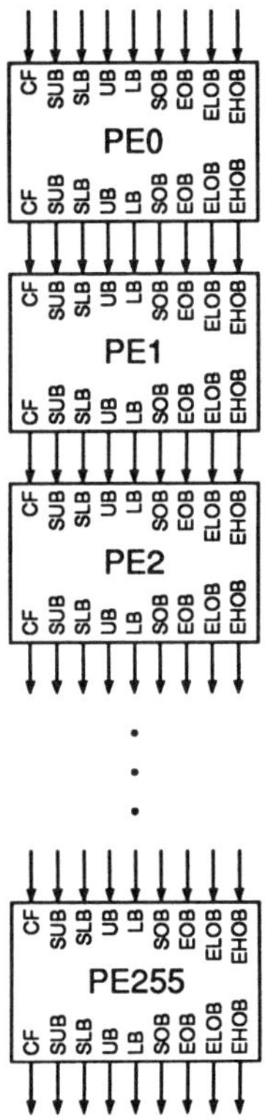

Figure 8 Array of 256 Processing Elements

There are actually two slightly different types of processing elements: the one in Figure 9 is used for all PE's except for PE0 and every 16th processing element thereafter—that is, for all PE's except PE 0, PE 16, PE 32, PE 48, PE 64, ... These PE's are identical to the one in Figure 9 except that the SR latch and associated multiplexer are omitted, and just Upper-Band pixels are fed to the absolute-difference module.

Table 1 Processing Elements & Associated Displacements

PE#		Displ.	PE#		Displ.
0	–	⟨-8,-8⟩	•		
1	–	⟨-8,-7⟩	•		
2	–	⟨-8,-6⟩	•		
3	–	⟨-8,-5⟩	237	–	⟨+6,+5⟩
4	–	⟨-8,-4⟩	238	–	⟨+6,+6⟩
5	–	⟨-8,-3⟩	239	–	⟨+6,+7⟩
6	–	⟨-8,-2⟩			
7	–	⟨-8,-1⟩	240	–	⟨+7,-8⟩
8	–	⟨-8,+0⟩	241	–	⟨+7,-7⟩
9	–	⟨-8,+1⟩	242	–	⟨+7,-6⟩
10	–	⟨-8,+2⟩	243	–	⟨+7,-5⟩
11	–	⟨-8,+3⟩	244	–	⟨+7,-4⟩
12	–	⟨-8,+4⟩	245	–	⟨+7,-3⟩
13	–	⟨-8,+5⟩	246	–	⟨+7,-2⟩
14	–	⟨-8,+6⟩	247	–	⟨+7,-1⟩
15	–	⟨-8,+7⟩	248	–	⟨+7,+0⟩
			249	–	⟨+7,+1⟩
16	–	⟨-7,-8⟩	250	–	⟨+7,+2⟩
17	–	⟨-7,-7⟩	251	–	⟨+7,+3⟩
18	–	⟨-7,-6⟩	252	–	⟨+7,+4⟩
	•		253	–	⟨+7,+5⟩
	•		254	–	⟨+7,+6⟩
	•		255	–	⟨+7,+7⟩

At the heart of both types of processing elements is an absolute-difference module and an accumulator. The circuitry to the left supplies the absolute-difference module with two bit-serial streams of pixels. The circuitry to the right resets the accumulator at the beginning of a new block, and at the end of the block places the accumulated error—both high- and low-order bytes[2]—on buses that transport the values to the end of the PE array.

The left-side circuitry assures that the proper pixels for the current block and search block are supplied to the absolute-difference module. There is no problem with the current-block pixels—they are constant for all PE's. The search-block pixels, however, must be synthesized from the Upper- and Lower-Band pixels. The key to making this happen is the *fast-track/slow-track* technique pioneered by Dick Lyon in his serial multiplier designs (see Lyon [7] and Denyer & Renshaw [8]). The SLB, UB and LB signals are on fast tracks—0 clock-cycle delay per PE—while the CF and SUB signals are on slow tracks—8 clock-cycle delay per PE.

2. The accumulator has high- and low-order bytes because although the maximum difference between two pixels is 11111111 (255), the maximum accumulated error between two blocks is 1111111100000000 (256X255=65,280).

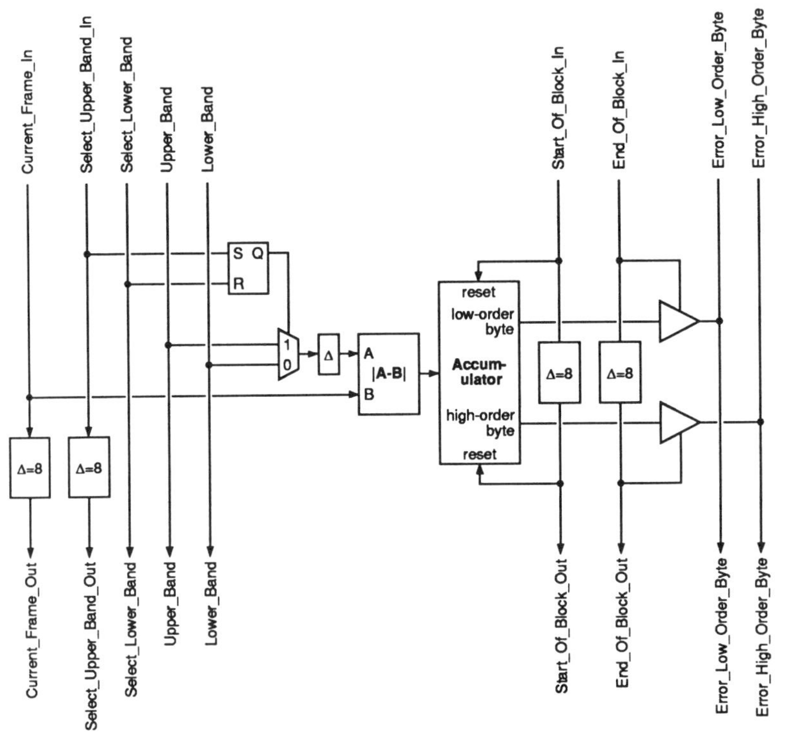

Figure 9 Processing Element N (N $\neq$ 0 mod 16)

To understand how the circuitry works, consider the pixel-alignments entering PE 0. The pixels in the Synthesized band, which are identical to the Upper-Band (UB) pixels, are displaced by $\langle -8,-8\rangle$ relative to the Current (CF) pixels.[3] This is consistent with Table 1 which indicates a displacement of $\langle -8,-8\rangle$ for PE 0.

Now consider the pixels entering PE 1. Because of the 8-clock-cycle delays on the Current Frame and the Select_Upper_Band (SUB) signal, the Upper and Lower Bands and the Select-Lower-Band (SLB) signal have been advanced by one pixel (8 clock cycles) relative to the Current pixels and SUB. SUB, which is always aligned with the Current pixels, sets the SR latch at the beginning of each new column of Current pixels. Setting the latch causes the Synthesized pixels to come from the Upper-Band until the latch is reset by SLB. SLB, which is coincident with SUB entering PE 0, is advanced by one pixel relative to SUB entering PE 1. The Synthesized Band, therefore, takes only the 16th (last) pixel in each

3. The unit delay just before the A input of the absolute-difference module is only for pipelining. Small delays on various array inputs, including CF, may have to be inserted to insure correct alignment.

column from the Lower Band. The result of all this is that the Synthesized pixels are now displaced by ⟨–8,–7⟩ relative to the Current pixels. In other words, the search block has dropped down one row. This pattern continues for PE 2 through PE 15 with each succeeding PE adding a row of pixels from the lower band.

The pattern is broken for PE 16 because we no longer want to move down one row but rather back to the top of the search window and to the right one column. This is accomplished by eliminating the Upper-Band/Lower-Band multiplexing logic, and simply reverting to the upper band. The pattern described above for PE's 0–15 now repeats for PE's 16–31, and for every succeeding group of 16 PE's.

On the right side of Figure 9, we again employ the fast-track/slow-track technique. The SOB and EOB signals are on slow tracks—8 clock cycles delay per PE—to match the speed of the CF signal on the left side. The ELOB and EHOB signals are on fast tracks—0 clock cycles delay per PE—so that the two errors from each PE fill unique *slots*.

The SOB signal is timed so that the low- and high-order registers in the accumulator are reset just as a new block begins[4,5]. The EOB signal is timed so that the values in these two registers are placed on the two buses just as the last pixel error for the current block is accumulated. Note that <u>no clock cycles are lost in the changeover from one block in the current frame to the next</u>. Except at the left and right edges of the frame, <u>the absolute-difference modules and the accumulators in all the PE's are always productively engaged</u>.

Finally, notice that because ERL and ERH are on fast tracks while EOB is on a slow track, errors placed on the ERL and ERH buses by the n'th PE immediately follow the errors of the (n–1)'th PE. The result is that errors appear at the output of the PE array in the order in which they are created. The 256 (high- and low-order) errors exactly fill the 256 slots available between Start-of-Block and End-of-Block.

6 Absolute-Difference Module and Accumulator

At the core of each processing element are a bit-serial absolute-difference module and a bit-serial accumulator.

The task of performing the absolute difference is actually split between a subtracter module (Figure 10) and the low-order stage of the accumulator (Figure 11). Within the subtracter module is a bit-serial subtracter consisting of a full adder, two inverters, AND gate and unit delay. Employing a technique known as *explicit* subtraction (see [8] for a discussion of bit-serial operations), it takes A, B and ~Start_Of_Word as inputs and produces A–B. The register

---

4. We are ignoring the latency in the absolute-difference module and the accumulator. Once these values are known, both the SOB and EOB signals will have to be delayed prior to PE 0 by an appropriate amount to compensate.
5. Because the accumulator is bit serial, the high-order byte is accumulated one word (8 clock periods) after the low-order byte.

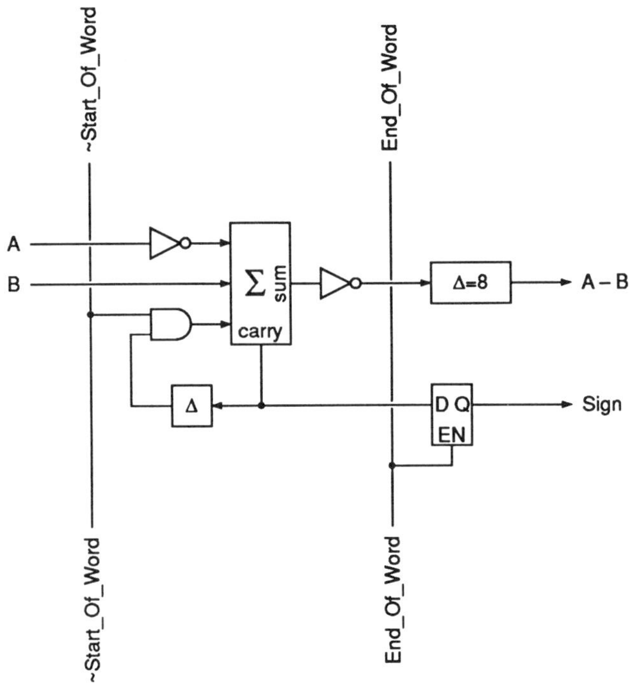

Figure 10 Bit-Serial Subtracter

enabled by End_Of_Word latches the carry (borrow) at the end of each subtraction. A borrow of 0 indicates that A–B is positive and may be accumulated without modification. A borrow of 1, on the other hand, means that A–B is negative and must be negated when accumulated. The 8-unit delay aligns A–B with Sign. The low-order stage of the accumulator (top part of Figure 11) uses a full adder and recirculated 8-bit sums to accumulate the results of the subtracter module. If the Sign input is 0, then A–B enters the full adder unaltered and the carry input at the start of a new addition is 0. If, however, Sign is 1, then A–B is inverted and a 1 is supplied on the carry input at the start of a new addition, effectively negating A–B.

Overflow of the low-order stage of the accumulator is handled by the high-order stage (bottom part of Figure 11) which employs a half adder and acts as an incrementer. Whenever there is a carry out from the low-order stage at the end of addition, the value in the high-order stage is incremented by 1. (Since the high-order stage never overflows, the carry input at the start of a new operation is always 0.)

Both the low- and high-order stages make their results available and are reset at the midpoints of their 8-stage feedback paths. (The location is arbitrary and is made for the convenience of layout.)

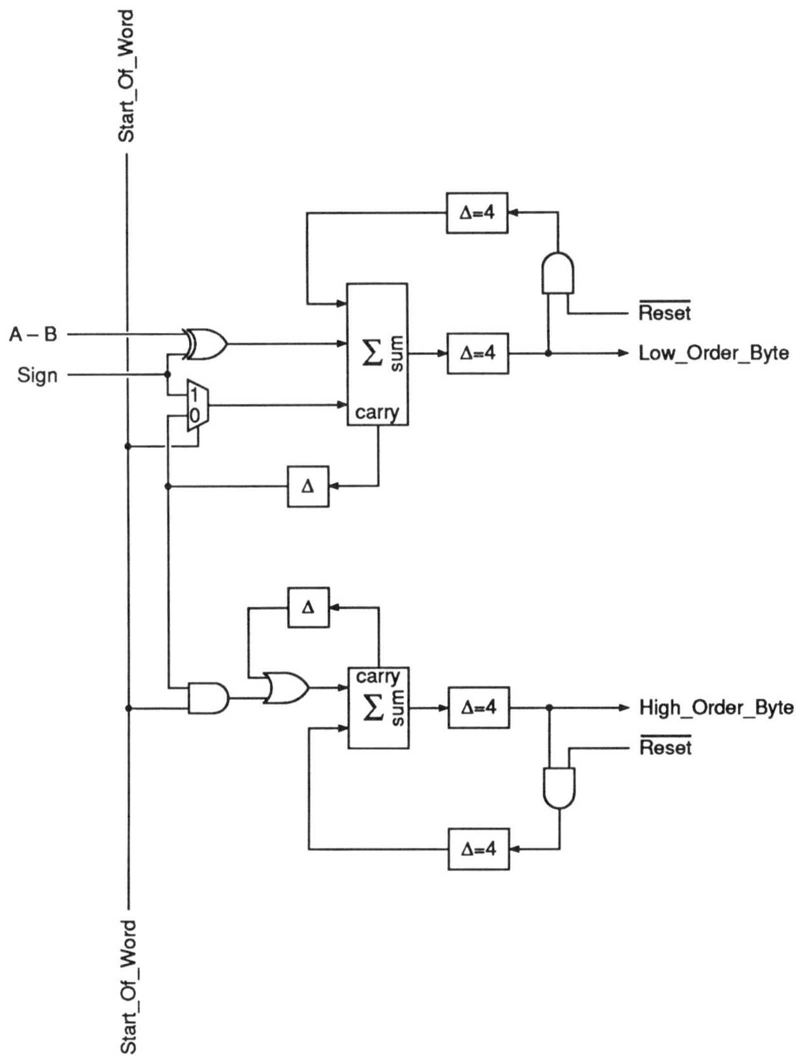

Figure 11 Bit-Serial Accumulator

7 Processing-Element Layout

At the heart of the CLi6000 implementation of the motion-estimation algorithm is the layout for the processing-element macro. For both PE types, the CLi6000 requires just 72 cells arranged in a 2X36 rectangle. And because inputs & outputs on the macro's bottom side mate perfectly with the corresponding outputs & inputs on the macro's top side, an array of PE's can be formed by simply *plugging* macros together.

Because it is more complex, we provide the layout (Figures 12 & 13) for the circuit (Figure 9) used for all PE's except those divisible by 16. Within each of these hard macros, there are approximately 285 equivalent gates contained in:

- 40 registers
- 5 NAND/XORs
- 3 Multiplexers
- 3 XORs
- 2 Tri-state drivers
- 2 NANDs
- 2 Inverters
- 1 SR Latch
- 1 AND

Symbols for various cell states are explained in documentation for the CLI design environment.

Figure 12 Overview of PE Layout

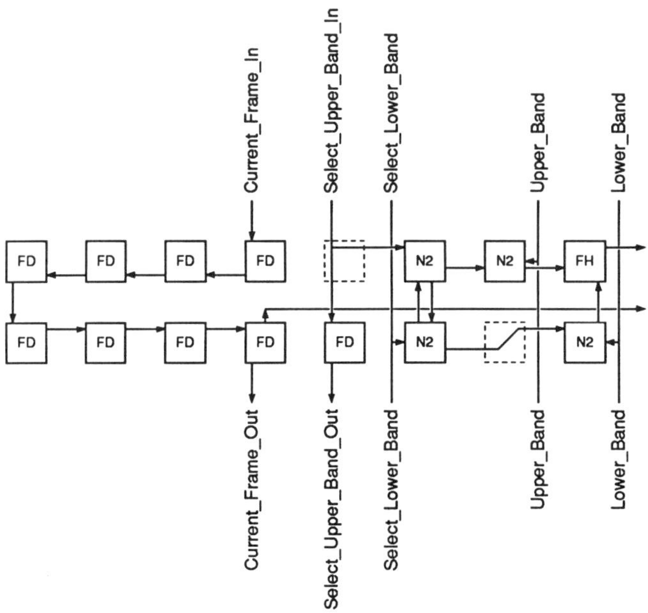

Figure 13(a) Input Section

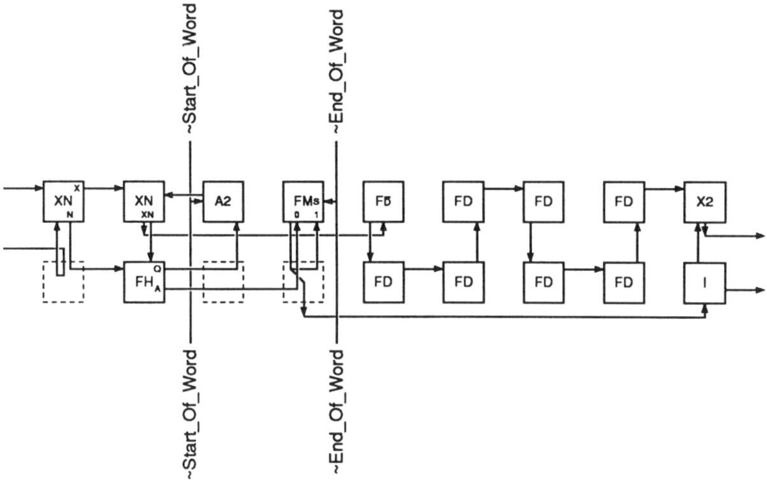

Figure 13(b) Absolute-Difference Section

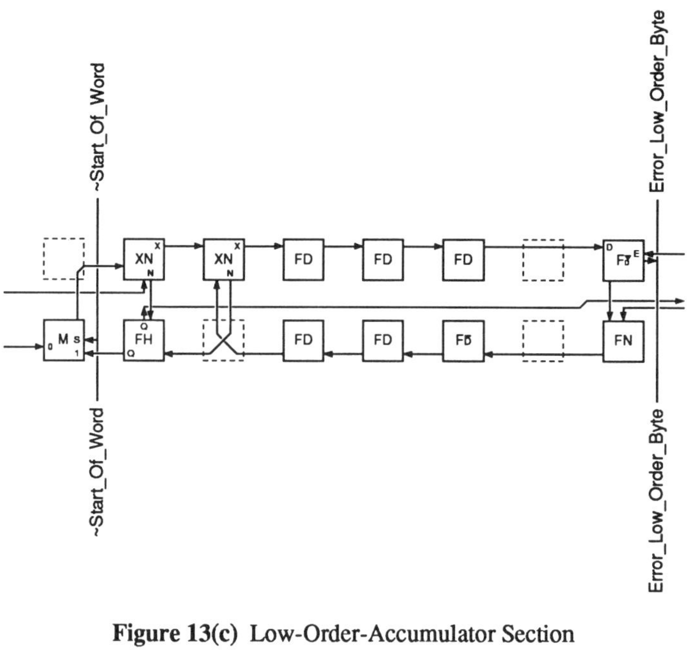

Figure 13(c) Low-Order-Accumulator Section

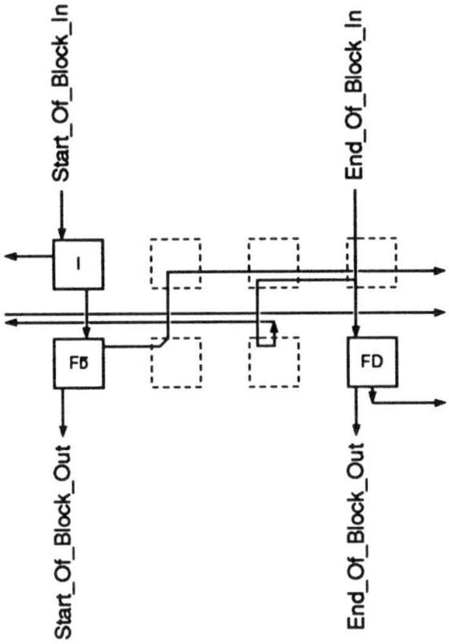

Figure 13(d) SOB & EOB Section

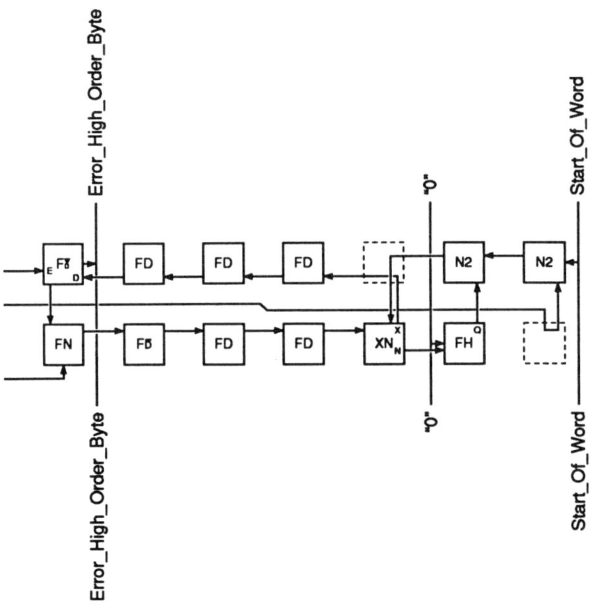

Figure 13(e) High-Order-Accumulator Section

8 Technology Comparisons

At 4,000 MIPS, motion estimation of a standard digital video signal easily surpasses the performance of even the most energetic uni-processors. In fact, the work reported on here began at Apple Computer where a Cray XMP could achieve only about 1/10 the performance needed for real-time motion estimation.

In Table 2, we provide comparisons of various technologies for implementing motion estimation. We make the following assumptions:

- We generously assume a performance level of 100 MIPS for the RISC microprocessor, but this is still a factor of 40 short of the 4,000 MIPS required.
- We assume that the FPGAs in the table implement the systolic array described above. Lower bounds for the Algotronix and Xilinx parts are based solely on register count.
- The CAL1024 SRAM FPGA from Algotronix requires two cells to implement a register and therefore at least three rows for each processing element, which has 40 registers.
- The XC3090 and XC4010 SRAM FPGAs from Xilinx provide two registers in each cell (CLB), and one 20-CLB row/column contains the 40 registers needed for a processing element.
- The CLi6005 fits one 2X36 processing element in two rows. The CLi6010 —available Spring 1993—fits two 2X36 processing elements in two rows.

Table 2 Technology Comparisons

Device	Size (Cells)	Devices Required
Cray XMP	—	≈10
RISC μP	—	≥40
CAL1024	32X32 = 1,024	≥24
XC3090	16X20 = 320	≥16
XC4010	20X20 = 400	≥13
CLi6005	56X56 = 3,136	11
CLi6010	80X80 = 6,400	4

9 Conclusions

The CLi6000 series of SRAM-based field-programmable gate arrays provides features supporting efficient implementation of pipelined arithmetic circuits. These features are illustrated through a systolic array for motion estimation, an especially compute-intensive algorithm used in digital video compression. The CLi6000's high register count, NAND/XOR cell states, multiplexer states, fine-grained architecture, array symmetry and high cell connectivity all work together in efficiently implementing the systolic array's heavily pipelined, bit-serial arithmetic operations.

While only one example, this exercise suggests the suitability of the CLi6000 as a flexible alternative to hard-wired systolic arrays.

Bibliography

[1] Gray, J.P. & Kean, T.A., "Configurable Hardware: A New Paradigm for Computation." Advanced Research in VLSI, Proceedings of the Decennial Caltech Conference on VLSI, March 1989.

[2] Gokhale, M. et. al., "Building and Using a Highly Parallel Programmable Logic Array," Computer, Vol. 24, No. 1, January 1991.

[3] Bertin, P., Roncin, D. & Vuillemin, J., "Programmable Active Memories: Performance Measurements," FPGA '92, February 1992.

[4] Furtek, F., Stone, G. & Jones, I., "Labyrinth: A Homogeneous Computational Medium," 1990 Custom Integrated Circuits Conference, May 1990.

[5] Furtek, F., "An FPGA Architecture for Massively Parallel Computing," 2nd International Workshop on Field-Programmable Logic and Applications, August 1992.

[6] "CLi6000 Series Field-Programmable Gate Arrays," Concurrent Logic, Inc., May 1992.

[7] Lyon, R., "Two's Complement Pipeline Multipliers", IEEE Transactions on Communications, April 1976, pp. 418-425

[8] Denyer, P. and Renshaw, D., VLSI Signal Processing: A Bit-Serial Approach, Addison-Wesley, 1985, pp. 303, 151 & 196.

Edge-Triggering vs. Two-Phase Level-Clocking

(Extended Abstract)

Marios C. Papaefthymiou
Keith H. Randall
MIT Laboratory for Computer Science
Cambridge, Massachusetts 02139

Abstract

Level-clocked designs that employ a two-phase, nonoverlapping clocking scheme have the theoretical potential to operate up to twice as fast as edge-triggered designs. We have run experiments that demonstrate, however, that edge-triggering is often just as fast as two-phase clocking, and that the speed potential of two-phase clocking is generally not obtained except when the delay between any two consecutive latches is roughly uniform and close to the maximum gate delay. Moreover, our experiments show that asymmetrical clocking of a two-phase circuit often does not provide any speedup over optimal symmetric clocking schemes.

Level-clocking can lead to substantial reductions in the number of storage elements in a circuit, however. Our experiments show that for edge-triggered circuitry that has been retimed to operate with the minimum possible clock period, by replacing each flip-flop by a pair of level-sensitive latches and subsequently retiming the resulting two-phase circuit, the number of storage elements can be reduced by up to 38% without increasing the clock period of the final design or affecting its I/O specification. Reductions of greater than 25% were achieved for more than one third of the circuits we tested.

Our experiments were performed using TIM, a timing optimization package currently under development at MIT. We ran our tests on MCNC benchmark circuits, AT&T communication circuits, and custom circuitry designed for MIT's Alewife machine.

1 Introduction

Level-clocking is becoming an increasingly popular alternative to edge-triggering as a clocking methodology for high-performance designs. In level-clocked circuitry, clocked storage elements are implemented using level-sensitive latches, which become transparent and transmit their inputs unimpeded to their outputs while their clock signals are asserted. Proponents of level-clocking argue that level-clocked circuitry can provide more flexibility in meeting a specific clock period and that it has the theoretical potential to operate faster than the more conventional edge-triggered circuitry. These arguments are based on the fact that in level-clocked circuitry computations are allowed to extend beyond a single clock period during the transparent phase of the level-sensitive latches, in constrast to edge-triggered circuitry, in which computations along every path must complete within a clock period. Advocates of edge-triggering, on the other hand, present simplicity and implementation ease as major advantages of edge-triggering, since edge-triggered flip-flops directly support the abstraction of a storage element that is synchronized by the ticking of a clock.

They also refer to the existence of powerful design tools as another major incentive for designing circuitry with edge-triggered flip-flops.

These arguments in support of edge-triggering and level-clocking are either theoretical or nonquantifiable. We wanted to make an empirical study of the two clocking methodologies. For this purpose, we have run experiments that compare edge-triggered implementations of synchronous circuitry and corresponding level-clocked implementations that employ a two-phase, nonoverlapping clocking scheme. Our empirical comparison focused on two specific quantitative measures: speed and number of storage elements. We ran our tests using TIM, a timing optimization tool currently under development at MIT [17].

Our speedup experiments show that edge-triggered circuitry often operates just as fast as two-phase circuitry, despite the theoretical advantage of two-phase clocking, and that the speed potential of two-phase clocking is generally obtained only when the combinational delay between any two consecutive latches is roughly uniform and close to the maximum gate delay. Our experiments also show that two-phase clocking leads to greater speedups when all gate delays in the circuit are roughly equal. Our experimentation with clock tuning suggests that asymmetric clocking schemes provide little or no speedup over optimal clocking schemes.

With respect to the number of storage elements, however, our experiments demonstrate that two-phase clocking can lead to substantial reductions in aggressive edge-triggered designs that operate at maximum speed. For two of our test circuits, we obtained reductions of 38%; for more than one third of our circuits, we obtained reductions of 25%; and for more than half of our circuits, reductions exceeded 18%. For low-performance designs that operate below their speed potential, however, our experiments show that two-phase clocking does not reduce the number of storage elements.

How can edge-triggering and two-phase clocking be fairly compared on an empirical basis? First, a fair experiment should compare competing circuit implementations that have the same functionality. It would be meaningless to compare two circuits that compute different functions. Second, a fair experiment should compare competing circuit implementations based solely on differences due to their storage elements. It would be unfair to compare two circuits which differ in their combinational elements, for example, because such a comparison would not depend only on the clocking methodology employed.

We did not have the resources to embark on designing pairs of competing circuits for various applications. We settled, therefore, on the strategy of taking edge-triggered circuits and using our timing tool TIM to produce equivalent two-phase circuits. We could have done the reverse, converting two-phase circuits to edge-triggered ones, but since edge-triggered designs are more popular, we were able

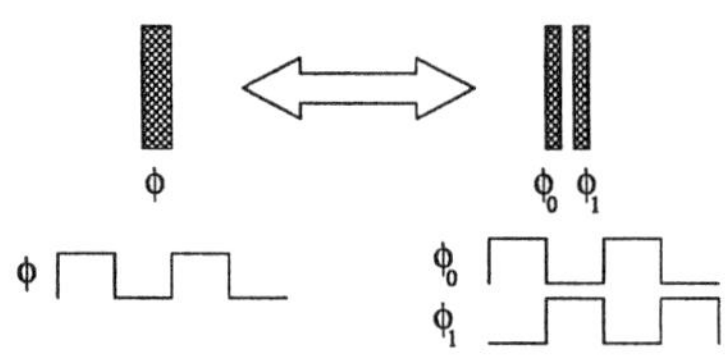

Figure 1: Replacement of an edge-triggered flip-flop by a pair of level-sensitive latches. The flip-flop is clocked by a single clock ϕ, and the two level-sensitive latches are clocked on the phases ϕ_0, ϕ_1 of a two-phase, nonoverlapping clocking scheme.

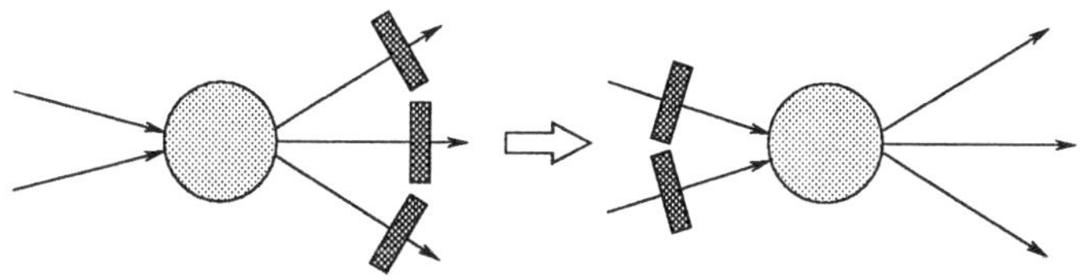

Figure 2: This figure illustrates retiming of a gate with lag 1. In this case, one storage element is removed from each output wire of the gate, and one storage element is inserted at each input wire of the gate. The total number of storage elements is reduced by 1. The critical paths in the circuit may also change as a result of the relocation of the storage elements.

to obtain several interesting edge-triggered circuits. (We hope to also obtain interesting level-clocked designs and remedy this situation before the full paper is published.)

We produced a two-phase circuit from an edge-triggered one by following a two-step procedure. The first step of this procedure was to replace each edge-triggered flip-flop by a pair of back-to-back level-sensitive latches that are clocked by a two-phase, nonoverlapping clocking scheme, as shown in Figure 1. (In fact, it is common in VLSI to implement edge-triggered flip-flops by a pair of back-to-back level-sensitive latches [3, 24].) The two-phase circuit that results after this conversion has the same clock period and the same number of storage elements as the original edge-triggered circuit, assuming that each edge-triggered flip-flop counts as two level-sensitive latches. Moreover, the placement of its latches is dictated by the original edge-triggered design, and the potential of two-phase clocking due to alternate placements of the latches in the circuit is not revealed. Thus, we needed a method to relocate storage elements and explore the space of possible placements in the circuit without changing its functionality or its I/O specification.

The second step of the procedure was to use the "retiming" transformation to relocate the storage elements of the two-phase circuit that resulted from the first step. Retiming relocates storage elements in both edge-triggered and level-clocked circuitry without changing its functionality [8, 9, 10]. In addition, retiming is a "universal" transformation for speeding up circuits, in the sense that any other functionality-preserving transformation that did better than retiming would depend on the functionality of the gates in the circuit [9]. Figure 2 illustrates the retiming operation for a gate in a circuit. Observe that retiming can change the clock period as well as the number of storage elements in a circuit.

Our experimental procedure compared an optimal edge-triggered implementation that we obtained from an original edge-triggered circuit with an optimal two-phase implementation of the corresponding two-phase circuit. The use of an optimal edge-triggered implementation as a reference point was essential to ensure that we did not penalize edge-triggering due to suboptimalities in the original edge-triggered circuit that depended on the placement of the storage elements by the circuit designer and were not intrinsic to edge-triggering.

We performed two kinds of experiments. The first kind of experiments compared edge-triggered and two-phase circuits with respect to speed. Our basic experimental approach was the following. First, we retimed a given edge-triggered implementation for maximum speed. Then, by using retiming in conjunction with tuning of the clocking schemes, we obtained the fastest possible implementation of the corresponding two-phase circuit, and we compared the speed of the two optimal implementations. The second kind of experiments compared edge-triggered and two-phase circuits with respect to their number of storage elements, when operating at some specified clock period. We first retimed a given edge-triggered implementation without changing its I/O specification, in order to achieve the specified clock period with the minimum number of flip-flops. We then retimed the corresponding two-phase circuit without changing its I/O specification, in order to achieve the same clock period with the minimum number of latches. We compared the number of storage elements in the two optimal implementations, under the reasonable assumption that each edge-triggered flip-flop is counted as two level-sensitive latches.

Timing verification and optimization of synchronous circuitry has been the subject of extensive study [1, 2, 4, 5, 6, 7, 8, 10, 11, 12, 18, 19, 20, 22, 23]. The concept of replacing each edge-triggered flip-flop by a pair of back-to-back level-sensitive latches, and then using retiming for speed optimization has been mentioned in [1, 6, 11]. The potential of level-clocking for reducing the number of storage elements has been mentioned in [11]. The idea of using latches instead of flip-flops has been also used in [21]. Retiming for speed has been studied in the context of single-phase level-clocked circuits in [19]. Despite the large amount of work in this area, our contribution is (we believe) the first attempt to quantify empirically the performance differences of edge-triggering and two-phase clocking.

The remainder of the paper has three sections. Section 2 describes our experimental methodology and reports our results on the relative speed of the two implementation approaches. In Section 3 we present our experimental results on latch-minimization. Section 4 concludes with directions for further research. Appendix A presents an upper bound on the relative speedup that can be achieved by two-phase clocking over edge-triggering and explains intuitively how two circuit characteristics, the maximum gate delay $d_{\max}$ and the critical ratio R of the circuit, determine the speedups that can be achieved by two-phase clocking. In Appendix B we show that the problem of retiming level-clocked circuitry in order to minimize the number of latches can be solved in polynomial time for symmetric clocking schemes.

2 Speedup experiments

In this section, we present our investigation of edge-triggering and two-phase clocking with respect to speed. First, we briefly refer to our tools and test circuits. We move on to describe and motivate our experimental methodology, and then we discuss our results.

Our experiments were performed using TIM, a timing optimization package currently under development at MIT [17]. TIM performs a variety of functions on two-phase circuitry, such as timing verification, retiming, clock tuning, and sensitivity analysis. It also performs timing verification, retiming, and sensitivity analysis on edge-triggered circuitry. TIM has been implemented in C, and most of its functions have been integrated in the SIS tools from Berkeley. Our test circuits were MCNC benchmark circuits and AT&T communication circuits, all of which were originally designed with edge-triggered flip-flops. The largest among these circuits had 290 gates, and the current version of TIM performed retiming and simultaneous tuning on that circuit in less than 2 minutes on a 16MB SPARCstation.

Experimental methodology

In our speedup experiments we employed the following three optimizations:

OP1 Retiming of edge-triggered circuitry for maximum speed of operation (minimum clock period).

OP2 Retiming of two-phase circuitry for maximum speed of operation with a symmetric clocking scheme.

OP3 Retiming and simultaneous clock tuning of two-phase circuitry for maximum speed of operation.

Using these three optimizations, we initially performed experiments SP1 and SP2.

SP1 We compared the speed of each original edge-triggered circuit that was optimized using OP1 with the speed of the corresponding two-phase circuit that was optimized using OP2.

SP2 We compared the speed of each original edge-triggered circuit that was optimized using OP1 with the speed of the corresponding two-phase circuit that was optimized using OP3.

The goal of our experimentation was not only to investigate whether two-phase clocking could speed-up the particular edge-triggered circuits in our test suite. We also wanted to determine specific design characteristics that may lead to faster two-phase circuits. To that effect, we performed experiments SP3 and SP4 on altered versions of our original test circuits that were obtained by modifying them in two ways.

The first modification changed the number of storage elements in the original circuits by pipelining, a transformation that increases the latency of a computation without decreasing its throughput. A degree-P pipelining of each circuit was obtained by multiplying the number of storage elements on each wire of the circuit by the integer P. The purpose of this transformation was to investigate which of the two implementation approaches is favored more when the number of storage elements is increased.

The second modification changed the gate delays of the original circuits. For each circuit G, we created four additional circuits G_i for $i = 0.8, 0.6, 0.4, 0.2$. Each G_i was topologically identical to G, but its gate delays d_i were modified. For each circuit G_i, each gate delay $d_i(v)$ was set equal to $d^i(v)$, where $d(v)$ was the delay assigned to v using the library corresponding to the circuit G. Thus, for smaller values of the exponent i, the gate delays in the circuits G_i became increasingly uniform. The objective of this modification was to see how uniformity of gate delays affects the speed of the two implementations.

Using the three optimizations on the modified circuits we performed experiments SP3 and SP4.

SP3 On each circuit G_i for $i = 1.0, 0.8, 0.6, 0.4, 0.2$, we applied the following procedure for $P = 1, 2, \ldots, 6$. We optimized the edge-triggered circuit using OP1, and we compared its speed with its corresponding two-phase circuit that was optimized using OP2.

SP4 On each circuit G_i for $i = 1.0, 0.8, 0.6, 0.4, 0.2$, we applied the following procedure for $P = 1, 2, \ldots, 6$. We optimized the edge-triggered circuit using OP1, and we compared its speed with its corresponding two-phase circuit that was optimized using OP3.

Note that for $i = 1.0$ and $P = 1$, experiments SP3 and SP4 were identical to experiments SP1 and SP2, respectively.

Experimental results

Remarkably, our initial experiments SP1 and SP2 indicated that two-phase clocking was no better than edge-triggering for any of our test circuits. The application of the three optimizations OP1, OP2, and OP3 on the original circuits, with gate delays assigned by their corresponding libraries, showed no speedup by switching to two-phase clocking. Although this result was surprising and unexpected, it could not have been a mere coincidence. Our subsequent empirical investigation with experiments SP3 and SP4 led us to the conclusion that there are two important circuit characteristics that determine the relative speed of the two implementation approaches: the maximum gate delay $d_{\max}$ and the *critical ratio* R, which is defined as the maximum ratio of total delay over total number of storage elements around the cycles in the edge-triggered circuit.

Our experimental results for SP3 are illustrated in the plots of Figure 3. Each plot gives data for an original test circuit $G_{1.0}$ and its four delay-modified versions G_i for $i = 0.8, 0.6, 0.4, 0.2$. For each of the five delay configurations of a test circuit and for degrees of pipelining up to 6, each plot gives two numbers: the speedup f_{lc}/f_{et} achieved by two-phase clocking over edge-triggering, and the ratio $d_{\max}/R$, where $d_{\max}$ is the maximum gate delay and R is the critical ratio of the circuit. For each circuit G_i, the value of the ratio $d_{\max}/R$ closest to 1 is boldfaced.

As is apparent from the graphs, for almost every delay configuration, the maximum speedup is achieved when the ratio $d_{\max}/R$ is closest to 1. In the five configurations of `mult16a`, for example, as the ratio $d_{\max}/R$ increases from small values and approaches 1 from below, the speedup constantly increases. When the ratio exceeds 1, the speedup soon drops down to 1. A similar pattern is revealed for almost all of our test circuits. This phenomenon can be justified as follows. The critical ratio R is a lower bound on the clock period of both the edge-triggered and the level-clocked circuit [6, 15]. Consequently, the longest combinational delay in the circuits is at least R under any transformation that does not change the number of storage elements around the cycles in the circuit. Retiming distributes the storage elements, however, so that combinational path delays are roughly equal across the circuit and close to the critical ratio R. When R becomes comparable to the maximum gate delay $d_{\max}$, then the longest combinational delay also tends to approach $d_{\max}$, and then the potential of two-phase clocking becomes apparent. Intuitively, when R approaches $d_{\max}$, level-clocking evens out differences among path delays more effectively than edge-triggering by letting the computations ripple through the transparent latches.

Let us examine more closely some characteristic graphs in Figure 3. Our initially surprising results from experiments SP1 and SP2 can be explained by looking at the ratios $d_{\max}/R$ of the original circuits, which correspond to $P = 1$ and $i = 1.0$. For every such circuit, the ratio $d_{\max}/R$ is smaller than 0.67. The only exceptions are `mult16b`, `ampseq2`, and `ampseq1`. `mult16b` has a ratio greater than 1, and consequently, it is already heavily pipelined. For higher degrees of pipelining, `mult16b` does not become any faster, which leads us to the conclusion that the original design of `mult16b` takes full advantage of any existing speed potential. The situation with `ampseq2` is similar. The original design has already no margin for improvement, and for higher degrees of pipelining there are sufficiently many storage elements for edge-triggering to be as fast as two-phase clocking. The situation with `ampseq1` is

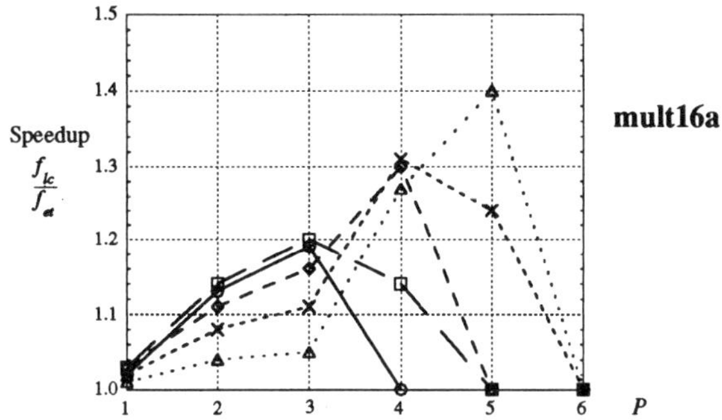

· △ · i=0.2	0.20	0.39	0.59	0.79	**0.98**	1.17	$\frac{d_{max}}{R}$
-×- i=0.4	0.23	0.46	0.68	**0.91**	1.14	1.37	
–◇– i=0.6	0.26	0.53	0.79	**1.05**	1.31	1.58	
–□– i=0.8	0.31	0.60	**0.90**	1.20	1.50	1.80	
–○– i=1.0	0.34	0.67	**1.00**	1.34	1.67	2.01	

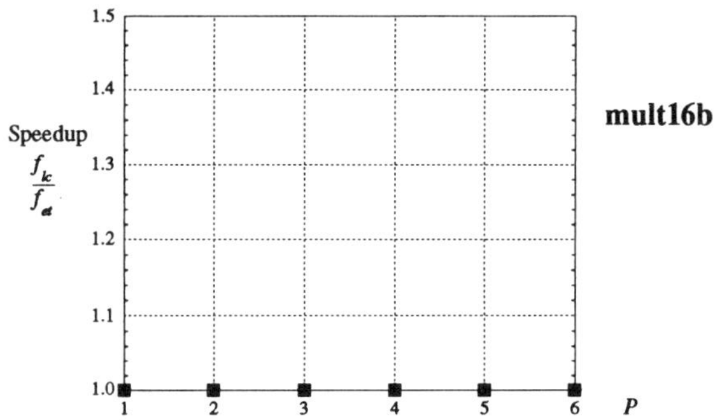

· △ · i=0.2	0.59	**1.18**	1.77	2.34	2.95	3.54	$\frac{d_{max}}{R}$
-×- i=0.4	**0.68**	1.36	2.04	2.68	3.40	4.02	
–◇– i=0.6	**0.76**	1.51	2.28	3.05	3.78	4.50	
–□– i=0.8	**0.83**	1.65	2.48	3.33	4.16	4.96	
–○– i=1.0	**1.02**	2.04	3.06	4.11	5.14	6.13	

Figure 3: Results of experiment SP3 on the MCNC benchmark and the AT&T circuits. Each plot corresponds to a test circuit. The first row of the horizontal axis gives the pipelining degree P. Each of the next five rows corresponds to a circuit G_i for $i = 0.2, 0.4, 0.6, 0.8, 1.0$, and it gives the ratio d_{max}/R for each pipelining degree. In each row, the ratio d_{max}/R closest to 1 is boldfaced. The vertical axis gives the speedup f_{lc}/f_{et} obtained for a specific i and a specific P. The clock frequency f_{et} was obtained by applying OP1. The clock frequency f_{lc} was obtained by applying OP2. For almost every test circuit, maximum speedups are achieved when the ratio d_{max}/R is closest to 1, or equivalently, when R is closest to d_{max}. Greater peak speedups are achieved as we move from $G_{1.0}$ to $G_{0.2}$, that is, as the gate delays become roughly equal across the entire circuit. The results of experiment SP4 have no significant differences from the results in this figure.

s344

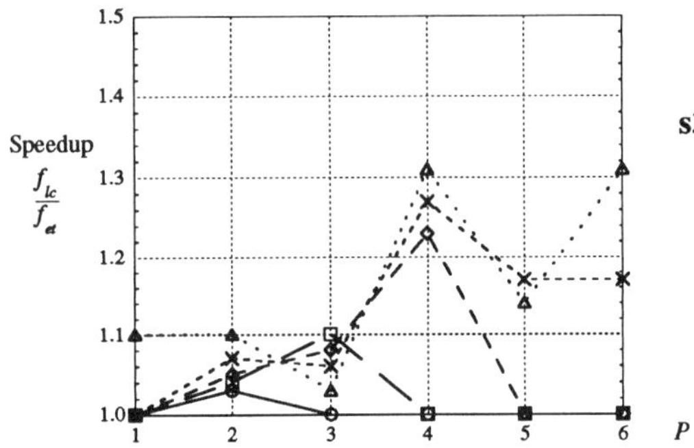

d_{max}/R	1	2	3	4	5	6
i=0.2	0.15	0.29	0.44	0.59	0.73	**0.88**
i=0.4	0.19	0.39	0.58	0.77	**0.96**	1.16
i=0.6	0.25	0.50	0.75	**1.00**	1.25	1.50
i=0.8	0.32	0.65	**0.97**	1.29	1.61	1.93
i=1.0	0.41	**0.83**	1.24	1.65	2.07	2.48

s382

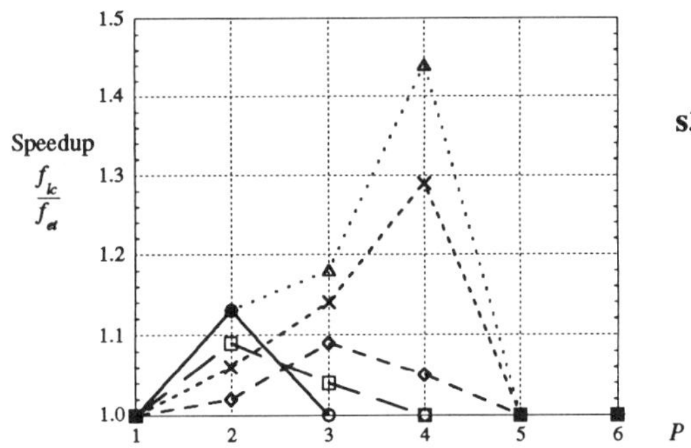

d_{max}/R	1	2	3	4	5	6
i=0.2	0.23	0.47	0.70	**0.93**	1.16	1.39
i=0.4	0.27	0.54	0.81	**1.08**	1.35	1.62
i=0.6	0.31	0.62	**0.93**	1.24	1.55	1.86
i=0.8	0.35	0.70	**1.05**	1.40	1.75	2.10
i=1.0	0.39	0.78	**1.16**	1.55	1.94	2.33

s400

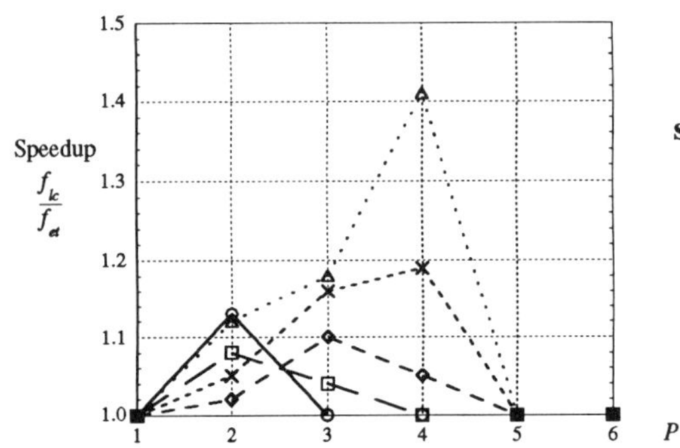

d_{max}/R	1	2	3	4	5	6
i=0.2	0.24	0.47	0.71	**0.95**	1.19	1.43
i=0.4	0.28	0.56	0.83	**1.11**	1.39	1.67
i=0.6	0.32	0.65	**0.97**	1.30	1.62	1.94
i=0.8	0.37	0.74	**1.11**	1.49	1.86	2.22
i=1.0	0.42	**0.84**	1.26	1.68	2.10	2.52

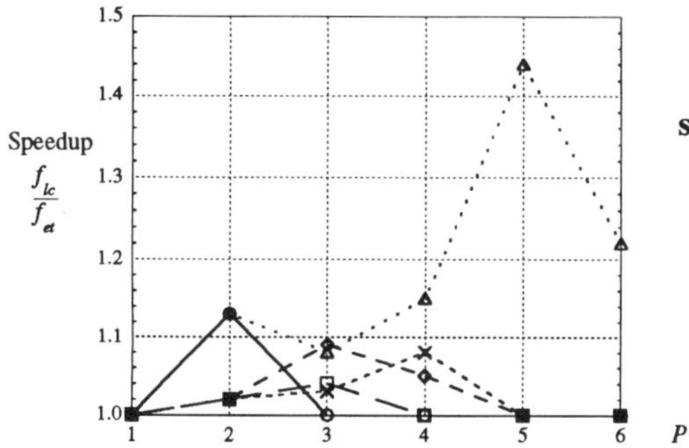

s820

	1	2	3	4	5	6	
· △ · i=0.2	0.17	0.33	0.50	0.67	0.84	**1.01**	$\frac{d_{max}}{R}$
-✕- i=0.4	0.24	0.48	0.72	**0.97**	1.20	1.44	
—◇— i=0.6	0.33	0.66	**1.00**	1.33	1.66	2.00	
—□— i=0.8	0.43	**0.87**	1.30	1.73	2.16	2.59	
—○— i=1.0	0.53	**1.07**	1.61	2.14	2.68	3.21	

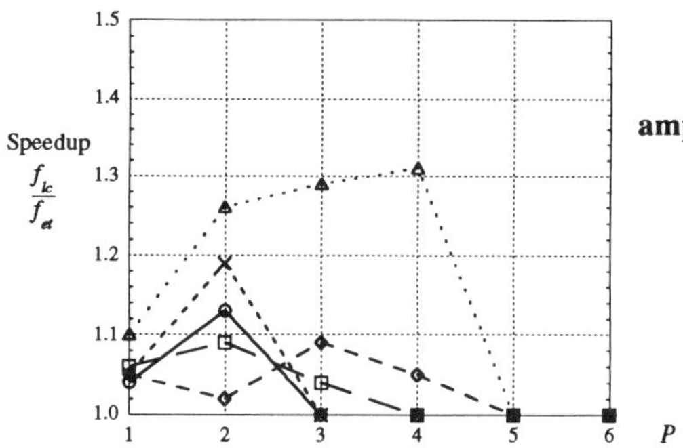

ampseq1

	1	2	3	4	5	6	
· △ · i=0.2	0.32	0.65	**0.97**	1.29	1.61	1.93	$\frac{d_{max}}{R}$
-✕- i=0.4	0.45	**0.90**	1.35	1.80	2.20	2.65	
—◇— i=0.6	0.61	**1.21**	1.82	2.43	3.03	3.64	
—□— i=0.8	**0.80**	1.60	2.40	3.19	4.00	4.80	
—○— i=1.0	**0.97**	1.93	2.90	3.87	4.83	5.80	

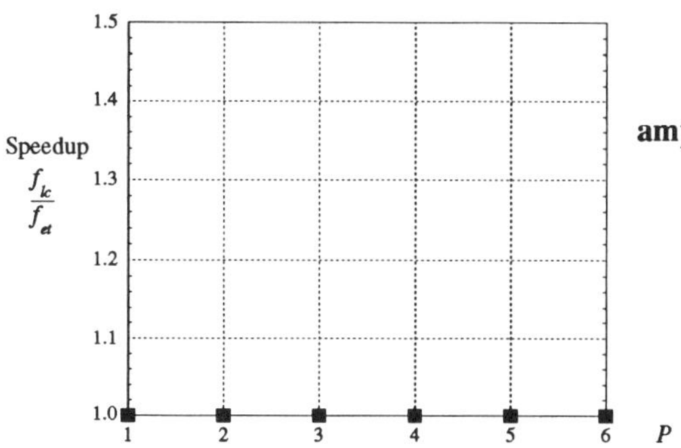

ampseq2

	1	2	3	4	5	6	
· △ · i=0.2	0.59	**1.17**	1.76	2.32	2.93	3.51	$\frac{d_{max}}{R}$
-✕- i=0.4	0.66	**1.34**	1.99	2.65	3.36	3.97	
—◇— i=0.6	**0.74**	1.47	2.20	2.94	3.70	4.40	
—□— i=0.8	**0.80**	1.59	2.39	3.20	3.98	4.73	
—○— i=1.0	**0.85**	1.69	2.54	3.40	4.23	5.07	

ampseq3

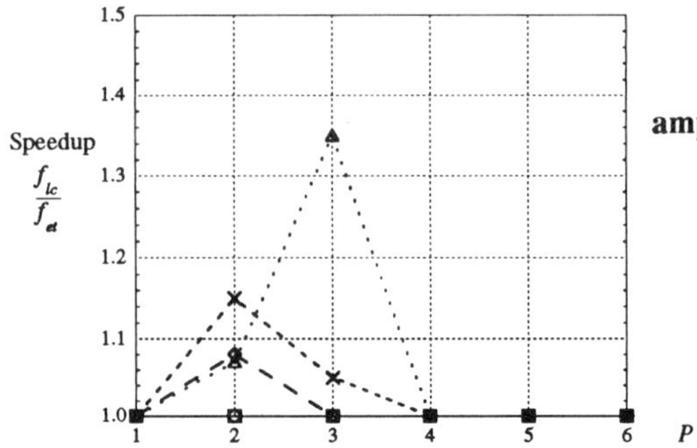

	1	2	3	4	5	6	
i=0.2	0.33	0.65	**0.97**	1.30	1.62	1.96	$\frac{d_{max}}{R}$
i=0.4	0.41	**0.82**	1.23	1.64	2.05	2.46	
i=0.6	0.51	**1.02**	1.54	2.04	2.56	3.07	
i=0.8	0.59	**1.19**	1.78	2.38	2.97	3.55	
i=1.0	**0.67**	1.35	2.03	2.69	3.38	4.04	

ampseq4

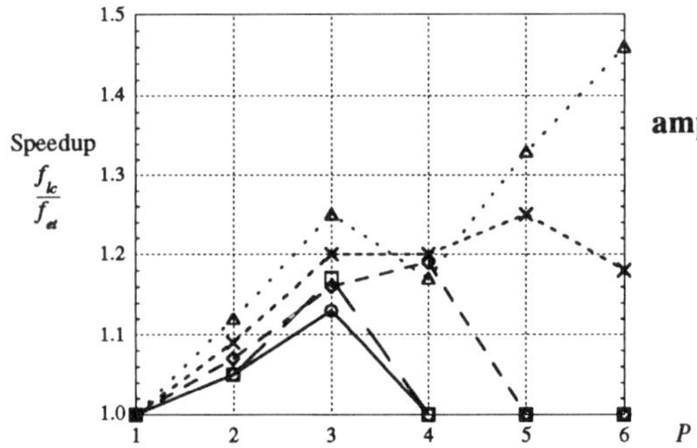

	1	2	3	4	5	6	
i=0.2	0.17	0.35	0.52	0.70	0.86	**1.05**	$\frac{d_{max}}{R}$
i=0.4	0.21	0.42	0.63	0.84	**1.05**	1.26	
i=0.6	0.26	0.51	0.77	**1.02**	1.28	1.54	
i=0.8	0.31	0.61	**0.92**	1.23	1.53	1.84	
i=1.0	0.37	0.73	**1.10**	1.46	1.83	2.19	

ampseq5

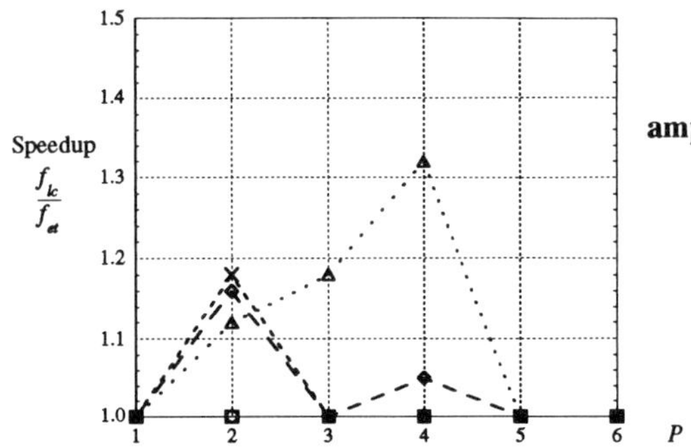

	1	2	3	4	5	6	
i=0.2	0.29	0.59	**0.88**	1.17	1.46	1.76	$\frac{d_{max}}{R}$
i=0.4	0.41	**0.82**	1.23	1.64	2.05	2.46	
i=0.6	0.50	**1.00**	1.50	2.00	2.50	3.00	
i=0.8	0.59	**1.17**	1.76	2.35	2.94	3.52	
i=1.0	**0.67**	1.34	2.02	2.68	3.36	4.03	

somewhat different. The ratio $d_{\max}/R$ is close to 1, but there is still room for improvement, since without any pipelining, that is, for $P = 1$, all versions of ampseq1 become faster by level-clocking.

Another conclusion that we can draw from the plots in Figure 3 is that two-phase clocking leads to greater speedups when the gate delays are more uniform. For every test circuit, peak speedups increase as the exponent i decreases, that is, as the gate delays become more uniform. This observation suggests that standard-cell designs in which gate delays are roughly equal are likely to benefit from two-phase clocking.

The data shown in Figure 3 are the results of experiment SP3, in which the two-phase circuits were clocked by symmetric clocking schemes. We also performed experiment SP4 that combines retiming with tuning of the clocking schemes. In all cases, however, OP3 did not provide any speedup greater than 2% over OP2. Thus, our experiments suggest that clocking with asymmetrical schemes often does not provide any speed advantage over symmetric schemes.

3 Latch-minimization experiments

In this section we present our experimental comparison of edge-triggering and two-phase clocking in terms of the number of storage elements required by each implementation approach. We first describe our methodology, and then we present and discuss our experimental results.

Experimental methodology

In our experiments, we employed retiming in order to minimize the number of storage elements in the circuits. We retimed both the original edge-triggered circuits and their corresponding two-phase circuits in order to achieve a given clock period with the minimum number of storage elements. In both cases, the retiming transformation was applied without relocating the I/O storage elements of the circuits, and thus the I/O specification remained unchanged.

We compared the two implementations of each circuit by performing experiments LM1 and LM2.

LM1 We retimed the original edge-triggered circuit, in order to achieve the minimum period possible with the minimum number of flip-flops. Then, we retimed the corresponding two-phase circuit, in order to achieve the same period with the minimum number of latches, and we compared the number of storage elements in the two circuits.

LM2 We retimed the original edge-triggered circuit, in order to achieve its original clock period specification with the minimum number of flip-flops. Then, we retimed the corresponding two-phase circuit, in order to achieve the same period with the minimum number of latches, and we compared the number of storage elements in the two circuits.

The motivation behind these two experiments was to investigate the impact of two-phase clocking on the number of storage elements under different conditions of operation. Experiment LM1 was aimed at the typical situation, where speed is the primary concern, and edge-triggered circuits are configured to operate at the maximum of their potential. It is often the case, however, that the clock period is dictated by external system considerations and cannot be changed easily. To

Circuit	clock period	e-t latch count	l-c latch count	reduction
mult16a	15.465	62	51	18%
s344	20.946	46	34	26%
s349	20.945	46	34	26%
s382	11.567	68	42	38%
s386	22.999	12	12	0%
s400	11.759	64	42	34%
s510	18.740	16	14	12%
s641	95.464	38	38	0%
s820	31.993	10	10	0%
ampseq1	11.773	86	74	14%
ampseq3	9.700	174	152	13%
ampseq4	15.004	98	76	22%
DRAM-ctl	8.692	52	32	38%

Figure 4: Results of experiment LM1. In this experiment, each edge-triggered circuit and its corresponding two-phase circuit operate at the minimum clock period that can be achieved by retiming the original edge-triggered circuit. For each circuit, the table gives its operating period, the minimum number of latches in the edge-triggered implementation after retiming, the minimum number of latches in the two-phase implementation after retiming, and the reduction in the number of storage elements with respect to edge-triggering. The number of latches in each edge-triggered circuit is twice its number of flip-flops.

that effect, we also performed experiment LM2, which compares the number of storage elements in the two implementations when the clock period equals that of the original edge-triggered circuit.

We performed our experiments using TIM. The latch-minimization algorithms in TIM run in polynomial time and take into account maximal sharing of storage elements. We ran our tests on MCNC benchmark circuits, AT&T communication circuits and custom circuitry designed for MIT's Alewife machine. The largest among these circuits had 340 gates, and the current version of TIM retimed that circuit to minimize its number of latches in less than 20 minutes on a 16MB SPARCstation.

Experimental results

Our experimental results for the two sets of experiments are shown in Figures 4 and 5. There is a striking difference betwen these two sets of results. When the operating period is the minimum clock period that can be achieved by retiming the original edge-triggered circuit, then two-phase clocking leads to substantial reductions in the number of storage elements. When the operating period is that specified for the original circuit, however, there are almost no gains in the number of storage elements when we switch from edge-triggering to two-phase clocking.

In the experimental results of Figure 4 the greatest reductions were achieved for two controller circuits. The number of latches in both s382 and the DRAM controller DRAM-ctl of the Alewife machine was reduced by 38%. Substantial reductions were also achieved for the multiplier circuits mult16a, s344, s349, for the controller circuit s400, as well as for the communication circuits ampseq1, ampseq3, and ampseq4. The two circuits s641 and s820 for which the number of storage elements did not decrease by two-phase clocking were PLD's.

Circuit	clock period	e-t latch count	l-c latch count
mult16a	66.987	32	32
s344	28.579	30	30
s349	28.579	30	30
s382	28.962	32	32
s386	22.999	12	12
s400	19.272	32	32
s510	20.369	12	12
s641	95.464	38	38
s820	31.993	10	10
ampseq1	17.041	70	70
ampseq3	12.041	144	141
ampseq4	23.087	64	64
DRAM-ctl	8.794	32	32

Figure 5: Results of experiment LM2. In this experiment, each edge-triggered circuit and its corresponding two-phase circuit are clocked at the original clock period specification of the edge-triggered circuit. For each circuit, the table gives its operating period, the minimum number of latches in the edge-triggered implementation after retiming, and the minimum number of latches in the two-phase implementation after retiming. The number of latches in each edge-triggered circuit is twice its number of flip-flops. Note that the level-clocked latch count decreases only for `ampseq3`.

Figure 5 shows that for all circuits except `ampseq3`, there was no reduction in the number of storage elements when the circuits were operating at the clock period specification of the original circuit. This seemingly negative result can be explained by comparing the clock periods of the original and the optimally retimed designs. In most cases, the original circuits operate substantially slower than the optimally retimed circuits. Most notably, the original `mult16a` is almost four times slower than its minimum-period retimed version. When the original clock period specification is so far from the minimum achievable, the placement of the storage elements in the edge-triggered circuit is as flexible as in the two-phase implementation, and thus no additional reductions are achieved by two-phase clocking. In the optimally retimed edge-triggered circuits, however, the minimum number of storage elements increases substantially, as it can be verified by comparing the columns that give the edge-triggered latch counts in Figures 4 and 5. Two-phase clocking can decrease this number without degrading circuit performance. In fact, as it is evident from the columns that give the level-clocked latch counts in Figures 4 and 5, the number of latches in more than half of the aggressive level-clocked implementations is not more than 15% higher than the number of latches in the low-performance implementations.

4 Conclusion – Work in progress

We compared edge-triggering and two-phase clocking in terms of speed and number of storage elements. Our methodology was independent of the functionality of the circuit and compared the two design approaches based on the effects of the storage elements in each one of them. In our speedup experiments, edge-triggering was often as fast as two-phase clocking, except when the average delay between any two consecutive latches was roughly uniform over the entire circuit and equal to the maximum gate delay, in which case the potential of two-phase clocking was generally obtained. Our experimental results suggest that circuits designed with standard cells of uniform delay benefit more from two-phase clocking. Moreover, symmetric clocking schemes seem to perform as well as tuned clocking schemes.

In terms of number of storage elements, two-phase clocking led to substantial reductions when the target clock period was set aggressively to the minimum that could be achieved by retiming the original edge-triggered circuit.

We are currently in the process of collecting experimental data from more circuits. We plan to perform experiments with circuits that were originally designed as two-phase circuits. We also plan to perform experiments with circuits in which the nonuniformity in gate delays is amplified, in order to check if, in this case, tuning of the clocking schemes provides any speedups. Another interesting question that remains open, and we plan to address, is whether asymmetric clocking schemes can decrease the number of storage elements further than symmetric ones.

Acknowledgements

We would like to thank Charles Leiserson of MIT for his ideas and suggestions throughout the course of this work. Thanks to Alex Ishii of NEC and Srinivas Devadas of MIT for helpful discussions and for providing us with test circuits. Thanks to John Kubiatowicz of MIT for his suggestions from the designer's perspective and for providing us with circuitry from the Alewife machine. Thanks also to Tom Szymanski of AT&T Bell Labs for enlightening discussions. Marios Papaefthymiou was supported in part by the Defense Advanced Research Projects Agency under Grant N00014-91-J-1698. Keith Randall was supported in part through the Undergraduate Research Opportunities Program at MIT.

A Upper bound on relative speedup

In this section we prove an upper bound on the relative speedup that can be achieved by two-phase level-clocking over edge-triggering. This bound is expressed in terms of the maximum gate delay $d_{\max}$ and the maximum ratio R of total delay over total number of storage elements around the cycles of the edge-triggered circuit.

The following two theorems give bounds on the clock period that can be achieved by retiming and tuning. The first theorem provides a lower bound and an upper bound for retiming edge-triggered circuitry.

Theorem 1 **([16])** *Let $G_{et} = \langle V, E, d, w \rangle$ be an edge-triggered circuit, with delay $d(v)$ for each gate $v \in V$ and $w(e)$ latches on each wire $e \in E$. Let*

$$R = \max_{C \in G_{et}} \frac{\sum_{v \in C} d(v)}{\sum_{e \in C} w(e)}$$

be the maximum ratio of delay over number of flip-flops around the cycles in G_{et}. Moreover, let d_{max} denote the maximum gate delay in G_{et}, and let $\Phi(G_{et})$ denote the minimum clock period that we can obtain by retiming G_{et}. Then

$$\max\{d_{max}, R\} \leq \Phi(G_{et}) \leq R + d_{max} \ . \qquad \square$$

The second theorem provides a lower bound for retiming and tuning two-phase circuitry.

Theorem 2 ([6]) *Let $G_{lc} = \langle V, E, d, w\rangle$ be a two-phase circuit, with delay $d(v)$ for each gate $v \in V$ and $w(e)$ latches on each wire $e \in E$. Let*

$$S = 2 \cdot \max_{C \in G_{lc}} \frac{\sum_{v \in C} d(v)}{\sum_{e \in C} w(e)}$$

be the maximum ratio of delay over number of latches around the cycles in G_{lc}. Then the minimum clock period $\Phi(G_{lc})$ we can obtain by retiming and tuning G_{lc} satisfies

$$\max\{d_{max}, S\} \leq \Phi(G_{lc}) \ . \qquad \square$$

Note that for a given edge-triggered circuit and its corresponding two-phase, level-clocked implementation that is obtained by replacing each edge-triggered flip-flop with a pair of level-sensitive latches, the lower bounds R and S are equal.

We use the bounds in Theorems 1 and 2 to prove the following upper bound on the speedup that can be achieved by switching from edge-triggering to level-clocking.

Lemma 1 *Let G_{et} be an edge-triggered circuit, and let G_{lc} be a two-phase circuit that is obtained by replacing each edge-triggered flip-flop in G_{et} by a pair of level-sensitive latches. Let $\Phi(G_{et})$ be the minimum clock period that we can achieve by retiming G_{et}, and let $\Phi(G_{lc})$ be the minimum clock period that we can achieve by retiming G_{lc} and simultaneously tuning its clocking scheme. Then*

$$\frac{\Phi(G_{et})}{\Phi(G_{lc})} \leq \frac{R + d_{\max}}{\max\{R, d_{\max}\}} \ . \qquad \square$$

The ratio $(R + d_{\max}) / \max\{R, d_{\max}\}$ can be used as a predictor of the speedup that may be achieved by two-phase clocking. The simple intuition behind this heuristic is that for greater values of the upper bound we have more space for improvement. There is no guarantee, however, on the actual improvement, since we have no lower bounds. The upper bound in maximized for $R = d_{\max}$. As we see in the experimental results of Section 2, it is when R approaches $d_{\max}$ that two-phase clocking becomes faster than edge-triggering.

B Retiming with minimum number of latches

In this section we consider the following problem: Given a level-clocked circuit G and a symmetric, k-phase clocking scheme π, compute a retimed circuit G_r that is properly timed by π and uses the minimum number of latches. We show that this problem can be solved in $O(V^3 \lg V)$ time by reducing it to the dual of an uncapacitated minimum-cost flow problem.

For the remainder of the section, we assume that along any path in a k-phase circuit, the phases that clock the latches appear in order modulo k. We also assume that in a symmetric clocking scheme with k phases, the duty cycle of every phase is ϕ, the rise times of any two consecutive phases are π/k units of time apart, and at least one of the k phases is low at any point in time.

The following lemma gives necessary and sufficient conditions for a k-phase circuit to be properly timed by a symmetric clocking scheme. Its proof is based on the notion of "rise-to-fall" times that has been introduced in [6].

Lemma 2 *Let $G = \langle V, E, d, w \rangle$ be a k-phase circuit, let π be a symmetric clocking scheme with k phases, and let ϕ be the duty cycle of each phase. Then the circuit G is properly timed by π if and only if for every path $u \overset{p}{\rightsquigarrow} v$ in G, we have*

$$d(p) \leq \frac{\pi}{k}(w(p)+1)+\phi \; . \tag{1}$$

□

A circuit $G_r = \langle V, E, d, w_r \rangle$ is a retiming of another circuit $G = \langle V, E, d, w \rangle$ if, for some retiming function $r : V \to \mathbf{Z}$, we have $w_r(e) = w(e) + r(v) - r(u)$ and

$$w(e) + r(v) - r(u) \geq 0 \tag{2}$$

for every edge $u \xrightarrow{e} v \in E$. The variable $r(v)$ counts the *lag* of vertex v, and indicates the number of latches that the retiming process moves from every output of v to every input of v. The following lemma must hold for a retiming G_r of G to be properly timed by a symmetric clocking scheme π.

Lemma 3 *Let $G = \langle V, E, d, w \rangle$ be a k-phase circuit, and let π be a symmetric clocking scheme with k phases. Then there exists a retiming $r : V \to \mathbf{Z}$ of G such that G_r is properly timed by π if and only if there exists an assignment of an integer number $r(v)$ to each vertex $v \in V$ such that*

$$r(u) - r(v) \leq w(e) \tag{3}$$

for every edge $u \xrightarrow{e} v \in E$, and

$$r(u) - r(v) \leq \left\lfloor w(p) - \frac{k}{\pi}(d(p) - \phi) + 1 \right\rfloor \tag{4}$$

for every path $u \overset{p}{\rightsquigarrow} v$ in G. □

The number of latches in the retimed circuit G_r is given by

$$\begin{aligned} \sum_{e \in E} w_r(e) &= \sum_{e \in E} w(e) + \sum_{e \in E} (r(v) - r(u)) \\ &= \sum_{e \in E} w(e) + \sum_{v \in V} (\text{indegree}(v) - \text{outdegree}(v))\, r(v) \; . \end{aligned}$$

Therefore, the problem of retiming a k-phase circuit that is clocked with a symmetric scheme, in order to minimize the number of latches in it, is expressed as follows.

Definition 1 (Latch-minimization problem) *Let $G = \langle V, E, d, w \rangle$ be a level-clocked circuit, and let π be a symmetric clocking scheme with k phases. The latch-minimization problem asks for a retiming $r : V \to \mathbf{Z}$ of G such that*

$$\sum_{v \in V} (\textit{indegree}(v) - \textit{outdegree}(v))\, r(v)$$

is minimized, subject to Inequality (3)

$$r(u) - r(v) \leq w(e)$$

for every edge $u \xrightarrow{e} v \in E$, and Inequality (4)

$$r(u) - r(v) \leq \left\lfloor w(p) - \frac{k}{\pi}(d(p) - \phi) + 1 \right\rfloor$$

for every path $u \overset{p}{\rightsquigarrow} v$ in G. □

The following theorem shows that the latch-minimization problem is solvable in polynomial time.

Theorem 3 *The latch-minimization problem can be solved in $O(V^3 \lg V)$ time by reduction to the dual of an uncapacitated minimum-cost flow problem on the graph defined by Inequalities (3) and (4). The cost of each edge equals the right-hand side of the corresponding inequality. The demand/supply of each vertex v equals the difference between the number of edges coming into v and the number of edges coming out of v.*

If every fanout wire of each gate in the circuit has the same output, then we can show that the latch-minimization problem with maximal sharing of latches can be solved even more efficiently in $O(V^3)$ time. If the objective is to achieve the minimum clock period with the minimum number of latches, we can find a solution in $O(V^2E + V^3 \lg^2 V)$ time. First, we compute a set of $O(V^3)$ possible clock periods using the $O(V^2E)$ algorithm from [6], and then we binary search this set for the minimum feasible clock period.

References

[1] T. Burks, K. Sakallah, and T. Mudge. Multiphase retiming using $minT_c$. *τ92 ACM Workshop on Timing Issues in the Specification and Synthesis of Digital Systems*, March 1992.

[2] M. R. Dagenais and N. C. Rumin. Automatic determination of optimal clocking parameters in synchronous MOS VLSI circuits. In *Advanced Research in VLSI: Proc. of the 5th MIT Conference*, pages 19–33, 1988.

[3] L. A. Glasser and D. W. Dobberpuhl. *The Design and Analysis of VLSI Circuits*. Addison-Wesley, Reading, Massachusetts, 1985.

[4] R. B. Hitchcock. Timing verification and the timing analysis program. In *Proc. 19th Design Automation Conference*, pages 594–604, June 1982.

[5] A. T. Ishii and C. E. Leiserson. A timing analysis of level-clocked circuitry. In *Advanced Research in VLSI: Proc. of the Sixth MIT Conference*, pages 113–130. MIT Press, April 1990.

[6] A. T. Ishii, C. E. Leiserson, and M. C. Papaefthymiou. Optimizing two-phase, level-clocked circuitry. In *Advanced Research in VLSI and Parallel Systems: Proc. of the 1992 Brown/MIT Conference*. MIT Press, March 1992.

[7] N. P. Jouppi. Timing analysis for NMOS VLSI. In *Proc. 20th Design Automation Conference*, pages 411–418, June 1983.

[8] C. E. Leiserson, F. M. Rose, and J. B. Saxe. Optimizing synchronous circuitry by retiming. *3rd Caltech Conference on VLSI*, 1983. R. Bryant, ed., pp. 87-116.

[9] C. E. Leiserson and J. B. Saxe. Optimizing synchronous systems. *Journal of VLSI and Computer Systems*, 1(1):41–67, 1983.

[10] C. E. Leiserson and J. B. Saxe. Retiming synchronous circuitry. *Algorithmica*, 6(1), 1991. Also available as MIT/LCS/TM-372.

[11] B. Lockyear and C. Ebeling. Optimal retiming of multi-phase, level-clocked circuits. In *Advanced Research in VLSI and Parallel Systems: Proc. of the 1992 Brown/MIT Conference.* MIT Press, March 1992.

[12] T. M. McWilliams. Verification of timing constraints on large digital systems. In *Proc. 17th Design Automation Conference*, pages 139–147, June 1980.

[13] J. B. Orlin. A faster strongly polynomial minimum cost flow algorithm. *Proc. of the 20th Annual ACM Symposium on Theory of Computing*, pages 377–387, May 1988.

[14] C. Papadimitriou and K. Steiglitz. *Combinatorial Optimization, Algorithms and Complexity.* Prentice-Hall, Inc., 1982.

[15] M. C. Papaefthymiou. Understanding retiming through maximum average-weight cycles. *3rd ACM Symposium on Parallel Algorithms and Architectures*, July 1991.

[16] M. C. Papaefthymiou. Understanding retiming through maximum average-delay cycles. *Mathematical Systems Theory*, 1992. To appear. A preliminary version of the paper appeared in the proceedings of the 3rd ACM Symposium on Parallel Algorithms and Architectures, July 1991.

[17] M. C. Papaefthymiou and K. H. Randall. TIM: a timing package for two-phase, level-clocked circuitry. Unpublished manuscript, October 1992.

[18] K. A. Sakallah, T. N. Mudge, and O. A. Olukotun. *check*T_c and *min*T_c: Timing verification and optimal clocking of synchronous digital circuits. In *Digest of Technical Papers of the 1990 IEEE International Conference on CAD*, pages 552–555, November 1990.

[19] N. Shenoy, R. K. Brayton, and A. Sangiovanni-Vincentelli. Retiming of circuits with single phase level-sensitive latches. In *International Conference on Computer Design*, October 1991.

[20] T. G. Szymanski. LEADOUT: A static timing analyzer for MOS circuits. In *Digest of Technical Papers of the 1986 IEEE International Conference on CAD*, pages 130–133, November 1986.

[21] T. G. Szymanski. Computing optimal clock schedules. In *Proc. 29th ACM/IEEE Design Automation Conference*, pages 399–404, June 1992.

[22] T. G. Szymanski and N. Shenoy. Verifying clock schedules. In *Digest of Technical Papers of the 1992 IEEE/ACM International Conference on CAD*, November 1992.

[23] S. H. Unger and C. J. Tan. Clocking schemes for high speed digital systems. *IEEE Transactions on Computers*, C-35(10):880–895, October 1986.

[24] S. A. Ward and R. H. Halstead, Jr. *Computation Structures.* McGraw-Hill, MIT Press, 1990.

Designing Salphasic Clock Distribution Systems

Vernon L. Chi
Microelectronic Systems Laboratory
Department of Computer Science
University of North Carolina
Chapel Hill, NC 27599-3175
chi@cs.unc.edu

Abstract

A Salphasic clock distribution method for globally synchronous systems was discovered, developed, and demonstrated in the Pixel-Planes 5 (PXPL5) graphics system. This method exploits the propagation properties of sinusoidal waves to create substantially constant phase regions in a propagation medium, thereby providing a simple means of distributing low-skew clock signals across a system of significant size. The engineering complexities associated with the clock distribution trees of conventional high performance systems are avoided at the expense of explicitly dealing with the electrical propagation behavior of the salphasic signal. Constraining the propagation medium to consist of an arbitrarily branching tree of transmission lines connected by lumped constant L-sections greatly simplifies analysis of the clock distribution system, and provides the basis for cview, a simulator used to design the PXPL5 clock system. This paper describes how to design salphasic clock distribution trees, discussing issues encountered in an actual digital system environment, and reports on the results achieved in the PXPL5 clock distribution tree.

1 Introduction

sal·pha·sic (săl-phā′zǐk) *adj.* Characterized by sudden jumps in phase between regions of relative constancy. [Lat., *salio*, to jump + NLat., *phasis*, phases of the moon.]

The engineering of large scale high-speed digital systems must deal explicitly with propagation delays of signals on their interconnects [9]. Two fundamentally distinct abstractions are used in the design of such systems, i.e., synchronous, and asynchronous signaling. Synchronous design is well understood and widely practiced, especially when the system consists of processing components encapsulated in a "single" machine. Asynchronous design is widely used in networking and distributed processing, but tightly-coupled asynchronous digital designs require delay-independent handshaking protocols [10,13,14] which are not widely understood or practiced today. Moreover, the hardware overhead for tightly coupled asynchronous protocols may not ever be economically justified for some designs. On the other hand, globally synchronous signaling becomes increasingly difficult to design with the ongoing trend towards higher length-frequency products of signal interconnects. Particularly problematical is distribution of the system clock [4,5,15,16], which is typically the highest frequency system signal, and is necessarily distributed across the entire extent of the system.

A novel means of transmitting low-skew global clock information on transmission lines cut to resonant lengths was disclosed by Peil and Raillard in 1963 [12], and applied to small but very high speed logic circuits implemented using tunnel diodes [11]. A substantially generalized low-skew transmission means, dubbed "salphasic", was independently discovered by the author in 1989 [1], and demonstrated in a fully functioning large scale system, Pixel-Planes 5 (PXPL5), in 1990 [6]. It was recognized by the author that salphasic behavior was not restricted to resonant transmission lines, and could be evoked through any sufficiently lossless signal propagation means, irrespective of geometry. Thus, salphasic clock signals can in principle be distributed across surfaces such as "clock planes" on circuit boards, or through an arbitrary topology of connected transmission lines such as an arbitrarily branching tree of printed circuit board runs.

This paper specifically addresses how to design a salphasic clock distribution network comprising an arbitrarily branching tree of connected transmission lines. The complexities of designing conventional low-skew clock distribution systems are largely avoided, but attention must be paid to a different set of issues. Fortunately, the necessary wave propagation analysis for salphasic distribution trees can be encapsulated in a software design tool and presented to the designer at an easy to use and understand level of abstraction. By using such a tool and observing a small number of simple rules, it becomes easy to design low-skew clock distribution networks.

Section II presents a review of general salphasic behavior as described in [3], focusing specifically on developing the case of an arbitrarily branching distribution tree implemented by recursive use of a canonical, reactively terminated transmission line segment. This section is not necessary, however, to understanding the remainder of this paper. Section III describes how the distribution tree formulas, encapsulated in the cview (acronym for "clock view") simulator, are used to design salphasic distribution trees. Rules-of-thumb for good design are presented and discussed to motivate their use and to provide an intuitive base for optimizing designs. Section IV describes the experience of designing the PXPL5 clock distribution tree, and presents measured results. Section V summarizes this work, and describes plans for future research in salphasic clock distribution.

2 Review of salphasic tree behavior

A fundamental property of a sinusoidal standing wave on a transmission line is that the temporal phase of the signal is a constant, independent of location. The polarity of the signal, however, reverses at each node (point of null voltage) of the standing wave, which is equivalent to a 180° phase reversal. Thus, a reasonable representation of a standing wave is in terms of a non negative magnitude, and a phase which which remains constant between adjacent nodes of the wave and which increments by 180° across each node. By convention, we let this increment be negative for increasing distance from the driving source. For a pure standing wave, this results in a staircase shaped phase function, motivating the term "Salphasic".

Another property of pure standing waves is that no net energy is propagated. If the standing wave is represented as the superposition of forward and reverse traveling waves, this is property is satisfied if and only if the amplitudes of the two waves are equal. If the load of a transmission line is lossless (i.e., non-dissipative), all of the energy incident on the load is reflected, which fulfills the equal amplitude condition at the load. If the line itself is lossless, the equal amplitude condition is fulfilled everywhere along the line, and a pure standing wave results.

Real systems are not lossless, but can be sufficiently low loss that nearly pure standing waves can be generated. In these cases, the corners of the staircase function become rounded and the constant phase regions develop a negative slope. In highly lossy systems, the phase function degenerates into a linear function, entirely losing its staircase properties.

Lossyness of transmission lines is both frequency and scale dependent. Good conductors such as copper or aluminum are sufficiently lossless at macroscopic scales, but not at IC scales. Superconductors have been proposed [7], however, to implement a 10 GHz clock with distribution system diameter of approximately 1 cm, relying on the salphasic properties of a multiple node standing wave. For PC boards and larger, sufficiently pure standing waves can generally be realized. MCMs are in a transition range, where the feasibility of generating sufficiently pure standing waves is still to be assessed.

Frequency-dependent "skin effect" losses will limit the frequency above which sufficiently pure standing waves can be realized for a given size transmission line. A system of several meters extent operating at 160 MHz has been found to be well within the maximum useful frequency-size limitations.

Most synchronous systems require a clock distribution network at least as complex as a tree; a single transmission line is rarely adequate. Consider a distribution tree of lossless transmission lines with lossless loads at its leaves. According to the above discussion, the transmission lines driving the leaves of the tree will support standing waves and thus exhibit salphasic behavior. It is also true that the transmission lines driving internal tree nodes will likewise experience standing waves, because the sub-trees they drive are lossless and are thus equivalent to lossless loads. Thus, the phase of the signal throughout the distribution tree is salphasic.

Since wavelength is inversely proportional to frequency, one limitation of multi-region salphasic systems is that they are not robust to variation in clock frequency. Except for very minor changes, at best they must be retuned, and in the worst case, re-designed. Thus, a machine that is to be manufactured with several different clock rates should use single region PC boards able to accommodate the maximum clock rate.

Of course for design purposes, it is necessary to treat the above concepts quantitatively. The following discussion summarizes the necessary formalism which, however, is encapsulated in the **cview** simulator [2] so that the designer need not explicitly deal with it.

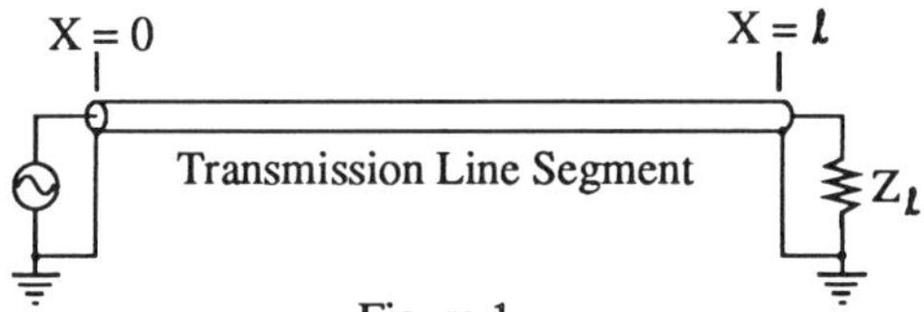

Figure 1

The salient parameters of a finite, linear, uniform transmission line driven at $x = 0$ and terminated by a load $Z_l = R_l + jX_l$ at $x = l$ (Figure 1) are,

$$Z_0 = R_0 + jX_0 = \sqrt{(R + j\omega L)/(G + j\omega C)}, \tag{1}$$

$$\gamma = \alpha + j\beta = \sqrt{(R + j\omega L)(G + j\omega C)}; \tag{2}$$

where the determining physical properties of the transmission line,

R = series resistance per unit length,
L = series inductance per unit length,
G = shunt conductance per unit length, and
C = shunt capacitance per unit length,

are all characterized by positive real numbers.

Z_0 is known as the *characteristic impedance*, and γ the *propagation constant* of the line. The propagation of sinusoidal signals along the line can be represented [8] as

$$V_x = V_A e^{-\gamma x} + V_B e^{\gamma x}, \tag{3}$$

where V_A and V_B represent the amplitudes of the forward and reverse traveling waves.

We define the *reflection coefficient* ρ [8] to be a complex voltage ratio of the reverse to the forward traveling wave at the load point $x = l$:

$$\rho = \frac{V_B e^{\gamma x}}{V_A e^{-\gamma x}}. \tag{4}$$

At the driving point $x = 0$ we observe that the terminal voltage is the sum of the forward and reverse traveling wave voltages:

$$V_0 = V_A + V_B. \tag{5}$$

Boundary conditions (4) and (5) are used to solve (3) for the voltage transfer function of the transmission line segment:

$$\frac{V_l}{V_0} = \frac{\rho + 1}{\rho e^{-\gamma l} + e^{\gamma l}}. \tag{6}$$

According to [8], the value of ρ depends on load impedance Z_l and characteristic impedance Z_0

$$\rho = \frac{Z_l - Z_0}{Z_l + Z_0}. \tag{7}$$

In the lossless limit, $R_l \to 0$ and from eq (1), $(R \to 0, G \to 0) \Rightarrow (X_0 \to 0)$. Eq(7) then becomes $\rho = -(R_0 - jX_l)/(R_0 + jX_l)$; thus, by inspection, $|\rho| = 1$, or equivalently, $\rho = e^{j\nu}$. Under these conditions, the right hand side of eq (6) is shown in [3] to be a real number; therefore any phase shift between V_0 and V_l must be $n\pi$ radians, i.e., the behavior is purely salphasic.

The driving point terminal impedance of the transmission line is shown in [8] to be

$$Z_{in} = Z_0 \frac{Z_l + Z_0 tanh(\gamma l)}{Z_0 + Z_l tanh(\gamma l)}. \tag{8}$$

In the lossless limit, $Z_l \to jX_l$ and $\gamma \to j\beta$, and simple substitution shows that the real part of Z_{in} vanishes. Thus, the driving point impedance of a lossless line having a lossless load is likewise lossless, i.e., it is purely reactive.

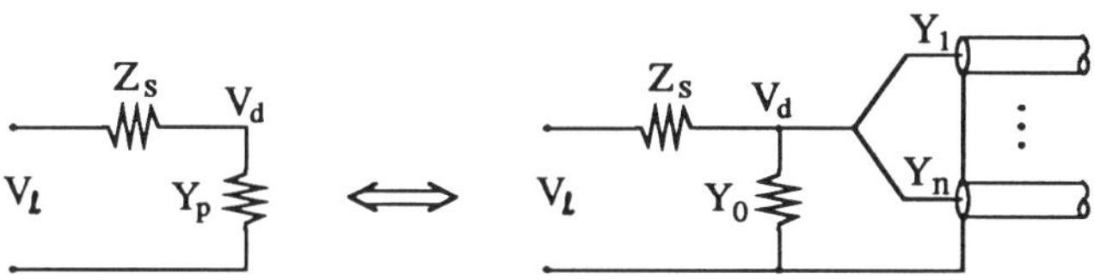

Figure 2

Now consider a load Z_l comprising an equivalent lumped constant L-section (Figure 2) with series leg of impedance Z_s and shunt leg of admittance

$$Y_p = \sum_{i=0}^{n} Y_i, \tag{9}$$

where Y_0 represents the admittance of a lumped shunt element, and Y_i, $i > 0$ represents the driving point admittance, $[1/Z_{in}]_i$ of the ith of n connected transmission lines. The transfer function of this L-section is

$$\frac{V_d}{V_l} = \frac{1}{Z_s Y_p + 1}. \tag{10}$$

In the lossless limit, $Z_s = jX_s$ and $Y_p = jB_p$, and substitution into eq (9) shows the transfer function of the L-section is real valued, hence its behavior is also salphasic.

We define a canonical transmission line segment with L-section load, behaving according to equations (6) through (10). Its overall transfer function is

$$\frac{V_d}{V_0} = \frac{(\rho + 1)}{(Z_s Y_p + 1)(\rho e^{-\gamma l} + e^{\gamma l})}, \tag{11}$$

which is easily shown to behave salphasically in the lossless limit.

Equations (7), (8), (9), and (11) are sufficient to characterize the amplitude and phase of a sinusoidal signal at every node in an arbitrary distribution tree recursively composed of such canonical transmission line segments. A software simulator, **cview**, was built using these equations, and used to study numerous tree configurations, and to develop the distribution tree for the PXPL5 clock system. Cview encapsulates all the necessary detailed wave propagation analysis of sinusoidal signals in linear distribution trees.

3 Designing salphasic distribution trees

Using the cview simulator, design of salphasic distribution trees can proceed without the need to understand the intricacies of the previous section. The cview simulator accepts a textual tree description as input, and provides outputs characterizing the amplitudes, phases, and driving point admittances of canonical transmission line sections composing the corresponding clock distribution tree. A detailed description of cview can be found in [2], and the source code is available from the author.

This section expresses a number of rules-of-thumb found by the author to be useful guides for configuring and optimizing a salphasic distribution tree. It is difficult in a paper such as this quantify these rules and their effects precisely. Therefore the author encourages interested designers to use **cview** to explore the behavior of salphasic trees and develop an intuition for themselves.

While cview insulates the designer from the details of signal propagation, it is nevertheless useful to develop a few key but simple insights to enable one to proceed. In the following, rules of thumb are presented, then motivated and discussed to help the designer incorporate some of these insights into a working intuition base.

> Minimize clock signal losses: select low loss components and materials; design to minimize radiation.

Salphasic behavior is critically dependent on maintaining nearly pure standing waves throughout the distribution tree. This is tantamount to reducing signal losses to the minimum possible. Referring to Figure 3, the knees of the "slightly lossy" S-curve become softer and the phase transition region longer as losses are increased. This degrades the near-constant phase properties of the "isophasic" regions, as well as diminishing their useful extent. It is therefore essential to select system components and materials to minimize signal loss.

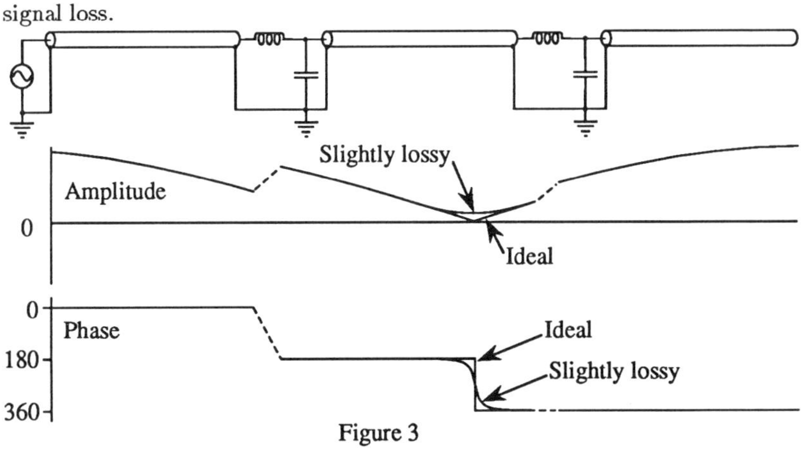

Figure 3

Typically, clock receivers having very high input impedances are preferred; in particular, the resistive component must be high. If comparators are used as clock receivers, most suitable ones have adequately low loss inputs unless the number of clock receivers is to be truly massive. PXPL5 uses up to 60 AD96685 high speed differential comparators, each having of the order of 10KΩ input resistance. The mismatch of a single clock receiver to the nominal 25Ω backplane clock line is 400:1.

High quality, well-shielded transmission lines with low attenuation figures are preferred, however even a modest 10 dB/100ft at the clock frequency will be adequate except for physically very large systems. PXPL5 uses feed lines each of about 1 meter length. Semi-rigid, hand-formable cable was used (Precision Tube Co. #BP50085) having 8.2 dB/100ft attenuation at 160 MHz. The dissipation in each feed line is thus 0.27 dB or about 3% voltage loss.

PC board dielectrics should have low dissipation factors, but except for very cheap boards, this is not a problem; e.g., FR-4 glass-epoxy has a dielectric dissipation of 0.020 at 1 MHz. PC board conductors should be copper, which is what is normally available. Available copper thicknesses are 1/2, 1, and 2 [oz/ft^2]. Increasing conductor thickness decreases losses until limiting at twice the skin depth of the clock frequency. Thus, for frequencies above 54 MHz, no benefit obtains from using more than 1/2 oz copper conductors. Circuit board runs should be designed not to lose energy through radiation. Thus, the safest strategy is to fully shield the clock distribution tree between ground and/or power planes.

> Clock receivers should not be located near phase transition regions.

Salphasic behavior is tied very closely to standing waves (Figure 3). In the ideal case, pure standing waves in a tree establish regions within which the signal is isophasic, i.e., its phase remains absolutely constant. Separating these regions are either nodes (points of null voltage) on transmission lines, or series legs of L-sections. The phase difference between any two adjacent isophasic regions is exactly 180°. The signal amplitude within a given region varies sinusoidally with distance along the transmission line. Actual designs, however, are at least slightly lossy, in which case, the isophasic regions are less clearly delineated by transition regions where the amplitude goes through a non-zero minimum, and the phase varies rapidly. Not only is the phase poorly controlled in the transition regions, but the amplitude of the signal is low. If a comparator is used as a clock receiver, the phase error ϕ_e due to its input threshold offset V_{io} depends on the input signal amplitude V_{is} according to $\phi_e = sin^{-1}(V_{io}/V_{is})$. Thus, for example, if a comparator input offset is specified as $\pm 3mV$ max, and the associated phase error is not to exceed $\pm 1°$, the signal amplitude should be at least $172mV$.

> Connect all clock receivers in mutually isophasic regions, modulo 360°.

Except under unusual circumstances, one desires all system clock receivers to be in phase. Since phase reversal occurs between isophasic regions, it is necessary to pay attention to which of these regions clock inputs are connected to. For smaller systems, it is often possible to connect all clock receivers within a single isophasic phase region, but for larger systems, this will usually not be possible. It can be seen from Figures 6 and 7 that each half backplane of PXPL5 contains an isophasic region. Moreover, these six regions are mutually isophasic, while being phase reversed from the isophasic region connected to the clock oscillator.

> On a line segment, a constant phase region is shifted towards the load by a capacitive load, and towards the source by an inductive load.

The wave position on a line segment is independent of everything except the characteristic impedance of the line and the value of its load. In particular, the wave position is independent of the the impedance of the driving source. Thus, the position of an isophasic region on a transmission line can be controlled entirely by tuning the load seen by that line. Figure 4 shows how the position of a standing wave can be controlled by use of an appropriate load. Recognizing that for an inductor, $L = \infty$ is equivalent to an open circuit, while $L = 0$ is equivalent to a short, we see that relative to an open circuit termination, the position of the wave can be shifted towards the source by up to 90° as the value of L decreases. Conversely, $C = 0$ is equivalent to an open circuit, while $C = \infty$ is equivalent to a short circuit. Thus relative to an open circuit, the wave can be shifted towards the load by up to 90° as the value of C increases.

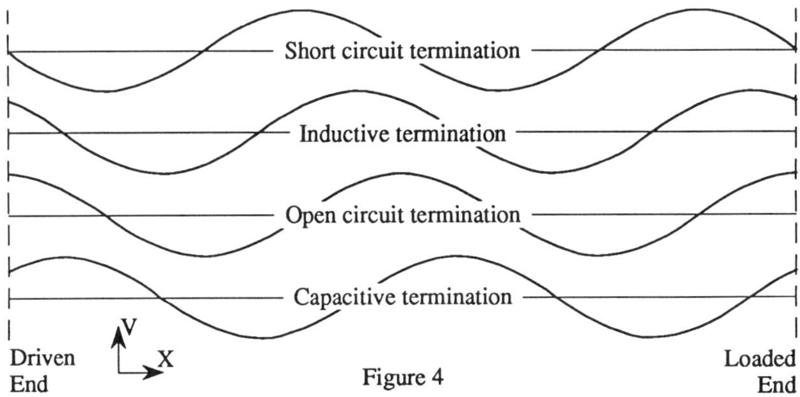

Figure 4

> The length of an isophasic region is extended by connecting inductance in shunt, and reduced by connecting capacitance in shunt.

The wavelength of a standing wave can be "artificially" lengthened by periodic inductive shunt loads (or shortened by capacitive ones) along the length of a transmission line. This can be useful in adjusting the sizes of isophasic regions, e.g. in cases where enough clock receivers won't fit comfortably within a single region. The following paragraph is intended to motivate an intuitive understanding of why this should occur.

If the spacing between these loads is sufficiently small relative to a wavelength (say $< \lambda/10$), the added loads can be treated as modifications to the transmission line parameters. Referring to eq (2), in the lossless case, note that $(R, G = 0) \Rightarrow \gamma = \omega\sqrt{-LC} = j\beta$. The phase constant β represents phase shift in radians per meter, so the wavelength $\lambda = 2\pi/\beta = 2\pi/\omega\sqrt{LC}$. In the periodically shunt-loaded case of interest here, the effective value $C_{eff} = C + C_a - 1/\omega^2 L_a$ is used to figure wavelength according to $\lambda_{loaded} = 2\pi/\omega\sqrt{LC_{eff}}$, where C_a and L_a are the average added capacitive and inductive loading per unit length of the line, and $\omega = 2\pi f_{clock}$.

Using cview, the above approximation is unnecessary. It is best to treat this situation as explicit loads in a chain of short canonical lines. Thus, it is possible to periodically load a long line with shunt inductors to extend the useful isophasic region on the line.

> The wave position on a multidrop chain should be adjusted so that the clock receiver nearest the source sees the same amplitude as the clock receiver nearest the load.

One useful configuration is a vine-form distribution tree, i.e., a chain of transmission lines connected in series, with a clock receiver also connected in shunt at each junction. Also known as bus-connected, or multidrop, this is the configuration of each half of a backplane in PXPL5 (Figure 7). While cview models a vine-form tree in terms of canonical segments, the overall behavior is well approximated by a single transmission line of lower characteristic impedance, tapped but not loaded at the clock receiver locations. It is easy to visualize the optimal tuning for a vine-form tree: the wave should be centered so that the clock receiver nearest the load sees the same amplitude as the clock receiver nearest the source. This provides the greatest worst-case margin between any clock receiver and a phase transition region. The tunable inductors depicted in Figure 6 provide adjustability for this purpose at system integration time.

> For plug-reconfigurable hardware, impedance maxima must be located at all points of disconnect.

A common system requirement is to accommodate pluggable hardware reconfiguration. Examples are addition, removal, or rearranging of boards plugged into a backplane, or addition or removal of card cages or racks from a system. To meet this requirement, a salphasic system should be designed to locate *current* nodes (corresponding to voltage maxima, or points of maximum terminal impedance) at all points of disconnect.

Referring to Figure 7, observe that the point of disconnect of the feed line for backplane 1 is at a current node, so when the feed is disconnected, the load on the oscillator does not significantly change. This tuning is achieved by selecting the length of the backplane feed lines, which for PXPL5 were cut to 98 cm. This length can be determined by solving equation (8) for l, given the terminal impedance presented by the backplane and Z_0 and γ of the feed line. In PXPL5, measurements were performed to be entirely sure of the backplane's terminal impedance; however the cplane simulation value was sufficiently accurate by itself. In practice, sensitivity to error in this tuning is low, as long as the clock oscillator impedance is reasonably low.

The system circuit boards are similarly designed to present current nodes at their backplane connectors. In the case of PXPL5, the clock receivers are very closely situated to their backplane connection, and while cview was used to model the short on-board connections, the results were equivalent to treating the on board circuitry as a lumped constant capacitor of about 6 pF in shunt with a resistor of about 10KΩ (due to the input impedance of the clock receiver, an AD 96685 ultrafast comparator). A parallel *LRC* circuit presents a maximum impedance (equal to R) at resonance, Thus, by selecting an inductor to resonate with the 6 pF equivalent capacitance at the 160 MHz clock frequency, and connecting it in shunt with the clock input, the circuit cards were designed to present a high terminal impedance at the backplane connector. As in the case of the feed lines, sensitivity to error in resonance is low. In PXPL5, we simply selected the closest valued surface-mount chip inductor, nominally 150 nH, for use on all circuit cards. Variation in clock skew as a function of number of populated slots was less than 90 pS.

Design the distribution tree to minimize coupled-in system noise

Noise, crosstalk, and other interfering artifacts are problems that cannot be ignored in designing a robust clock system. One criticism of salphasic methodology is that the necessary sinusoidal signal does not have sharp leading and falling edges, and is thus more sensitive to system noise. In fact, when pushing any logic family to its limit, the rise and fall times are not that much better than a sine wave. Nevertheless, it is good design practice to minimize coupling between the clock network and the rest of the system. Thus, on a circuit board, shielding the clock distribution lines on both sides is advisable. This was noted above as also being desirable to *keep the clock signal in*, i.e. to keep signal losses to a minimum. Common mode current noise in the shielding planes should be minimized where feasible. A complete study of this has not been conducted, but in PXPL5, the backplane ground planes were split to provide an isolated pair of clock system shielding planes (IsoGND). The IsoGND planes of each backplane are connected through the shields of their feed lines to a single point clock system ground at the clock oscillator, and are otherwise isolated from the system ground. This strategy was successful, as indicated by the measured data shown in Figure 9.

Design the system to minimize harmonic distortion.

In salphasic systems, any non-linearities will scatter energy from the fundamental clock frequency into harmonics, and can cause problems similar to coupled-in noise. It is therefore important to select components having minimal non-linearities. While the initial choice of a μA6685 comparator for PXPL5 clock receivers was excellent for its low loss, it was found that the input impedance was substantially non-linear, primarily in its capacitance. This is due to the Miller capacitance of its input stage which exists only when this stage is not limiting. Normal comparator operation causes limiting during substantial but not all portions of the signal wave, causing the input capacitance to vary. A subsequent choice of comparator, an AD96685, utilizes a differential cascode input stage, which largely eliminates the Miller effect, with dramatic improvement of linearity.

Back-match the clock tree to the clock source as best possible to encourage dissipation of noise and distortion products.

In spite of best efforts to the contrary, some harmonics will be generated, and some system noise will be coupled into the clock tree through cross-talk and related means. Some mechanism is needed to minimize this unwanted interference. This can be accomplished by allowing the interference to effectively couple back to the clock driver which is the only part of the clock system which may be substantially lossy, and can therefore be used to dissipate the interference.

In an arbitrarily branching tree, perfect back-matching of each canonical line to its driving point is in general not possible. Nevertheless, it is desirable to reduce the reverse direction reflection coefficient (i.e. back-match) of canonical line segments, especially at the higher levels of the tree. Fortunately, if a system is otherwise well designed for noise and distortion, it requires only a nod to back-matching to obtain very good clock signal integrity.

4 Designing the PXPL5 clock distribution tree

The PXPL5 system [6] built in the Microelectronic Systems Laboratory of the Department of Computer Science at UNC Chapel Hill, was used as a test bed for evaluating salphasic clock distribution tree technology. PXPL5 is a heterogeneous 3D raster graphics supercomputer having a massively parallel SIMD subsystem comprising hundreds of thousands of processors, a MIMD subsystem comprising some 50 Intel i860 based processors, and various other system functions. It is currently packaged in three 19″ equipment racks with three side-by-side backplanes each having 20 slots.

This system is globally synchronous with a 160 MHz clock being distributed across all backplanes from a Vectron model CO-233 crystal oscillator through semi-rigid coaxial transmission lines to the centers of the backplanes (Figure 5). The clock circuit on each backplane branches at the driving point with each branch driving half of the backplane in a bus style multidrop configuration. As shown in Figure 6, each branch is terminated by a variable shunt inductor, which is used to adjust the position of the standing wave on the branch.

The clock receivers on the daughter boards are AD96685 high speed comparators, the ECL outputs of which drive conventional clock distribution trees with MC100E111 clock buffers and equal length distribution lines. On each

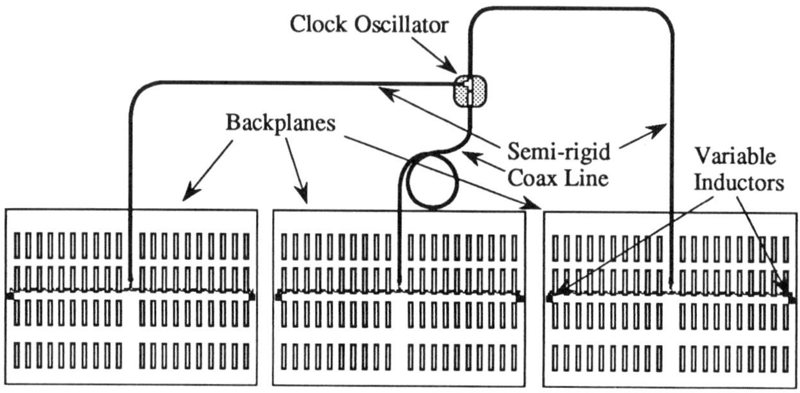

Figure 5: Pictorial of Pixel-Planes 5, three rack configuration

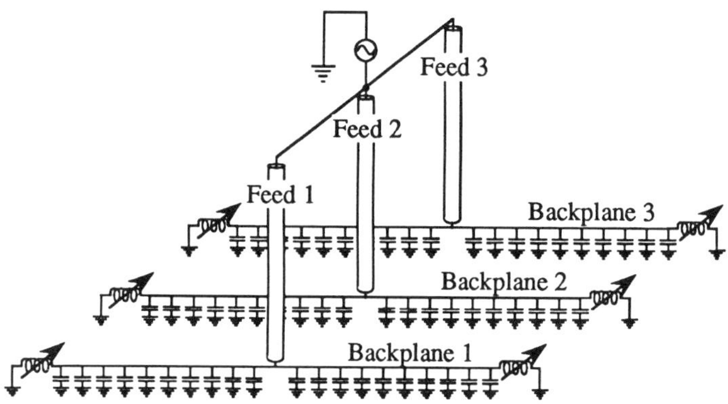

Figure 6: Schematic of Pixel-Planes 5, three rack configuration

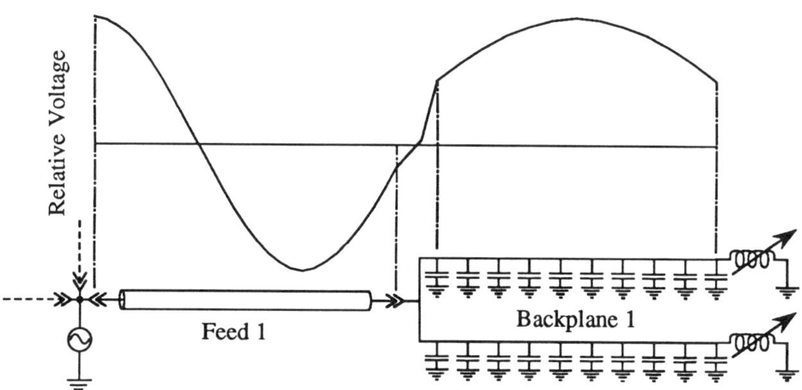

Figure 7: Wave distribution in one rack of Pixel-Planes 5

board, the comparator is located immediately adjacent to the backplane connector, and the effective parasitic capacitance presented to the backplane is approximately nulled by a surface mount chip inductor in shunt with the input. This minimizes the change of the standing wave distribution on the backplane with number of populated slots. Both the clock oscillator and the receiving comparators are ground-referenced with no threshold level adjustments, resulting in detection of sinusoidal zero crossings within a tolerance determined by the input offsets of the comparators.

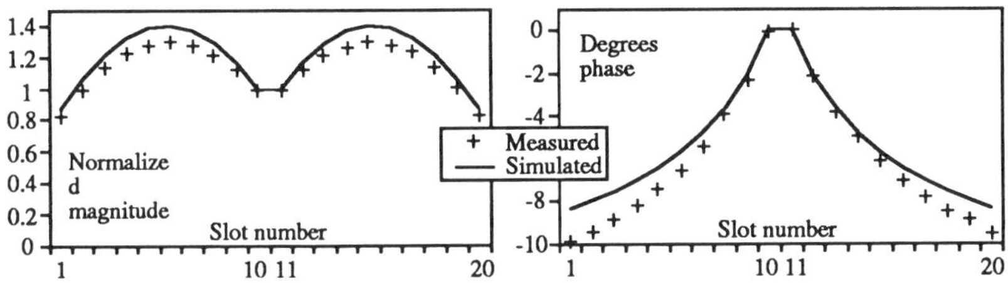

Figure 8

Detailed simulations were performed in the development of the salphasic backplane clock distribution system, and measurements were performed on the running system. A comparison between simulated and measured results for a single backplane is shown in Figure 8. Phase and amplitude measurements were performed using a Hewlett-Packard Model 4195A Network Analyzer with a pair of Tektronix P6201 1 GHz active probes with 10× attenuators to reduce probe capacitances to 1.5 pF. A measured 200 pS skew across all three backplanes is an order of magnitude improvement over the 2 nS propagation delay induced skew which would be experienced in conventional technology using the same conductor geometry.

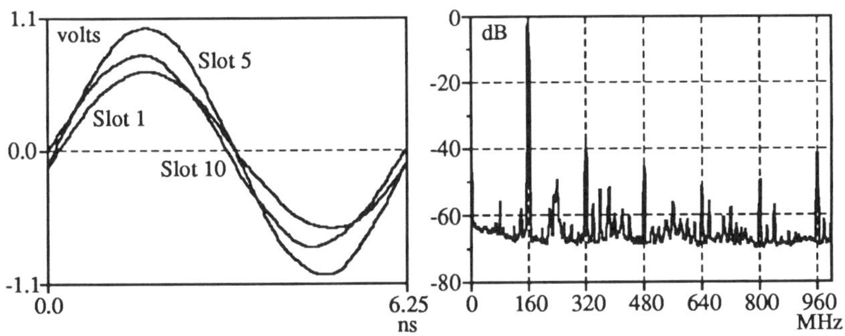

Figure 9

Figure 9 shows oscilloscope waveforms of the clock signal taken at selected slots on a fully populated backplane. These measurements were taken with a Tektronix Model DSA602 Digital Signal Analyzer with an 11A72 1 GHz preamp, using P6201 active probes with 10× attenuators. These waveforms demonstrate the near-sinusoidal nature of the clock signal. A Discrete Fourier Transform of the slot10 waveform using a Blackman window is also shown in Figure 9. The main spectral components are harmonics of the master clock frequency, with the second harmonic being about 2% and all others below 1%. Other spectral components appear at multiples of the 20 and 40 MHz daughter board clocks. These are all well below 1/2% of the master clock amplitude.

It is finally worth noting that as of this writing, the PXPL5 clock system has provided over two years of continuous service. It came up on the first attempt, and after a simple tuning procedure using the variable inductor terminations, has performed continuously without attention since August 1990. During this time, there have been many physical reconfigurations of the overall system, from minimally populated single rack through fully populated three racks. Moreover, of all problems that have occurred with PXPL5 to date, none have ever been attributable to the system clock.

5 Summary and future research plans

This paper attempts to codify the useful insights encountered in the conception, development and application of salphasic clock distribution, including lessons learned in the building of the Pixel-Planes 5 clock system. The term "building" is used advisedly here, as we experienced no debugging phase. Fielding a totally novel technology without undergoing extensive debugging is a rather extraordinary experience in the view of the author, and is testimony to the relative ease with which salphasic systems can be designed. To be sure, many false starts and misconceptions occurred in the conception and "on paper" reduction to practice of the salphasic concept, but once the key insights were understood, and the mathematical analysis was encapsulated in the cview simulator, the actual design of the PXPL5 clock system was quite straightforward. It is hoped that the rules of thumb presented herein will provide sufficient guidance to designers who wish to try their hand in salphasic design.

There are many as yet unexplored aspects of salphasic design, and thus its limits and the degree of necessary precaution are not well characterized. For example, it is not known how the PXPL5 clock would have performed had we not split ground planes to provide IsoGND for the clock system. The use of high speed differential comparators as clock receivers was a convenient off-the-shelf choice, but noise robust phase-locked-loop receivers may offer significant advantages. A major barrier to the rapid acceptance of salphasic technology is that no off-the-shelf logic family can directly accept salphasic signals. It's worth noting, however, that a differential input clock receiver suitable for salphasic inputs would be level compatible with virtually all other logic families. Thus, should a salphasic-compatible logic family ever be developed, it's clock inputs could automatically be cross compatible with other logic families, and with conventional clocking technology.

The most exciting possibility for salphasic distribution rests on the fact that salphasic behavior does not depend on the specific geometry of the distribution system, merely on its low loss properties. Thus, the possibility exists of distributing clock signals on a circuit board "clock plane", similar to power and ground planes. Theoretically, such a scheme should be far more robust to clock input loading because of the extremely low characteristic impedance of the plane. If this possibility can be reduced to practice, it is believed salphasic distribution techniques could revolutionize clock distribution on PC boards.

6 References

[1] V. L. Chi, "Salphasic Distribution of Clock Signals", UNC Chapel Hill Department of Computer Science Technical Report number 90-026, June 1990.

[2] V. L. Chi, "Cview – A Salphasic Clock Distribution Tree Simulator", UNC Chapel Hill Department of Computer Science Technical Report number 92-035, October 1992.

[3] V. L. Chi, "Salphasic Distribution of Clock Signals for Synchronous Systems", to appear in *IEEE T-C*.

[4] A. L. Fisher and H. T. Kung, "Synchronizing Large VLSI Processor Arrays", *IEEE T-C*, vol. C-34, pp. 734–740, August 1985.

[5] E. G. Friedman and S. Powell, "Design and Analysis of a Hierarchical Clock Distribution System for Synchronous Standard Cell/Macrocell VLSI", *IEEE J-SC*, vol. SC-21, pp. 240–246, April 1986.

[6] H. Fuchs, J. Poulton, et.al., "Pixel-Planes 5: A Heterogeneous Multiprocessor Graphics System Using Processor-Enhanced Memories", *Computer Graphics*, vol. 23, no. 3, pp. 79–88, July 1989.

[7] M. Hosoya, et.al., "Quantum flux parametron: A single quantum flux device for Josephson supercomputer", *IEEE T-ASC*, pp. 77-89, June 1991.

[8] R. E. Matick, "Transmission Lines for Digital and Communication Networks", New York, NY: McGraw-Hill, 1969, ch. 2.

[9] C. A. Mead and L. Conway, *Introduction to VLSI Systems*. Reading, MA: Addison-Wesley, 1980, ch. 7.

[10] C. E. Molnar, "Introduction to Asynchronous Systems", in *Proc. New Frontiers in Computer Architecture Conf.*, pp. 83–94, March 1986.

[11] W. Peil and R. Marolf, "Computer Circuitry for 500 Mc", in *1962 International Solid-State Circuits Conference Digest of Technical Papers*, February 1962.

[12] W. Peil and H. Raillard, "Clock Power Distribution Arrangement for High speed Logic Systems", United States Patent Number 3,267,394, filed 13 February 1963, issued 16 August 1966.

[13] C. L. Seitz, "Self-timed VLSI Systems", in *Proc. Caltech Conf. on VLSI*, pp. 345–355, January 1979.

[14] I. E. Sutherland, "Micropipelines", *Communications of the ACM*, vol. 32, pp. 720–738, June 1989.

[15] S. H. Unger and C-J. Tan, "Clocking Schemes for High-Speed Digital Systems", *IEEE T-C*, vol. C-35, pp. 880–895, October 1986.

[16] K. D. Wagner and E. J. McClusky, "Tuning, clock distribution, and communication in VLSI high speed chips", Stanford University CRC Tech. Rep. 84-5, June 1984.

Practical Implementation of Charge Recovering Asymptotically Zero Power CMOS

Saed G. Younis younis@ai.mit.edu
Thomas F. Knight, Jr. tk@ai.mit.edu
Artificial Intelligence Laboratory
Massachusetts Institute of Technology
545 Technology Square
Cambridge, Massachusetts 02139, USA

Abstract

As clock and logic speeds increase, the power requirement of CMOS circuits are rapidly becoming a major concern in the design of personal information systems and large computers. In this paper we present a practical implementation of a new CMOS logic family, Charge Recovery Logic (CRL), with a power dissipation that falls with the square of the operating frequency, as opposed to the linear drop of conventional CMOS circuits. We show that such power saving is achieved with moderate increases in area and circuit complexity.

The technique relies on constructing an explicitly reversible pipelined logic gate, where the information necessary to recover the energy used to compute a value is provided by computing its logical inverse. Information necessary to uncompute the inverse is available from the subsequent inverse logic stage.

Depending on the Q of available inductors, we anticipate energy savings of over 99% per logic operation for logic which is clocked sufficiently slowly.

1 Introduction

In principle, a computing engine need not dissipate any energy, as shown in the work of Landauer, Bennett, and Feynman [8] [1] [3]. Although these authors approach the problem from different disciplines and use different physical as well as theoretical models, they all conclude that the transfer of energy through a dissipative medium dissipates arbitrarily small amounts of energy if this transfer is made sufficiently slowly. This should not come as a surprise: to eliminate the dissipation we have to make the potential applied to the medium zero, and from thermodynamics we know that the only way to transfer energy with zero potential is to allow the transfer to happen over an infinite amount of time.

Power dissipation in conventional CMOS primarily occurs during device switching. One component of this dissipation is due to charging and discharging the gate capacitance through a conducting, but slightly resistive, device. We note here that it is not the charging or the discharging of the gate that is necessarily dissipative, but rather that a small time is allocated to perform these operations. In conventional CMOS, the time constant associated with charging the gate through a similar transistor is RC, where R is the ON resistance of the device and C its input capacitance. However, the cycle time can be, and usually is, orders of magnitude longer than RC. An obvious conclusion is that power consumption can be reduced by spreading the transitions over the whole cycle rather than "squeezing" it all inside one RC.

To successfully spread the transition over periods longer than RC, we forbid any device in our circuit from turning ON while a potential difference exists across it. Further, once the device is switched ON, the energy transfer through the device occurs in a controlled and gradual manner. While the latter is relatively easy to control through the use of inductors, the former leads to some interesting restrictions on the way we usually perform computations. We can always determine and control the potential on one side a CMOS device since it is usually connected to a power supply rail. The potential on the other side, however, depends solely on the result of the computation. To perform a non-dissipative transition of the output, we must know the state of the output *prior to and during* this output transition. Stated more clearly, to non-dissipatively reset the state of the output we must at all times have a copy of it. The only way out of this circle is to use *reversible logic.* It is this observation that is the core of a number of proposals to construct low power electronic computing engines. By low power, or non-dissipative, we mean that the energy per computational step can be made arbitrarily small by spreading the computation over a longer period.

Fredkin and Toffoli [4] demonstrated one realization of a low power universal gate using conservative logic. In conservative logic, the information content as well as the number of 1's and 0's are conserved throughout the computation. One property of a computation using conservative logic is the production of unwanted intermediate outputs. Unfortunately, discarding these outputs results in energy dissipation. Recycling it however, does not. It is this operation of recycling that requires the use of reversible logic [5]. The CMOS gate proposed by Fredkin and Toffoli could not be easily integrated, however, as it requires the use of inductors internal to the computational network. The sizes and numbers of these inductors are well beyond what can be easily accommodated on a silicon substrate.

Aware of the requirement to spread the energy transfer over a longer period of time, Seitz et al. [11] proposed a new reduced-power CMOS design style. The authors elected to use only N-Channel devices and therefore depended on bootstrapping action in order to eliminate the V_T voltage drop through their devices. The authors having successfully fabricated and operated numerous circuits using this style still warned of the importance of device sizing to achieve enough bootstrapping for proper operation. More recently, Koller and Athas [7] proposed a form of dual rail logic that achieved adiabatic switching. Not using reversibility, they were unable to completely remove the power dissipation of their latches and commented that "to return the energy to the power supply without dissipating it, we must remember which output line we are charging. However, the only record we have of that is the very piece of information we are erasing". In addition, attempting to string multiple stages of this gate without intervening latches required a separate clock phase for each stage. The beauty of their implementation however is that their latch dissipation was about $\frac{1}{2}CV_T^2$ per bit, which is independent of V_{dd} and much less than the usual $\frac{1}{2}CV_{dd}^2$.

Recent and independent work by Hall [6] and Merkle [10] showed how to connect *Retractile Cascade* stages to eliminate the power dissipation in the latches. Merkle's proposal requires the use of intermediate voltage boosters made out of varactors and needs a relatively large number of independent clocks. In addition, the proposal calls for the explicit use of latches between the successive *Retractile Cascade* stages thus adversely affecting both throughput and latency.

In this paper we present a new technique for constructing non-dissipative CMOS circuits. We feel that this technique has a number of distinct advantages which warrant its use in future circuits. First, we require only two clocks and their inverses, for a total of four swinging rails, regardless of the number of stages. Second, each stage has an implicit latching action, and thus explicit latches between successive stages are not needed, simplifying pipelining and improving throughput and latency. Third, the technique, through proper sizing, allows the effective RC of the circuit as seen by each rail to be independent of the data values used in the computation. Since we use inductors to non-dissipatively move the charge, keeping RC constant is important for maintaining constant and data-independent rise and fall times. We note however, that correct functional operation of our gates does not depend on proper sizing. Fourth, we use CMOS pass gates and dual-rail logic throughout the circuit avoiding the need for bootstrapping, reversible varactor amplifiers, etc. Fifth, the topology and design rules of these circuits are identical to those of conventional CMOS except for a redundancy factor. Finally, we use reversible logic techniques to non-dissipatively reset each stage without ever reversing the direction of the pipeline.

The major disadvantage of our technique is the requirement for sixteen times as many devices as conventional CMOS. Area increases for large circuits may be far less, because of the similarity in the wiring patterns in duplicated logic. Four times as many signal wires are used, which may represent a more accurate estimate of the area ratios. The gates provide true and complement output forms of each logic value, eliminating the logical need for inverters, and partly compensating for the explosion in transistor count. We believe that for many applications, trading even a factor of sixteen in silicon area is quite appealing for a power reduction of literally orders

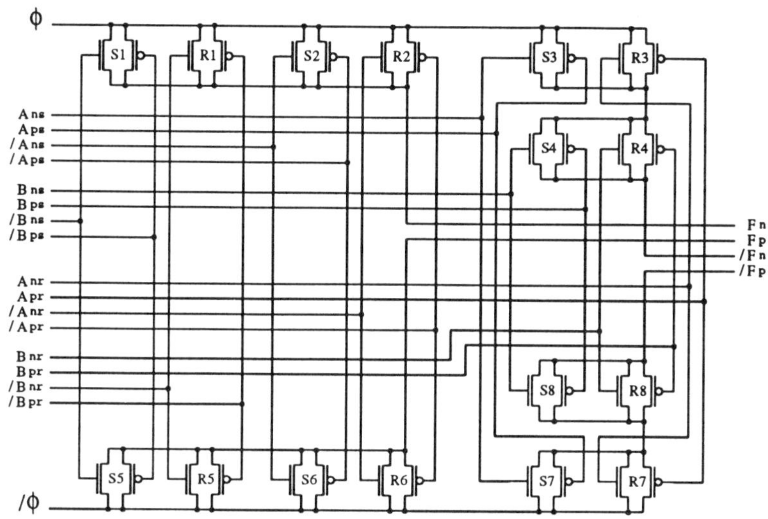

Figure 1: CRL 2-input NAND gate.

of magnitude.

2 Charge Recovery Gate

In this section we describe the topology and operation of CRL gates. We use the 2-input AND/NAND gate implementation shown in Figure 1 for our example. Because CRL uses pass gates instead of single transistors, we need to generate and distribute the *electrical* complement of each of our data lines in order to control the P-Channel side of the pass gates. We identify the lines intended to drive the N-Channel devices by the letter "n" in the subscript, and those for the P-Channel devices by the letter "p". Furthermore, our CRL circuit requires us to generate and distribute the *logical* complements of the data lines. We identify those in the figure by the forward slash character, /. We will clarify why we need these logical complements when we discuss the universality of this logic family.

From the above, we see that CRL uses four wires to represent each data line. An active data line A, representing the logic value 1, has $A_n = 1, A_p = 0, /A_n = 0, /A_p = 1$ for example. The reason for the apparent redundancy will become apparent shortly. In addition to the above, each CRL gate accepts two input groups. The first is the SET group, and we identify the lines belonging to this group by the letter "s" in the subscript. The second is the RESET group, and for it we use the letter "r" in the subscript.

CRL gates have two stable states. They are the SET state and the RESET state. While in the SET state, the output lines of a CRL gate are valid and can be sampled by a subsequent gate. While in the RESET state, the output lines of a CRL gate must be forced to levels that will turn off any pass gates they are driving. Note that while *electrically* complementary lines are always at opposite levels, *logically* complementary lines are at opposite levels only when the gate driving them is in the SET state, thus the distinction between the two and the need for four wire signalling.

In going from the RESET state to the SET state, the gate uses the SET inputs to compute its valid output. At that time the gate assumes, and we require, that all the RESET inputs are idle. We define an idle input as one that is at a level that will turn off any pass gate it controls. In going from the SET state to the RESET state, the gate uses the RESET inputs to non-dissipatively restore itself to the RESET state.

In Figure 1, we have 16 pass gates labeled S1 thru S8 and R1 thru R8. The S gates are associated with the set inputs, and the R gates are associated with the reset inputs. Note that every SET pass gate is paralleled by a corresponding RESET pass gate. It is important that the set and reset pass gates be paralleled in the same order and with connections between intermediate pass gates in a series chain.

We are now ready to describe the operation of a CRL gate. We start with the gate in the RESET state. In the RESET state, the following is true in our example NAND gate:

- The top rail and all the nodes connected to it are at V_{ss}.
- All the sources and drains of the transistors in S1 thru S4 and R1 through R4 pass gates are at V_{ss}.
- All the outputs driving N-Channel devices are at V_{ss}.
- The bottom rail and all the nodes connected to it are at V_{dd}.
- All the sources and drains of the transistors in S5 thru S8 and R5 through R8 pass gates are at V_{dd}.
- All the outputs driving P-Channel devices are at V_{dd}.

We further assume that both the SET and RESET inputs are idle. We now wait until the SET inputs become valid. Note that since the voltage across every pass gate is now zero, turning the SET inputs active from their idle levels creates no dissipative transients. Recall that to go to the SET state, all the SET inputs must be valid and all the RESET inputs must be idle. Thus only the SET pass gates are active during this transition.

To SET the gate, we gradually swing the top rail from V_{ss} to V_{dd} and swing the bottom rail from V_{dd} to V_{ss}. Examining the circuit diagram and following the effects of these swings, we see that F_n will go high and F_p low if and only if the input data lines are such that $/(A \wedge B)$ is true. Otherwise, each will remain at its idle level while $/F_n$ swings high and $/F_p$ low. Therefore F, when active, computes $/(A \wedge B)$.

While in the SET state, we must guarantee that every pass gate that was off during the transition remains off for as long as the circuit is in the SET state. This is because a pass gate that was off during the SETting has a potential difference across it. On the other hand, turning off a pass gate that was on during the transition has no adverse effect.

We could trivially restore the circuit to its original state non-dissipatively by gradually returning the top and bottom rails to their RESET levels. Note that the SET inputs must remain valid while the circuit is reset in order to completely restore the state of the circuit, including the voltage levels of internal nodes. Unfortunately, it is not possible to build pipelined circuits using this simple scheme. A pipelined circuit accepts a new data value as soon as the first stage finishes processing the previous data point. However if every stage in the pipeline must wait until the subsequent stage resets before it can reset and accept a new value, then this "pipeline" can only accept a value after the previous value propagates to the end of the pipeline and after all the subsequent stages reset themselves in a reverse order. This circuit has the same throughput as a non-pipelined circuit.

We *can* however pipeline CRL gates if we do the resetting in the following manner. After SETting the circuit, we provide an identical copy of the SET inputs on the RESET lines. We will show how to generate these RESET inputs in Section 3. At the same time we allow the SET inputs to go idle, i.e. the previous pipeline stage can reset itself so as to accept new data inputs. Since every SET pass gate is paralleled by an identical RESET pass gate, we can now non-dissipatively restore the circuit by gradually restoring the top and bottom rails to their RESET levels. After RESETting the circuit, we allow the RESET inputs to go idle. Now, being in the RESET state, the circuit can accept new inputs.

The energy stored in a CRL circuit includes the energy stored in the internal nodes in addition to the energy stored on the output lines. By providing exact duplicates of the set inputs on the reset lines, and by connecting reset and set pass gates in parallel everywhere, we guarantee that the energy on these internal nodes is recovered.

Generating and distributing the logical complement of every line doubles both the wire and device counts. The device count doubles because of the complementary networks that are needed to generate them. In the circuit of Figure 1, S1, S2, R1, R2, S5, S6, R5 and R6 belong to the networks generating true outputs. The remaining pass gates belong to the networks generating the logical complements. Just as in conventional CMOS gate where the P-Channel network is the complement of the N-Channel network, in CRL gates, the network that generates a complementary output is the complement of the network that generates the true output. We pay this price of generating and distributing the logical complement to retain a crucial property of any logic family, *universality.* Since when a line is idle, the pass gates it controls are off, a logical 0 input cannot produce a logical 1 output. We need the complementary outputs in order for CRL family

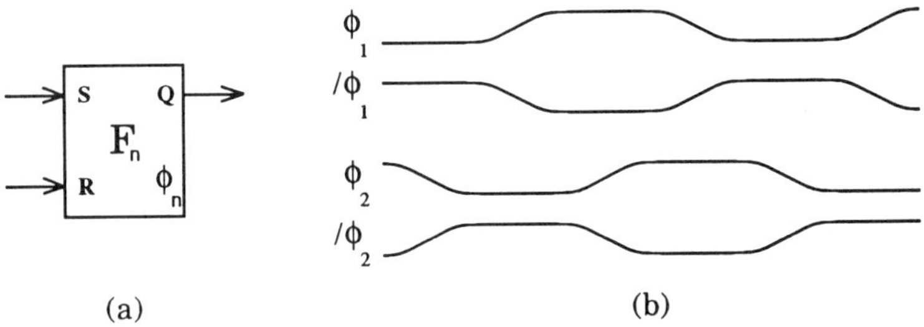

Figure 2: (a) Abstraction box. (b) Timing of the four clock rails.

to support negation. By demonstrating the implementation of a 2-input AND/NAND gate, we equivalently have shown that CRL is universal. It should be obvious how we can build any and-or logic function based on the above techniques in a way similar to conventional CMOS.

So far, our CRL gate requires eight times as many devices as conventional CMOS as illustrated by the 2-input AND/NAND gate example. For circuits that require complementary outputs, such as address decoders, the redundancy factor may be somewhat less.

3 Network Connection

In this section we show how to connect CRL gates, or stages, in a non-dissipative pipeline. The main purpose of this method of interconnection is to provide the RESET inputs to each gate at the correct time. We build the pipeline out of copies of an abstraction box shown in Figure 2a.

We think of this box as containing a parallel set of CRL gates performing any logical function of an arbitrary number of inputs. Symbolically, the output of the box represents a bundle containing the outputs of the CRL gates internal to the box. The box has two input branches. One is the SET branch and identifies a bundle containing all the SET inputs of the gates internal to the box, the other is the RESET branch and identifies a bundle containing all the RESET inputs. The function computed by the box is identified by the letter in the center of the box. Finally, indicated at the lower corner of the box is the clock phase used to control all the CRL gates internal to that box. A clock of ϕ_1 indicates that the top rail is connected to ϕ_1 and the bottom rail to $/\phi_1$, while a clock of $/\phi_1$ indicates that the top rail is connected to $/\phi_1$ and the bottom rail to ϕ_1.

Using this abstraction, Figure 3 illustrates how CRL gates are connected to produce a non-dissipative pipeline. The timing of the four clock lines is shown in Figure 2b. Note that the box with a function F^{-1} performs the inverse operation of the box with a function F. To SET a box, all the SET inputs must be valid and stable and all the RESET inputs must be idle, i.e.

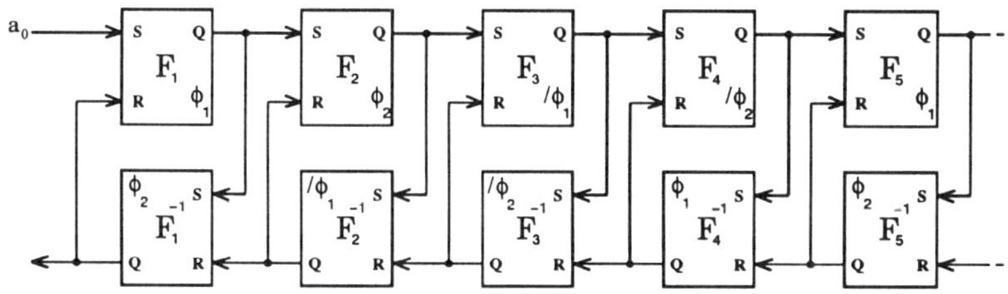

Figure 3: Non-dissipative multi-stage pipeline connection.

they come from a box that is currently in RESET, so that all the RESET pass gates are OFF. With these inputs, swinging the clock rails of the box away from their rest level will SET the box. To RESET the box, the rails are returned to their rest levels while the SET inputs are idle and the RESET inputs are active and stable.

To follow the operation of the circuit we start with ϕ_1 and ϕ_2 at their rest state and assume that the pipeline has been operating for some time. We follow the propagation of the input a_0 only, even though other parallel activity is going on. From the states of the clocks we see that box F_1 is RESET and its RESET inputs are idle as well. Swinging ϕ_1 SETs F_1 and computes $F_1(a_0)$. Swinging ϕ_2 now SETs F_2 and F_1^{-1} and produces $F_2(F_1(a_0))$ and $F_1^{-1}(F_1(a_0)) = a_0$ respectively. Now swinging ϕ_1 to its rest level RESETs F_1, produces $F_3(F_2(F_1(a_0)))$ and $F_2^{-1}(F_2(F_1(a_0))) = F_1(a_0)$. The circuit is now ready to safely RESET boxes F_2 and F_1^{-1}. One can see that we can continuously drive a new input into the network every ϕ_1 and successfully operate the pipeline in a non-dissipative fashion. In addition, this pipeline can have any arbitrary number of stages and still be driven entirely by ϕ_1, ϕ_2 and their inverses only.

There remains one problem however. At the end of the pipeline the RESET input to F_5^{-1} is not available and hence resetting this box is dissipative. Furthermore, it could not be generated, as this is the place where reversibility is broken. We can however, restore reversibility here through brute force by connecting to the end of this pipeline a mirror, and an inverse, image of itself. The missing input at the end of this extended pipeline that is needed to reverse the last inverse box is now simply a_0. With this topology, we can proceed without any dissipation by continually supplying delayed copies of the input to the pipeline at the inverse input on the far right. The technique of connecting an inverse network to the forward network was previously used in [5] and [3] to eliminate dissipation through recycling the intermediate *garbage* that results in conservative logic.

Admittedly, the above solution is more of theoretical than practical interest. If reversibility needs to be broken, that is, when information loss cannot be avoided, then some dissipation will occur for every lost bit of information. For these situations, we can reduce the dissipation by ending the pipeline

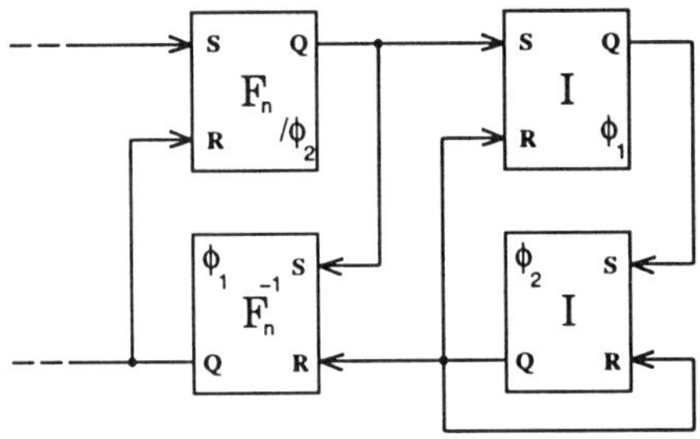

Figure 4: Last pipeline stage connection for dissipation reduction.

with two identity boxes, $I(a) = a$, and use the output of the lower identity box to reset itself as shown in Figure 4. Closer examination shows that the dissipation is $\frac{1}{2}CV_T^2$ per bit per cycle as opposed to $\frac{1}{2}CV_{dd}^2$ for conventional gates. Since the output of an identity box is the same as the input, the resetting swing proceeds normally until the output levels are insufficient to keep the appropriate pass gates on. Because of this, some internal nodes will have a potential that is one V_T away from their reset levels. The next input to the gate will short this potential difference resulting in $\frac{1}{2}CV_T^2$ dissipation per bit. Note that that we only pay this penalty at the last stage of a long pipeline.

4 Reversibility

The reason for choosing a NAND gate for the implementation example was to demonstrate the universality of CRL . Obviously, the NAND function is not reversible. To be reversible, a function must minimally have as many output as it has inputs. In addition, a non-conservative function generally requires more outputs than inputs to be reversible. If in addition to generating $/(A \wedge B)$ we pass thru a copy of A and a copy of B, then we would have enough information to make this augmented NAND gate reversible. So long as the information entropy of the system is kept constant, making every functional block of the system reversible is always possible and usually simple. However, an operation resulting in an increase in the informational entropy of the system is fundamentally irreversible and hence dissipative. With infinite storage we can always maintain enough information history to reconstruct the past regardless of the nature of the computation. Fortunately, we can be clever about trading off a little dissipation at critical places in the computation where reversibility would be complex or costly and use the system's *finite* resources to maintain reversibility and asymptotic zero dissipation elsewhere. CRL in essence provides the link between power dissipation and information content.

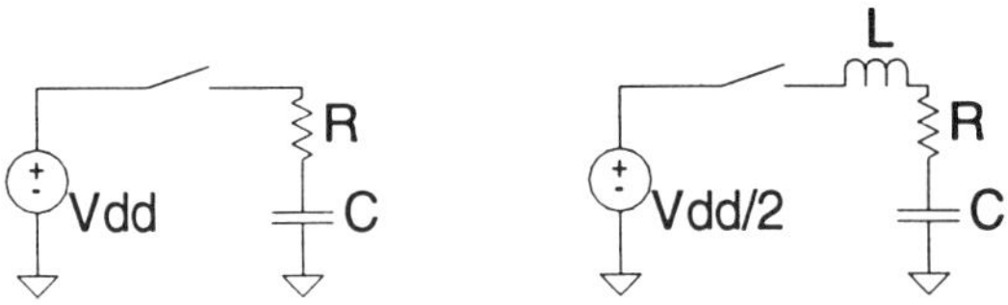

Figure 5: Conventional and non-dissipative circuit analysis models.

5 Power-Delay Product

The power-delay product is used as a figure of merit to compare logic families. In this section we compare the performance of a conventional CMOS gate with that of our gate on the basis of power dissipation per period. We use the circuits shown in Figure 5 as our models. These models are based on a lumped element approximation where the R's and C's are the effective R and C as seen from the supply terminals of the chip. For appropriately sized circuits we can assume that the gate voltages have roughly the same rise time and therefore we can approximate R by R_{device}/N and C by $C_{device} \times N$ for an N gate chip. Note that R and C depend only on N and the fabrication process technology. In conventional CMOS, to go through a single period of operation, we discharge C from an initial voltage of V_{dd} to ground and then charge it back to V_{dd}. For our CRL gate to go through a single period we use inductors connected to $V_{dd}/2$ on one side to swing the rails to the opposite polarity at the appropriate time. To swing a rail that rests at V_{dd} we

1. disconnect the rail from the positive supply and connect it to the inductor,
2. we keep the inductor connected until the current reaches zero signaling a complete polarity inversion,
3. we disconnect the rail from the inductor and connect it to the negative rail,
4. and finally we reverse the above steps to go back to the positive state.

The inductor in the above circuit acts as an electrical "flywheel" that inverts the polarity of internal nodes connected to the rail. For now, assume the switch connecting the rail to the inductor is external to the chip. We will revisit this decision in a later section.

As is widely known, the internal energy dissipation of conventional CMOS circuits is attributable to three major components. The first is due to the static leakage current, which we will assume to be negligible. The second is due to the transient current associated with charging and discharging the gate capacitance C through the ON resistance R and is equal to CV_{dd}^2 per period for a rail-to-rail voltage equal to V_{dd}. The third is due to the switching current which is caused by both N-Channel and P-Channel devices being simultaneously ON for a brief time during a swing of $V_{dd} \geq 2V_T$ [2]. We can approximate this switching dissipation by

$$V_{dd} \times \frac{1}{2}\frac{V_{dd}}{2R} \times \frac{V_{dd} - 2V_T}{V_{dd}} \times 2RC$$

which for a typical $V_{dd} = 4V_T$ simplifies to $\frac{1}{4}CV_{dd}^2$. As V_{dd} drops below $2V_T$ this switching dissipation becomes negligible. Unfortunately, the transfer curve begins to exhibit hysteresis thus limiting the utility of the gate at these power supply voltages.

Since the switching dissipation has a form similar to the transient dissipation, and is usually smaller in magnitude, we concentrate on the dissipation due to the transient dissipation associated with charging the gate capacitances of the devices, and approximate the total dissipation per period of conventional CMOS, E_{CMOS}, as

$$E_{CMOS} = CV_{dd}^2$$

We now examine the dissipation of our proposed circuit. To simplify the algebra, we let $V_{dd} = +V_0$ and $V_{ss} = -V_0$ so that V_0 is equal to half the rail-to-rail voltage V_{dd}. Our R,L,C circuit is described by

$$\frac{d^2V_C}{dt^2} + 2\alpha\frac{dV_C}{dt} + \frac{V_C}{{\omega_0}^2} = 0$$

where V_C is the capacitor voltage, $\alpha = R/2L$, and $\omega_0 = 1/\sqrt{LC}$. For the solution to oscillate, the circuit must be *underdamped*, requiring that

$$2\sqrt{LC} > RC$$

and we find that the frequency of oscillation, ω_d, is given by

$$\omega_d = \sqrt{{\omega_0}^2 - \alpha^2}$$

Since R and C are fixed for a given chip, we can only adjust L to affect ω_d. Examining the formula for ω_d we discover that ω_d steadily increases as L decreases up to a maximum, ω_{dmax}, and then sharply decreases as the circuit approaches the *critically damped* point. We find that

$$\omega_{dmax} = \frac{1}{RC} = \frac{1}{R_{device}C_{device}}$$

and the smallest inductance we would ever need, L_{min}, is found by

$$L_{min} = R^2C/2 = R_{device}^2C_{device}/2N$$

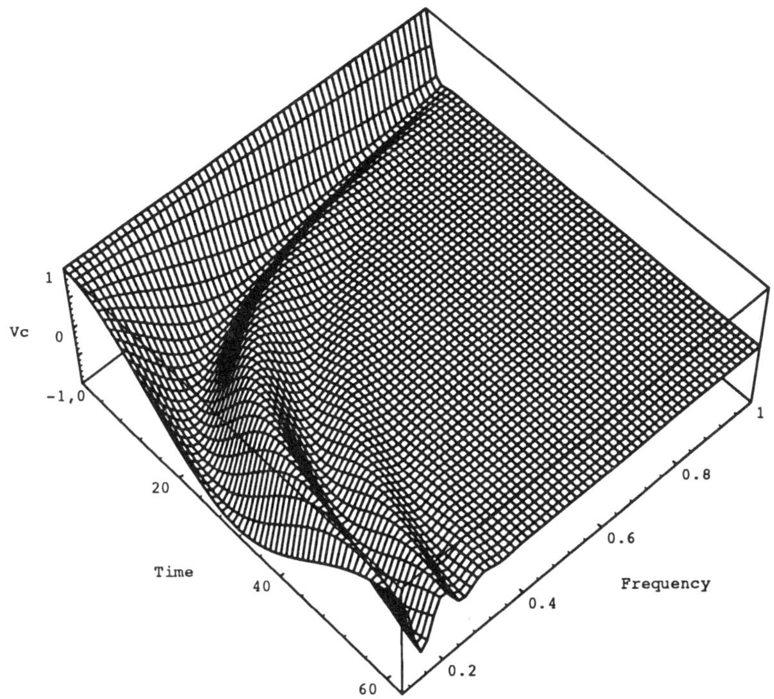

Figure 6: Plot of $V_c(t, \omega_d)$ showing the effect of fast rise and fall times on energy loss.

For $V_C(0) = V_0$ and $i_C(0) = 0$ we find

$$V_C(t) = V_0 e^{-\alpha t}(cos\omega_d t + \frac{\alpha}{\omega_d} sin\omega_d t) \tag{1}$$

and

$$i_C(t) = V_0 C \frac{{\omega_0}^2}{\omega_d} e^{-\alpha t} sin\omega_d t \tag{2}$$

Figure 6 shows a plot of normalized V_c as function of time and normalized ω_d. Initially $V_C(0) = 1$. With ω_d close to ω_{dmax} the voltage drops rapidly, dissipating most of the capacitor energy on the way down. The reclaimed energy is insufficient to fully charge the capacitor back to the negative rail. As ω_d moves away from ω_{dmax} we in effect spread each oscillation over a longer period of time. We see from the plot that the capacitor retains most of its energy since V_C comes close to the bottom rail. With ω_d only an order of magnitude lower than the maximum, the capacitor recovers most of its energy.

The power dissipated in a single rail-to-rail swing for our gate consists of two components. The first is the power dissipated in R during the swing,

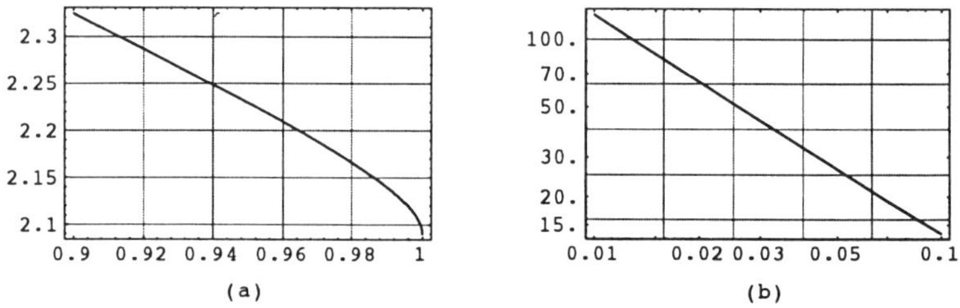

Figure 7: (a) Linear plot of F_{saving} factor vs. ω_d. (b) Log-Log plot of F_{saving} factor vs. ω_d. In both plots ω_d is normalized to ω_{dmax}.

E_s, which results in a final voltage that is lower than $|V_0|$. Using equation 1 we find that

$$E_s = \frac{1}{2}C\,V_0^2\,(1 - e^{\frac{-2\pi\alpha}{\omega_d}})$$

The second is the energy lost in R while charging C to the rail to compensate for the lower peak voltage due to E_s. We will do this by simply connecting the line to the rail and losing some energy in R, E_c, during this process. We find that

$$E_c = \frac{1}{2}C\,V_0^2\,(1 - e^{\frac{-\pi\alpha}{\omega_d}})^2$$

adding these two terms and multiplying by two to allow for both directions of swing, we find that

$$E_{loss} = 2C\,V_0^2\,(1 - e^{\frac{-\pi\alpha}{\omega_d}}) \tag{3}$$

per period.

Using Equation 3, we compute the ratio of the power consumption per period of conventional CMOS to that of our gate, F_{saving}, as

$$F_{saving} = \frac{2}{(1 - e^{\frac{-\pi\alpha}{\omega_d}})}$$

Figure 7a shows a Linear plot of F_{saving} near ω_{dmax}. In this region, the circuit's performance is close to conventional CMOS but improves rapidly with lower ω_d. As we get away from ω_{dmax}, the graph attains a nearly constant slope as shown in Figure 7b. Note that we only need to slow down a little in order to see substantial power saving. For $R_{device} = 10\text{K}\Omega$ and $C_{device} = 0.1\text{pF}$, $\omega_{dmax} = 1\text{GHz}$. Operating at a speed of 40 MHz we will consume 33 times less power and at a speed of 4 MHz we achieve an impressive 319 times power saving. We emphasize that these numbers reflect the power saving advantage per period and are therefore in addition to the linear power drop as a function of frequency found in conventional CMOS.

6 Dynamic Considerations and Nonlinearities

The above analysis assumes that we can, in theory, lump the gate capacitances of MOS devices into one equivalent linear capacitance. In practice however, we need to be more careful. Each rail in CRL feeds a number of branches. Ideally, the effective RC time constant of each branch as seen by the rail is data independent and equal to the RC time constant of the entire rail circuit. A branch with a longer time constant would lag behind during the transition. This would create a potential difference across pass gates that are switched on, leading to dissipation. The effect is minimized by the symmetry of CRL . Because of the existence of the true and complement networks for every output line, a swinging rail is always connected to one and only one output line. Therefore, regardless of the output level, the rail will always drive the same output capacitance. The only difference in RC comes from the fact that the true and complement networks are not identical and as such could contribute different RC's depending on the data. Properly sizing the devices so that the two networks exhibit the same time constant independent of the state of the inputs will eliminate this problem.

Another point of consideration is the nonlinearity of the capacitances of MOS devices. An enhancement mode MOS device has a higher gate capacitance while in inversion, i.e. conducting, than when it is off. At the beginning of a SETting swing, all the outputs are at idle and all the devices driven by these outputs are off. At the completion of a SET, devices controlled by a swinging output are on. For this reason, a rail that is SETting a gate sees lower effective capacitance at the start of the swing than at the end. With the inductor anchored to $V_{dd}/2$, the rail will not reach the opposite voltage at the end of a swing. This leads to dissipation when the rail is connected to V_{dd} or V_{ss} after completing the swing. Fortunately, a rail SETting a gate is at the same time RESETting another. We feel that as the number of gates connected to a rail increases in a balanced way, the adverse effects of this nonlinearity is minimized. Intuitively, the effective RC time constant is now equal to $R \times C_{average}$.

In addition to the above, there is also the effect of capacitive coupling. Take an output wire that carries a logical true. The devices that it controls in a subsequent gates are on. Because of the gate-to-channel capacitance, when the subsequent gates SET, the varying voltage in the channels of turned on devices capacitively couples to the gate. This dumps, or extracts, charge from the output wire of the previous stage. Again, the effect is minimized due to symmetry. Since each output drives identical devices that are capacitively coupled to the top and bottom rail swinging in opposite directions, this capacitive coupling is almost entirely eliminated by the symmetry. We say almost because due to the capacitance nonlinearity, the symmetry will cancel the coupling when integrated over the entire swing and not instantaneously.

We want to stress here that while minimizing the effects of the above phenomenon improves the power saving factor of CRL circuits, none of the above effects, even when extreme, jeopardizes the logical functionality of CRL circuits. HSpice simulations of the 2-input NAND gate, as well as simulations of other CRL gates and circuits, have demonstrated proper operation of the CRL circuit in the presence of these effects. This is important in simplifying the design of CRL logic. If a designer incorrectly sizes a branch in a CRL circuit, the worst he can expect is higher power dissipation in that part of the circuit and not a disfunctional chip.

7 Practical Issues

We have presented a logic family in which each transistor internal to the logic is turned on only when there is zero voltage across it, thus eliminating energy dissipation in the switch. There remain, however, a number of practical implementation issues.

First, while we have minimized the use of external inductors by requiring only four external clock rails, there remains potential dissipation associated with the switch in series with each inductor. This switch is now the only component with significant voltage across it when it is turned on. This is where we expect to see significant power loss. Yet, we are switching the transistor at a time when we can guarantee, as a result of the series inductance, that there is near zero current flow. If we can fully turn on the transistor prior to significant current build up, then dissipation can be minimized.

Turning this switch transistor on requires charging and discharging its gate capacitance. If we do this with a dissipative switching operation, then much of our potential power savings will be thrown away here.

However, we can recursively apply the power recovery techniques discussed above to recover the energy stored in the switch transistor gate. Presumably, applying a single level of such recursive power recovery will be adequate for most applications.

A second difficulty in fabricating circuits with high power recovery is the Q of the inductor. While our derivation assumed an infinite inductor Q, realistic high frequency inductors have Q's limited to the range of 100-200, placing an upper limit on the achieved power recovery.

In some applications, the use of high temperature superconducting high Q inductors, in combination with liquid nitrogen cooled CMOS might be an attractive combination. Achievable Q's with superconducting technology are in the several hundred thousands, and the major limitation might become leakage currents in CMOS devices. These dark currents will also be drastically reduced by low temperature operation. Lowering the temperature also increase the carrier mobility. At liquid nitrogen temperatures, the mobility of carriers in MOS devices increases 4 times under low field conditions and 1.7 times under high field conditions. Operating under both high and low field conditions, the measured mobility increase in conventional CMOS averages to about 2.5 times [9]. In contrast, CRL operation is by design limited to low field operation and would therefore retain the full benefit of 4 times mobility increase at LNT.

8 Reliability

The reliability of CRL circuits should be excellent. First, the low dissipation of CRL circuits ensure operation at near ambient temperatures with minimal attention to cooling. Second, during conventional CMOS switching, the circuit temperature can rise above the average temperature for a short time. Since the failure rate is related to a high power of temperature, the overall failure rate is typically higher than that predicted by the average operating temperature. As CRL circuits disallow sudden transients, the failure rate of CRL devices will be lower than that of conventional CMOS operating at the same temperature. Third, the elimination of transistor conduction in the presence of high drain to source voltages removes hot electron trapping as a failure mechanism. This may allow finer transistor geometry, higher performance, and process simplification by elimination of LDD and sidewall spacer techniques used to control hot electron behavior in conventional technologies. Finally, the peak currents in CRL are minimized thus further reducing other failure effect such as metal migration.

9 Conclusion

We have demonstrated a set of techniques for constructing a reversible, energy efficient family of computing devices, with theoretically near-zero dissipation. Key ideas include the reduction in the number of external swinging rails through the use of our phased reverse computation technique and the paralleling of all forward computing pass gates with a second pass gate driven by the recomputed logic values.

An analysis of the available energy savings demonstrates the asymptotically zero dissipation available. We presented a discussion of practical issues and considerations in the construction of these circuits.

Acknowledgments

This research is supported in part by the Defense Advanced Research Projects Agency under contract N00014-91-J-1698. We acknowledge useful discussions with Carl Feynman, Norman Margolis, Tom Simon, and Tom Toffoli.

References

[1] Bennett, C., "The Thermodynamics of Computation – a Review", *International Journal of Theoretical Physics*, Vol. 21, No. 12, 1982, pages 905-940.

[2] Calebotta, S., "CMOS, the Ideal Logic Family", *National Semiconductor CMOS Databook,* Rev. 1, AN-77, pp. 2-3, 1988.

[3] Feynman, R., "Quantum Mechanical Computers", *Foundations of Physics,* Vol. 16, No. 6, 1986.

[4] Fredkin, E., and Toffoli, T. (1978),"Design Principles for Achieving High-performance Submicron Digital Technologies," Proposal to DARPA, MIT Laboratory for Computer Science.

[5] Fredkin, E., and Toffoli, T., "Conservative Logic", *International Journal of Theoretical Physics*, Vol. 21, Nos. 3/4, 1982, pages 219-253.

[6] Hall, J. S., " An Electroid Switching Model for Reversible Computer Architectures," in *Proceedings of Physics of Computation Workshop,* Dallas Texas, October 1992.

[7] Koller, J. G., and Athas, W. C., "Adiabatic Switching, Low Energy Computing, and the Physics of Storing and Erasing Information," in *Proceedings of Physics of Computation Workshop,* Dallas Texas, October 1992.

[8] Landauer, R., "Uncertainty Principle and Minimal Energy Dissipation in a Computer", *International Journal of Theoretical Physics* , Vol. 21, Nos. 3/4, 1982, pages 283-297.

[9] Laramée, J., Auburn, M.J., and Cheeke J.D.N., "Behavior of CMOS Inverters at Cryogenic Temperatures," *Solid-State Electronics*, vol. 28, no. 5, pp. 453-456, May 1985.

[10] Merkle, R. C., "Reversible Electronic Logic Using Switches", Submitted to *Nanotechnology*, 1992.

[11] Seitz, Charles L. *et al.*, "Hot-Clock n*MOS*," in *Proceedings of the 1985 Chapel Hill Conference on VLSI,* Computer Science Press, 1985.

Multi Module Focal Plane Processing Sensor with Parallel Analog Support for Computer Vision

Marc Tremblay, Denis Laurendeau and Denis Poussart
Computer Vision and Digital Systems Laboratory
Department of Electrical Engineering, Laval University,
Québec, Canada, G1K 7P4
marct@gel.ulaval.ca

Abstract

Computational sensors offer interesting system design opportunities by the integration of dedicated low-level image processing at the sensor level. These sensors are able to achieve higher performance by exploiting the synergy of combining transduction and analog processing on the same solid state substrate or on closely associated subsystems. This novel kind of smart image acquisition system should provide significant performance improvement compared to the conventional use of frame grabbers. In this paper, the Multiport Access photo-Receptor (MAR) architecture is described including its high resolution hexagonal image sensor, a parallel analog processing module, a microcoded system controller/interface and a set of digital peripheral processing modules. The sensor uses a multiport addressing architecture of the pixel array in order to apply external multiresolution analog spatial convolution followed by edge detection, scale-space integration and interactive edge segment extraction. The implementation of each part of the system requires different technologies and then favors a multi-chip module assembly. This multi module approach allows the implementation of high resolution sensors with a very powerful computational capability tailored to computer vision applications. A 256 x 256 pixels version of the MAR sensor over a 0.8 cm x 0.8 cm die using 1.2 μ CMOS technology is currently operational. Experimental results are presented for this prototype along with the corresponding multiresolution edge maps.

1 Introduction

Hardware for computer vision has often been oriented on high performance digital processor without considering the image sensor as a possible computing element. Basic low-level image processing tasks transform digitized illuminance data obtained from a CCD camera into a more dense representation which is required by further segmentation and recognition processes. The high-level of integration offered by current VLSI technologies allow the designers to develop very efficient architectures by exploiting the regularity and the locality of computer vision algorithms. The most well-known approaches are related to massively and distributed parallel processing or pipelined architectures. While these recent developments on parallel computer are quite exciting in a machine vision perspective, it is definitively more relevant to use their computing capabilities for medium and high-level recognition tasks leaving the low level image processing for dedicated hardware at the sensor level. A major consequence of this computational sensing approach is the significant reduction of data between the sensing unit and the recognition processes.

VLSI technology provides opportunities for the design of a variety of specialized sensors. Previous work on smart sensing has considered the design of complex photo-sensitive elements [11] with emphasis on the communication between neighbors [6] [7]. Other approaches use the implicit access of parallel row data at one end of the photo-sensitive array for SIMD computation [4], or a complete numerical sequential processor which is implemented on the same substrate [1]. A common goal of computational sensing approaches is to integrate photosensitive elements and analog processing on CMOS technology [12] [16]. Biological vision is a major source of inspiration in the development of artificial retinas [8]. Some conclusions of the research on biological systems suggest that focal plane processing strategies extract contour features at different spatial scales as a means to compress data. While the neural network architecture of a biological retina is generally known, the implementation of a similar architecture on a 2D silicon substrate remains a major challenge and is limited by the maximum area that is available. This implied a basic trade-off between the computational power (or complexity) integrated at each pixel site and the spatial resolution that is allowed by a given technology for maximum die size.

A multi module hexagonal image sensor is proposed with an associated satellite analog processing module in order to perform multiresolution edge computation at the focal plane level. By the addition of a microcoded digital controller and potential dedicated co-processors, the proposed system architecture is a typical candidate for a multi module chip assembly and defines a smart integrated image acquisition device for computer vision applications. An hexagonal multi-port addressing strategy of the pixel illuminance on the 2D sensor coupled to an external parallel analog computing unit is at the heart of the system and is detailed on the first part of the paper. This is followed by an overview of the open digital architecture of the system controller allowing independent development an integration of digital signal processing units such as image co-processors and signal conditioning modules. In the following, Section 2 presents the concept of parallel analog extraction which facilitates the fabrication of higher resolution sensors. Section 3 discusses the motivation for the hexagonal tessellation of pixels, while Section 4 describes the pixel architecture and its operating mode. The parallel analog spatial filtering capability of the MAR system is presented in Section 5 with a typical application for multiresolution edge extraction. An overview of the global system architecture is described in Section 6. The paper concludes with a presentation of typical results obtained with the 256 x 256 prototype of the MAR sensor.

2 Parallel Analog Extraction and Satellite Processing

Focal plane processing implies a delicate balance between pixel complexity, spatial resolution, and data flow. It is clear that the small cell size of a 2D photo-sensing array leaves only a limited area available for computation. In the case of large arrays, until technology allows much denser circuits (or 3D structures), most of the non-photosensitive area will has to be dedicated to communication, with little built-in processing. Furthermore, there exist serious I/O limitations. For instance, even if edge information could be computed in parallel at each pixel location, a simultaneous read-out of such data over large image regions shall remain challenging. Current technologies favor simple operators with great spatial homogeneity. The MAR architecture recognizes such a trade-off between the simplicity of the basic pixel design and the complexity and penalty of rapid communication by making use of external, but tightly coupled, processing support.

This so-called *satellite processing* is illustrated in Figure 1, which shows the symbolic representation of a sensor with relatively high resolution (256 x 256) and its associate off-chip parallel analog processing module. The dark circle on the sensor delineates a region of interest (ROI) centered on the pixel of interest (POI) from which illuminance data is retrieved and routed to a conditioning module. These channels are commonly used by a set of *N* different analog filters *H* which may implement multiresolution edge extractors, Gaussian filters, etc. The Section3 presents the sensor architecture and illustrates how several illuminance values may be extracted simultaneously from a ROI.

3 Multi-port Access Architecture

The tessellation used for the design of a specialized planar sensor is not limited to conventional rectangular grids. However, regular grids are restricted to rectangular or hexagonal maps. An hexagonal grid has been selected for the MAR sensor for two main reasons. Firstly, it is relatively easy to extract analog data by a set of parallel (vertical) buses as shown in Figure 2. Secondly, the distance between a given POI and its immediate neighbors is the same along any of the six main directions of the

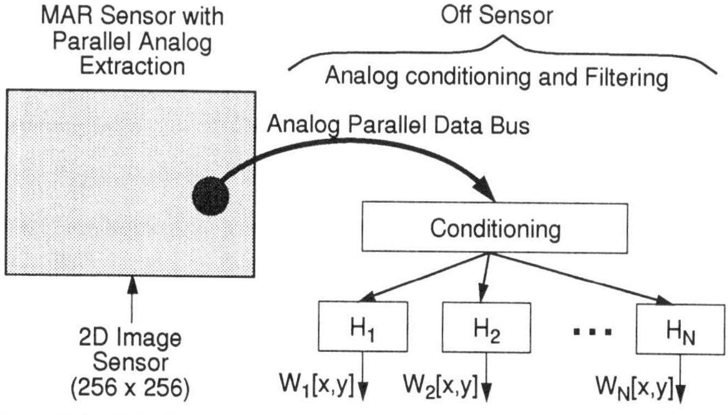

Figure 1 Satellite Processing approach. The sensor implements multi-port addressing of photo-sensors while signal conditioning is performed off chip. The set of analog filters uses the output analog signals for parallel spatial convolution purposes.

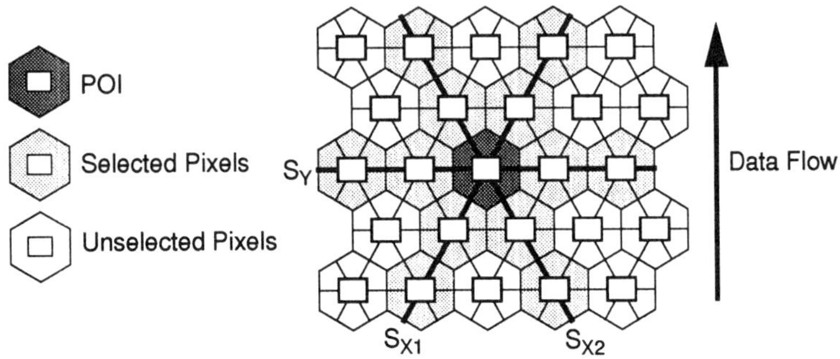

Figure 2 Hexagonal tessellation for circularly symmetric operators. Active selection lines are traced in thick lines and are kept in lock with the pixel of interest (POI). The analog data extraction is made vertically by three sets of data busses.

hexagonal tessellation, a condition which considerably facilitates the implementation of circularly symmetric kernels. As shown in Figure 2, selection buses (S_i) are routed at relative orientations of 60 degrees and define the overall tessellation on the hexagonal grid. This topology allows for simultaneous access to the illuminance data of the selected pixel (intersection of the three active selection lines) together with the illuminance of all neighbors located on the three axes of symmetry of the sensor array (corners of the concentric hexagon). It yields a natural compatibility with circularly symmetric operators since all pixels located at the corner of a given hexagon have identical radial distance from the POI.

In Figure 2, a small array of 5 by 5 pixels (centered on rectangular boxes) is addressed by a set of three selection busses labelled S_Y, S_{X1} and S_{X2}. Any pixel for which one of its selection line is active (or addressed) outputs the illuminance value recorded by its photosensitive area on a vertical analog data bus. This signal is routed out the sensor on an single channel and is fed to an external analog module for spatial filtering operations as shown in Figure 1. A block diagram of the entire sensor is shown in Figure 3. Each set of selection busses (Y, X_1 and X_2) are activated by an individual bidirectional shift register. A fourth bidirectional shift register (named T) is used to control the parallel analog multiplexor which is located at the upper edge of the sensor. While every pixel which is intersected by an active selection diagonal (thin black lines) is readout, the size of the resulting extraction kernel (white star) is defined by the size of the analog multiplexor.

The POI may be moved along any of the axes of the underlying hexagonal structure. The direction code uses 3 bits and is labeled according to the trigonometric circle (counterclockwise) as shown in Figure 3. An internal decoder, which holds the parity of the active line, interprets the direction command in order to control shift registers for both motion and direction. The three active selection lines track the POI and the output shift register stores the spatial combination required to insure the proper routing of analog information. The *Init* Signal is used to

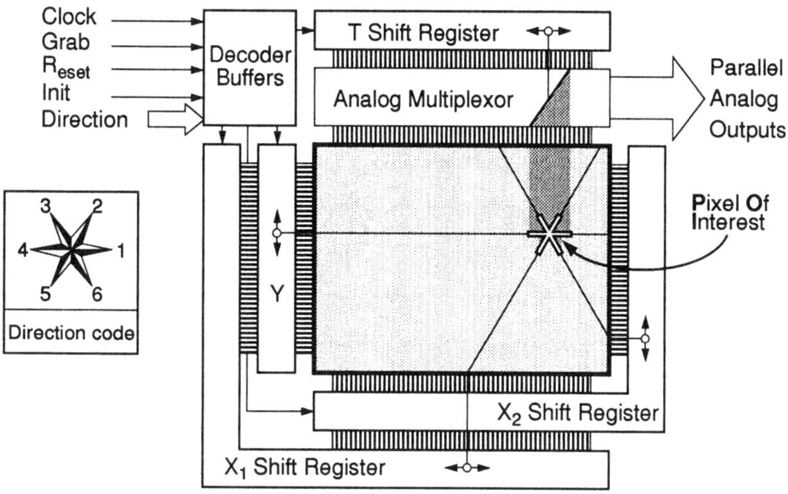

Figure 3 Block diagram of the MAR Sensor. A set of four bidirectional shift registers drives selection lines which converge upon the POI. A decoder module uses a direction code of three bits in order to drives each shift register. The analog data path is shown in the shadowed area until it reaches the analog multiplexor.

initialize the four shift registers, while the R_{eset} signal is used to reset the integration process, which is discussed in Section 4, as well as the function of the global *Grab* signal.

Figure 4 describes the electronic structure of each pixel. The single current I_S generated by the integration of photo-current I_E is retrieved through a set of three N-transistors (M_Y, M_{S1} and M_{X2}) with individual selection lines. This is similar to a multi-port memory access with individual data buses (D_Y, D_{X1} and D_{X2} respectively). Each signal is routed out via one of these data buses, according to the status of the selection lines. Figure 3 shows a typical state of each register with its associated active selection lines which concurrently intersect the POI. The dark area on top of the convolution kernel represents the extracted analog data flow until it reaches the parallel analog multiplexor. Pointer *T*, which tracks the location of the POI, controls a set of transmission gates in order to put each analog signal on the proper output channel.

Figure 5 shows the layout of a 256 x 256pixel array of a MAR sensor. The four shift registers and the analog multiplexor are located as in Figure 3. The upper left-hand corner consist in a 2-phases clock generator and control signals buffering. The direction decoder may be seen on the lower left-hand. The upper right-hand corner of the circuit collects the analog signals from the multiplexor, merges them in a single channel per radius, and routes the resulting output current to an individual analog output pad. The sensor provides only two digital outputs. The first one flags an overflow on one of the four shift registers (also used as a digital test output). The second one identify the parity of the selected line in the hexagonal grid (Odd Y indices represent unshifted sensor lines while Even ones are right shifted).

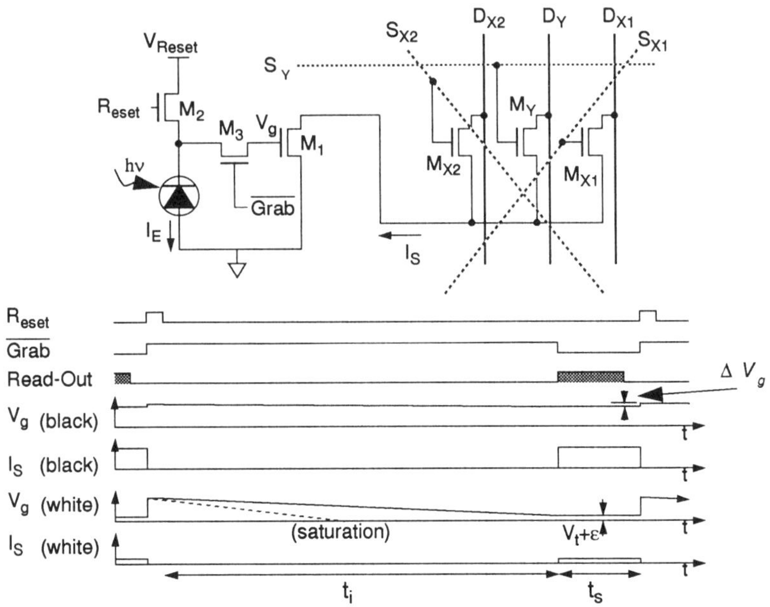

Figure 4 Multi-port addressing architecture of each pixel and non-destructive read-out of illuminance data. The timing diagram shows a typical illuminance integration cycle and its associate read-out cycle.

4 Pixel Architecture and Operating Mode

The circuit of Figure 4 is based on a photo-current I_E generated by a photo-diode which drives the gate capacitance of transistor M_1 over an adjustable integration interval t_i. The integration process is initialized ($R_{eset} = 1$ and $\overline{Grab} = 1$) by placing a positive voltage (V_{Reset} -V_{SN}) through M_2 and M_3 on the gate of M_1. As shown in the timing diagram of Figure 4, transistor M_3 is disabled to stop the integration process during the scanning of the sensor t_s. The scanning time is usually a fraction of the integration time and transistor M_3 could be omitted if the application was not critical in terms of the acquisition of absolute values (because the integration process continues during the scanning time). It is also possible to circumvent this feature by adding an electro-mechanical shutter in the optical path of the sensor which is controlled by the signal *Grab*.

Transistor M_1, which converts gate voltage into a current signal, operates in saturated mode ($V_{D1} > V_g - V_t$) according to the following relation:

$$I_S = \frac{\beta}{2}(V_g - V_t)^2 \tag{1}$$

In equation (1), β is defined by the geometric shape of transistor M_1 and V_t is the threshold voltage of the N transistor. We then derive the analytic expression of the output current I_S as a function of the illuminance E_{in}.

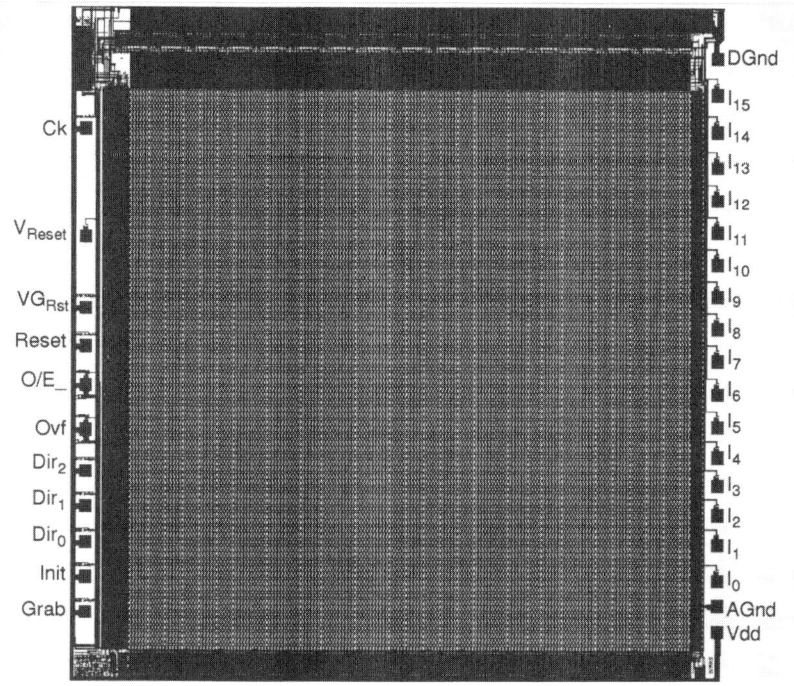

Figure 5 The layout of the 256 x 256 version of the MAR sensor. For display purposes, each pixel in this plot is replace by a small dot centered on the photo-diode (since it looks like a black square). This layout may be related to Figure 3 for the functional blocks.

The photo-current I_E is defined as a linear function of the thickness of the region of space charges w [13] which is given by:

$$I_E = E_{in}Aw = E_{in}A\sqrt{\frac{2\varepsilon(V_g + V_\gamma)}{|q|N_d}} \tag{2}$$

where V_γ represents the forward voltage drop of a diode (approx. 0.6 V), ε is the dielectric constant (11.9 ε_0 for the silicon), q is the electron charge and N_d is the dopant concentration of the substrate region (~1.5x10^{16}/cm^3 for NT 1.2 μ process) [3]. The value of A represents the photo-diode area and the proportionality constant. Equation (2) may be rewritten for this particular case as:

$$I_E = E_{in}K\sqrt{V_g + V_\gamma} \tag{3}$$

with

$$K = A\sqrt{\frac{2\varepsilon}{|q|N_d}} \tag{4}$$

the sensitivity constant of the photo-diode which is uniform throughout the sensor plane.The integration of photo-current I_E on the gate of transistor M_1 thus controls gate voltage V_g as:

$$V_g = \left(-\frac{1}{C_g}\right)\int I_E dt \tag{5}$$

$$I_E = -C_g \frac{dV_g}{dt} \tag{6}$$

It is possible to write the final gate tension V_g following integration interval t_i of the photo-current I_E by combining equations 5 and 2. This yields the following differential equation:

$$\frac{dV_g}{dt} = (-\frac{1}{C}) I_E = (-\frac{K}{C}) E_{in}\sqrt{V_g + V_\gamma} \tag{7}$$

$$\int_{(V_{Reset} - V_{SN})}^{V_f} \frac{dV_g}{\sqrt{V_g + V_\gamma}} = (-\frac{K}{C}) E_{in}\int_0^{t_i} dt \tag{8}$$

The final gate voltage V_f after the integration time t_i is then a function of the illuminance value E_{in} and is given by:

$$2(\sqrt{V_f + V_\gamma} - \sqrt{V_{Reset} - V_{SN} + V_\gamma}) = (-\frac{K}{C}) E_{in} t_i \tag{9}$$

$$V_f = (\sqrt{V_{Reset} - V_{SN} + V_\gamma} - \frac{K}{2C} E_{in} t_i)^2 - V_\gamma \tag{10}$$

By replacing equation (8) into equation (1) with $V_g = V_f$ we get the final relation between the output current I_S and the illuminance E_{in}:

$$I_S = \frac{\beta}{2}\left(\left[(\sqrt{V_{Reset} - V_{SN} + V_\gamma} - \frac{K}{2C} E_{in} t_i)^2 - V_\gamma\right] - V_t\right)^2 \tag{11}$$

Equation (11) shows the non-linear behavior of the MAR sensor response since some of its terms are functions of E_{in}^4 and in E_{in}^2. It is interesting to note that the particular case of zero illuminance signal ($E_{in} = 0$) gives the expected response of maximum saturated current defined by the initialization voltage:

$$I_S = \frac{\beta}{2}([V_{Reset} - V_{SN}] - V_t)^2 \tag{12}$$

Equation (11) is valid only if the value of V_{DS} is high enough to ensure saturated mode for transistor M_1. This is possible by sizing the selection transistors (M_Y, M_{X1}, M_{X2}) with a larger channel than transistor M_1.

Figure 4 shows the two extreme cases of the pixel response in a bright and in a dark region when only one pixel is selected during the scanning interval t_s (for proper biasing of output transistor M_1). For the dark case, photo-current I_E is limited to the reverse leak current of the photo-diode which causes a small deviation ΔV_g on the gate of M_1. In this condition the output current is maximum. The bright case is illustrated for an unsaturated pixel which sinks a small current (near zero) until a minimum threshold voltage V_t is applied at the gate of the output transistor M_1. When an isolated region has a very high level of illuminance (caused by a light source in the scene or by specular reflections) pixels in this region should saturate. In this particular case (see dotted line in Figure 4), the gate voltage decreases rapidly, resulting in a zero current output signal in the entire saturated region without any effect on the unsaturated neighbor pixels.

While a CCD sensor resets its illuminance integration process in a line-by-line sequence following a destructive readout, the MAR sensor has a global *Reset* signal for the entire array. The integration process is thus uniform for each pixel of the sensor and the readout of the output current I_S is non-destructive, with voltage V_g remaining unchanged irrespective of the frequency or duration of the access to a pixel element. This property is essential since, after a complete scan of the sensor, each pixel is addressed several times due to the intrinsic property of parallel analog signal extraction.

It is clear that the integration process of the MAR sensor is not limited to proceed at a standard video rate (1/30 sec). There is a very useful characteristic for this type of sensor. The integration interval t_i may be set by the operating system depending on the illumination condition of the scene. This parameter is then defined as an equivalent aperture control (or the light source intensity thereof) for a fixed lens but varying lighting conditions. A simple procedure which uses the histogram of a previous image could dynamically adjust the level of the white saturation of pixels by changing the integration interval t_i. The range of adjustment for this parameter is only limited by the voltage deviation due to the dark current of the back-biased PN junction. In fact, best results are obtained with rapid integration and scanning intervals.

The relatively high current level which is sunk by each selected pixel element provides a favorable Signal to Noise Ratio (SNR). In fact, transistor M_1 operates in saturated mode and typically sinks a current between 0 and 0.1 mA, which corresponds to a considerable dynamic range. As in biological vision systems [2], the output response is not a linear function of the illuminance value at a particular point in the scene. It may be appropriate to trade off some of the dynamic range of the sensing device for an improved linearity of the output response. A smaller aperture or a shorter integration interval t_i could be used, thus avoiding the nonlinear behavior of the pixel response when V_g is near the threshold value V_t. While a reduced dynamic range of the pixel improves its linearity, it is clear that it also reduces the SNR.

5 Parallel Multiresolution Edge Extraction

The most interesting property of the proposed architecture is that it allows parallel analog filtering for edge detection at multiple scales. As previously discussed, the hexagonal tessellation is naturally compatible with circularly symmetrical operators. The Laplacian Of Gaussian (LOG) operator has been identified by Marr *et al.* [8] [9] [10] as a most natural

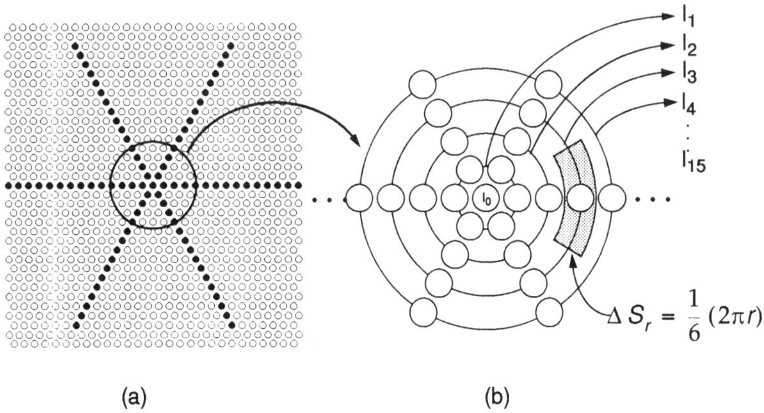

Figure 6 Complete convolution Kernel of the MAR sensor on (a). The information from each set of six pixels which are located at the same radial distance from the POI (b) is routed out from the sensor on individual data buses in order to apply circularly symetrical filters.

isotropic edge detector. Unfortunately, their computational cost on sequential computers is quite high. On the MAR architecture, the computation of 10 or 20 simultaneous filtered images is straightforward. It is the most relevant property of the MAR architecture, especially when scale-space integration is planned as an edge reinforcement procedure.

Figure 6(a) gives an overall view of the star-shaped convolution kernel of the MAR sensor which has a maximum radius of 16 pixels from the POI (I_0). Each extracted current I_r is put on an individual analog output channel, where r defines the radial distance of the pixel from the POI. Since each analog signal is retrieved via an individual bus, it is not desirable to access all the 91 pixels as individual sensor outputs. Rather, each set of six pixels with equal radius are grouped and dedicated for circularly symmetrical operators. The discrete spatial convolution of a given image I with an operator H is expressed for the general case as:

$$W[x,y] = \sum_{i=-X}^{X} \sum_{j=-Y}^{Y} H[i,j]\, I[x-i, y-j]\, \Delta S \tag{13}$$

with ΔS, the effective area of sampled pixel (usually constant and equal to 1). For the particular case of the MAR sensor, where a sub-sampling of the image is made by the convolution kernel, we need to compute the effective area of each selected pixel before applying any weight $H[i,j]$. This area is shown in Figure 6 (b) (dark area surrounding a pixel) and has a value of one sixth of the effective ring ($\Delta S_r = \pi r/3$). The effective size of the center pixel I_0 is the area of a small disk with radius of *1/2* ($\Delta S_0 = \pi/4$). Equation 13 may be corrected for the particular case of the MAR sensor with a circularly symmetric operator *F[r]*:

$$W_k[x,y] = \frac{\pi}{3}\left[\frac{3}{4}F_k[0]I_0 + \sum_{r=1}^{15} rF_k[r]I_r\right] \tag{14}$$

In equation 14, the value of $F_k[r]$ corresponds to the sampling value of the continuous function $F_k(r)$ for r ranging from 0 to 15 pixels. For LOG filters, F is a function of parameters σ and r and is given by:

$$F_\sigma[r] = \nabla^2 G_\sigma[r] = \frac{-1}{2\pi\sigma^4}\left(2 - \frac{r^2}{\sigma^2}\right)e^{-\left(\frac{r^2}{2\sigma^2}\right)} \tag{15}$$

For multiresolution LOG filtering, we implement several filters computed by (15) for different values of σ. As shown in Figure 7, each filter is implemented by a simple resistive network connected as an analog adder (see detailed implementation formulation in [5]). The resistor values $R_\sigma[r]$ are chosen such that their value is proportional to $1/(rF_\sigma[r])$. The left-hand part of Figure 7 holds the signal conditioning module as a current-voltage converter. Each filter output W_σ is routed to a zero-crossing detector which transforms the corresponding analog output into a two-bit digital signal: (i) the sign of the analog signal and (ii) the thresholded value of its amplitude. This set of binary values (2N bits) are stored in an image memory and is used by the digital controller. Details on the implementation of the zero-crossing and edge tracking algorithms are available in [15].

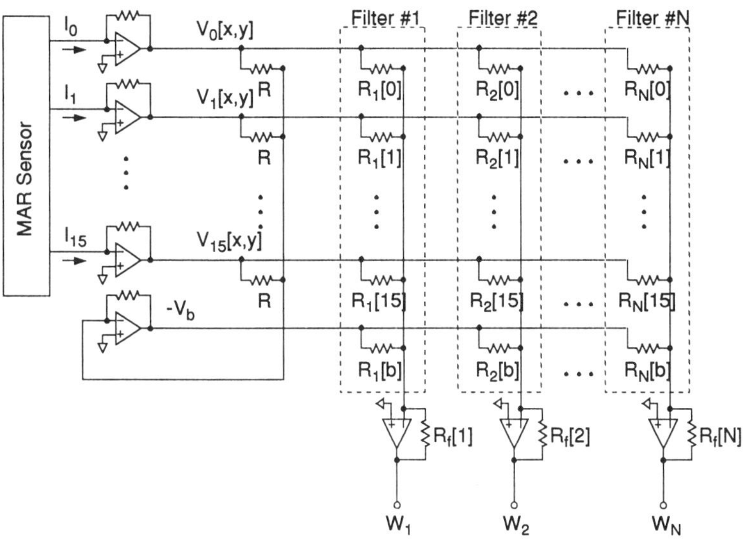

Figure 7 Parallel analog filtering module. Each output of the sensor is converted to a voltage signal and processed by a set of resistive networks which implement N different filters.

6 Global System Overview

The MAR sensor has been developed with great care in order to allow system integration, modularity, and flexibility for future enhancements. Figure 8 shows the general architecture of the system with the microcoded controller, which is responsible for the scanning of the sensor, the addressing of local memory, the control of some peripheral co-processors and the interfacing with an host computer. The controller also performs the edge tracking algorithm on hexagonal tessellation (which may be found in [14] and [15]) and provides interrupt signals in order to allow interactive operation of the MAR system during feature extraction. The co-processor modules are defined to process hexagonal data which are pointed into memory by the MAR controller. The address must be computed by special hardware that decides whether the addressed pixel is on a right-shifted line (Even line) or not. Co-processing may be related to the *scale-space integration, histogram calculation*, or *shape from shading*. A two port co-processor is also planned for *stereo disparity evaluation* using a pair of MAR systems. The hexagonal mapping may be advantageously exploited in many computer vision algorithms in order to implement very simple and isotropic operations.

A typical processing sequence includes the following operations: 1) scanning the sensor and store edge maps and raster image into the memory (may include some pipelined co-processing operation during the sensor scan), 2) restarting the integration process for the next acquisition, 3) running the rest of the co-processing operations (which may require 2 or 5 complete scans at the digital rate of 30 MHz) and 4) edge tracking, the interactive extraction of relevant information from the host computer. The edge tracking reefers to the extraction of strait lines from regular contours

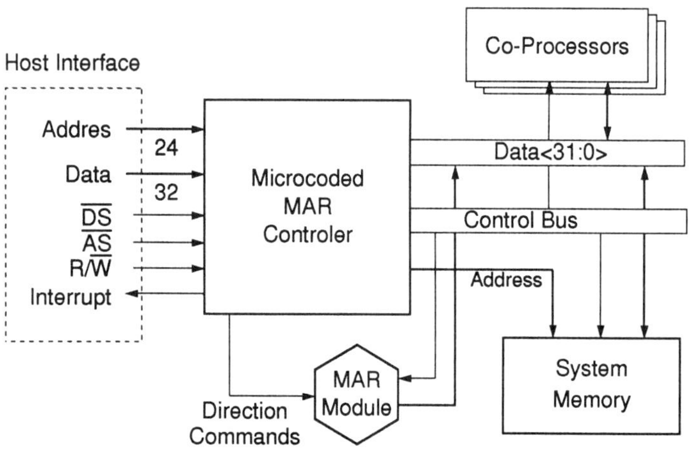

Figure 8 General architecture for the MAR system. The MAR module includes the MAR sensor, the 16 channels analog computing module, the zero-crossing detectors and A/D converter for the illuminance data acquisition. The entire system is accessed from a host computer with interrupt generation capabilities.

in the scene. At every line segment, only the start and end points are stored along with the corresponding properties previously computed by the co-processors. The final data-base representation of the extracted line segments may significantly compress the amount of data to be extracted from the sensing element. It may also increase the efficiency of the recognition process if this data-base includes several pre-computed properties.

7 Results

The 256 x 256 pixel version of the MAR sensor is currently installed in a custom camera case with an opto-electric shutter. This version implements sixteen different filters with their associated zero-crossing detector for multiresolution edge detection. Several image acquisitions have been performed on different scenes. A typical result is shown in Figure 9 for a photographic enlarger along with the resulting edge data. Four of the sixteen edge maps are presented as well as the digitized illuminance data (on hexagonal tessellation) of the enlarger. The high frequency rejection is particularly relevant on the small vertical slots which are correctly detected for high resolution filters (σ=.8 and σ=1.1) but are almost completely eliminated and grouped in a single feature by low resolution filters (σ=2.4 to σ=6.9). It may be also observed that every extracted edge segment is oriented according to one of the three main diagonals of the hexagonal structure. An other relevant property, which is related to Marr's theory [9], is the uncertainty of the localization of edges for low resolution filters. This is particularly visible on the knobs of the enlarger.

These edge images are used as the input of a scale-space integration algorithm which tracks edges in scale space. The resulting integrated image (obtained by software simulation from edge maps) is presented in the lower right corner of Figure 9. This final image is used for edge tracking where localization of edges (or accuracy) is well defined but noisy edges are attenuated. It is clear that this image may be used in the generation of relevant contours for scene description and will be supplied as a compact representation of the scene. This information will be used for medium and high-level algorithms for pattern and object recognition. This integrated edge map of the scene is generated in a single frame period and is updated every 100 ms.

8 Conclusion

A new concept of focal plane processing sensor has been presented. It is designed to link with tightly integrated sub-systems. The prototype implements Marr's theory based on the convolution of an image with circularly symmetric operators followed by zero-crossing edge detection. It defines a reasonable resolution sensor with multiresolution edge extraction capabilities within a compact package and with flexible and high performance operating modes. The feasibility of a sensor with a resolution up to 500 x 500 pixels is an important consequence of this concept. The integration of the microcoded controller as well as the

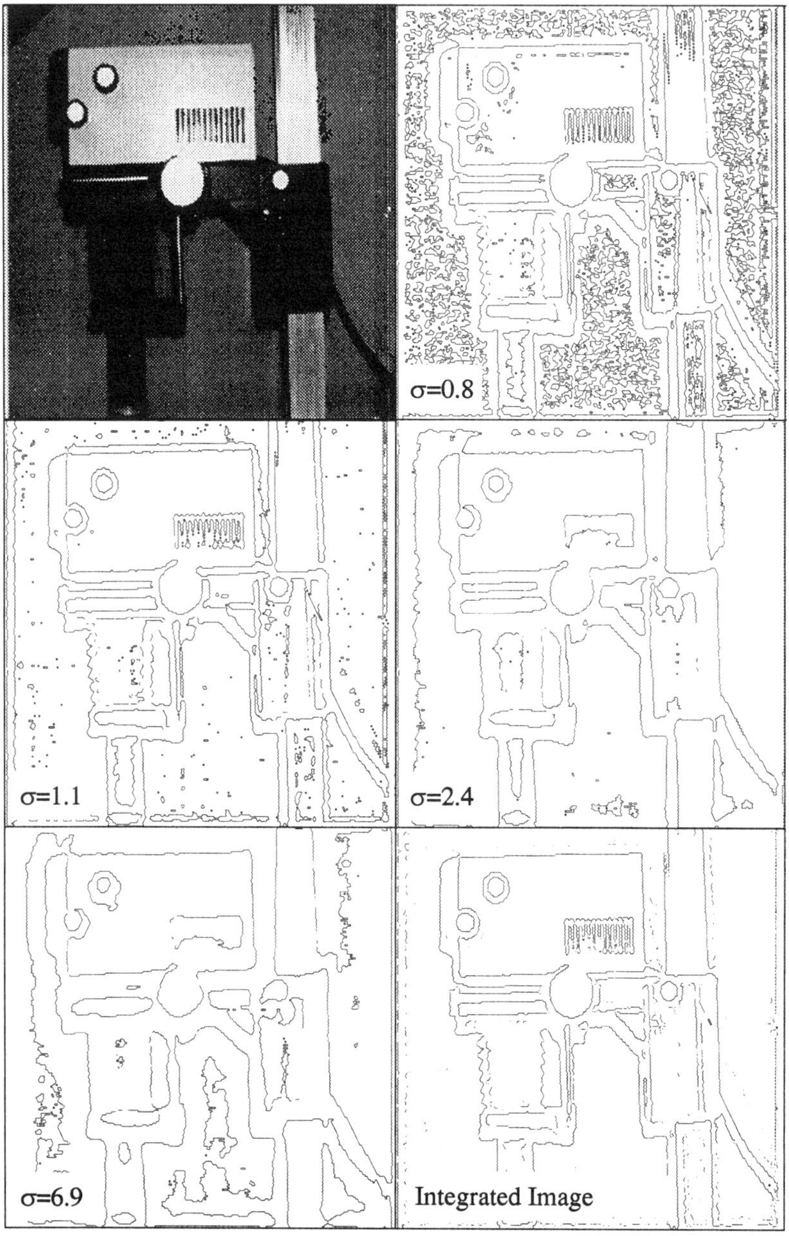

Figure 9 Results from the 256 x 268 MAR sensor for the image of a photographic enlarger . The illuminance image is shown in the upper left corner along with four of its sixteen edge maps at different spatial resolutions which are generated simultaneously in a single frame period. The resulting representation of the scene following scale space

operating software which will support intelligent and automatic features extraction are being carried out. Further work is in progress and include the development of post-processing VLSI modules for scale-space integration, stereo matching with two MAR sensors, and shape-from-shading. It is believed that the early and medium-level processing capabilities of the architecture discussed above could contribute to the development of low-cost and efficient implementations of intelligent sensing devices capable of generating a compact scene description for subsequent segmentation and recognition processes.

9 Acknowledgments

This work was performed with financial support made available through the Institute for Robotic and Intelligent Systems of Canada (projects A5 and NP-1) and through grants FCAR 92-ER-0380 of the Government Québec and NSERC A5274 of the Government of Canada. The Canadian Microelectronics Corporation (CMC) provided software, hardware and fabrication support through the Northern Telecom foundry.

10 References

[1] Anderson, S., W.H. Bruce, P.B. Denyer, D. Renshaw, G. Wang, "A Single Chip Sensor & Image Processor for Fingerprint Verification", *IEEE 1991 Custom Integrated Circuit Conference*, San Diego, California, May 1991.

[2] Baylor, A., "Photoreceptor Signal and Vision", *Invest. ophtamologiy vision SCI.*, Vol. 28, pp. 34-49, 1987.

[3] Brown, D., A. Scott, "*Design Rules and Process Parameters for the Northern Telecom CMOS4S Process*", Report IC90-01, Société Canadienne de Microélectronique, février 1990.

[4] Chen, K., A. Aström, P.-E. Danielson, "PASIC. A Smart Sensor for Computer Vision", *Proc. 10th International Conference on Pattern Recognition*, V.2, pp. 286 - 291, Atlantic City, juin 1990.

[5] d'Anjou, M., M. Tremblay, D. Laurendeau, "A VLSI Implementation of a Parallel Analog Laplacian of Gaussian Filter with Multiple Resolution levels", *Proc. of Canadian Conference on Very Large Scale Integration*, Halifax, pp. 204-211, Octobre 1992.

[6] Knight, T.M., "Design of an Integrated Optical Sensor with On-Chip Preprocessing", *Ph.D. Theses, Massachusetts Institute of Technologie*, June 1983.

[7] Mahowald, M.A., C. Mead, "The Silicon Retina", *Scientific American*, Vol 264(5): pp. 76-82. 1991.

[8] Marr, D., "Early Processing of Visual Information", *Philosophical Transactions of the Royal Society of London* B, Vol 245: pp. 483-519. 1976.

[9] Marr, D., E. Hildreth, "Theory of edge detection", *Proceedings of the Royal Society of London*, Series B, Vol 207: p. 187-217. 1980.

[10] Marr, D., "Vision - A Computational Investigation into the Human Representation and Processing of Visual Information", *San Francisco: Freeman*. 1982.

[11] Mead, C.A., M.A. Mahowald, "A Silicon Model of Early Visual Processing" *Neural Networks*, Vol 1: pp. 91-97. 1988.

[12] Spiegel, J., F. Kreider, C. Claiys, I. Debusschere, G. Sandini, P. Dario, F. Fantini, P. Belluti, G. Soncini, "A Foveated retina-like sensor using CCD technology" Analog VLSI Implementation of Neural Networks, C. Mead and M. Ismail editors, Kluwer, Boston, 1989.

[13] Sze, S.M., "Physics of Semiconductor Devices", Second Edition, Wiley-Interscience Publication, 1981.

[14] Tremblay, M., D. Poussart, "MAR: An Integrated System for Focal Plane Edge-Tracking with Parallel Analog Filtering and Built-in Primitives for Image Acquisition and Analysis", Proc. 10th Int. Conference on Pattern Recognition, V.2, pp. 292 - 298, Atlantic City, juin 1990.

[15] Tremblay, M., D. Laurendeau, F. Parent, D. Poussart, "Smart 2D Hexagonal Sensor with Integrated Analog Processing for Multiresolution Edge Extraction", Proc. of *International Conf. on Intelligent Autonomous Systems*, Pittsburgh, February 1993.

[16] Wang, G., D. Renshaw, P.B. Denyer, M.Lu, "CMOS Video Camera", *EURO ASIC '91*, 1991

Low-Power DSP Circuit Design Using Bit-level Pipelined Maximally-Parallel Architectures

Phillip J. Duncan, Shobana Swamy, Rajeev Jain

UCLA Electrical Engineering
56-147 Engineering IV
Los Angeles, California 90024 USA
duncan@janet.ucla.edu
(310) 206-3280

Abstract

Low power design techniques are becoming increasingly important with the rapidly expanding market for portable devices requiring sophisticated DSP functions. Previously published papers have shown that low-power designs can be achieved through the use of parallelism and pipelining by over-designing for performance and then operating the circuits at reduced supply voltages. This paper presents results which indicate that maximal parallelism and bit-level pipelining are most appropriate for low-power design of real-time DSP systems. These concepts are explained using a power measure normalized to the data sample period. Measured design results are then presented which clearly show the benefits which can be gained from bit-level pipelining.

Two versions of an 11-tap FIR filter have been fabricated in 2 μm pwell technology; one with bit-level pipelining and one with word-level pipelining. At a sample rate as high as 35 MHz, the bit-level pipelined circuit allows a 36% reduction in voltage which leads to a power savings of 66% compared to the word-level pipelined version. The corresponding increase in area is 16%, and two stages of latency are required.

I. Introduction

Low-power design techniques for digital CMOS systems are presented in a paper by Chandrakasan, Sheng, and Brodersen[1]. Many issues are discussed, including device technology, circuit design style, transistor sizing, and architecture design. The main conclusion reached is that the best method of achieving low-power designs is to use reduced supply voltages.

The authors discuss the technique of using increased circuit complexity to maintain the required throughput while operating at reduced voltages. Once the real-time constraints of the system are met, there is no advantage to increasing the computational throughput. The authors thus suggest using the increasingly large numbers of transistors available with shrinking feature sizes to over-design for performance so as to allow operation at reduced supply voltages. The specific techniques suggested for performance over-design are parallelism and pipelining. For a given DSP function there are several alternative methods to achieve parallelism and pipelining — the question is which of these is optimal for low-power.

This paper investigates architectural approaches to low-power and shows that maximally-parallel architectures with bit-level pipelining are the best design technique for real-time DSP when power consumption is the overriding design constraint.

Section II presents a new expression for power measure which is normalized to the data sample period. This new expression represents a more useful method of analyzing the architectural issues of low-power design. In Section III the tradeoff between time multiplexed and maximally-parallel architectures is considered. Pipelining using the technique of retiming is investigated in section IV. It is shown that over-design for performance is best achieved using fine grained or bit-level pipelining. Section V presents two fabricated designs which demonstrate the benefits of bit-level pipelining for low-power applications.

II. Power Measure for Real-Time DSP Design

In order to evaluate low-power architecture design techniques, it is necessary to understand the factors which contribute to power dissipation. The following simplified expression for total average power dissipation is used in [1]:

$$P_{total} = C_{eff} \cdot V_{dd}^2 \cdot f_{clk}$$

In real-time DSP applications, the sample period is almost always a fixed design constraint. Therefore, for real-time DSP design, the power measure should be normalized to the sample period as follows:

$$P_{samp} = C_{eff} \cdot V_{dd}^2 \cdot f_{clk} \cdot T_{samp}$$

C_{eff} = average capacitance switched per clock cycle

$f_{clk} \cdot T_{samp}$ = clock cycles per data sample

For most real-time DSP systems there is a relatively constant number of computations that must be performed during each data sample period. From a power use perspective it is irrelevant whether these computations occur in one clock cycle or in many cycles. The impact on power consumption is due to the number of computations, not the sequence in which they are performed. By quantifying the number of computations needed for each input data sample for a particular architecture in terms of the number of circuit node transitions that occur, an even more useful expression for power consumption can be defined:

$$P_{samp} = \Pi \cdot C_{\Pi} \cdot V_{dd}^2$$

Π = average nodes transitioning per data sample

C_{Π} = average capacitance switched per node transition

The above equation factors the average data sample power consumption into terms which roughly correspond to the traditional phases of ASIC design. The first term Π represents the contribution of architecture and logic design to power consumption. After architecture and logic design are completed, the goal is to perform layout generation while minimizing C_{Π}. Finally, V_{dd} is the only parameter that can be varied after the circuit is fabricated.

For low-power design this paper shows how maximal parallelism can be used to minimize Π (Section III), and how bit-level pipelining can be used to provide a performance margin that allows a reduced V_{dd} level even at high sample rates (Section IV).

III. Architecture Considerations

One method which might be considered for producing low-power implementations is the use of time multiplexed architectures. A time multiplexed architecture uses a small set of hardware functional units to implement a larger set of required algorithm operations. In this section we show that this is not a good approach when low-power is required for real-time DSP applications.

A possible time multiplexed architecture for an FIR filter is shown as figure 1. This implementation has a single adder/subtracter and a single shifter. Since a single adder/subtracter must perform multiple additions/subtractions per input data sample, the time multiplexed architecture will require multiple clock cycles per data sample. In general, more functional units (adders, shifters, etc.) are added until the necessary throughput is achieved or the power budget is exhausted. Several useful synthesis systems have been developed which can automatically produce such time multiplexed architectures from algorithm descriptions [6][7].

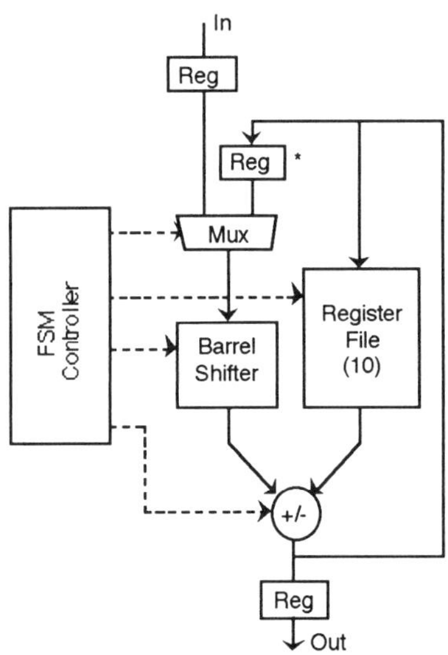

Figure 1: Time multiplexed architecture for FIR.

Using time multiplexed architectures one can trade off throughput for lower power consumption. For real-time DSP applications, the throughput is almost always a fixed design constraint, thus preventing throughput to be used in trade for reduced power consumption. Since the data sample period is a fixed constraint, distributing the per sample computational requirements, represented by Π, among several clock cycles does nothing to reduce the average power consumed per sample P_{samp}.

Time multiplexed architectures actually have a negative impact on power consumption since they increase the amount of computation per data sample due to the overhead of using hardware for multiple algorithmic operations within the same data sample period. Figure 1 shows the overhead functions (FSM controller, extra register, and multiplexor) which perform computations auxiliary to the operations present in an FIR algorithm. The overhead functions produce a large number of node transitions beyond those strictly required by the algorithm description, thus increasing Π and P_{samp}.

Another factor which results in a higher value of Π for time multiplexed designs is the increased complexity that often results when a time multiplexed functional unit must perform tasks that are similar but not identical. For instance, in the example a combination of additions and subtractions are being performed on a single functional unit; this unit must have additional circuitry to allow a control signal from the FSM to determine which operation is performed on each clock cycle. The extra circuitry results in further node transitions which are not strictly required by the FIR algorithm.

The conclusion to be drawn from the above discussion is that when power considerations are more important than area considerations, time-multiplexed architectures should not be used. For a low-power design it is therefore desirable to implement every algorithmic operation appearing in the signal flow graph as a separate circuit element. An architecture using this strategy is defined as a maximally-parallel architecture.

After reducing the value of Π as much as possible by using a maximally-parallel architecture and employing careful logic design, low-power design methodologies should focus on reducing the supply voltage. In the next section we illustrate how bit-level pipelining can be applied to achieve this objective [5].

IV. Pipelining

Maximally-parallel architectures succeed in reducing the Π factor, however, they do not always yield better performance than time shared architectures due to long delay free logic paths which limit the frequency at which the circuit can be clocked. The maximally-parallel architecture for an FIR filter shown in figure 2A demonstrates this fact (registers are represented by thick bars intersecting the signal lines). Four adders have been included so that all four taps will be computed in parallel. The additions will occur during the same clock cycle, but due to the propagation delay through the adders they will still occur sequentially.

Higher performance can be achieved by pipelining the design so that long delay free paths are broken up. Figure 2B shows the word-level pipelined circuit corresponding to figure 2A. In the word-level pipelined form, each of the additions begins in parallel. The critical path has now been reduced to the propagation time of a single word-level adder.

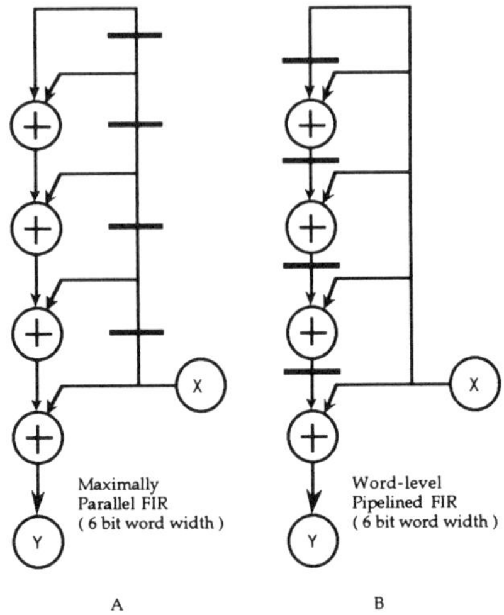

Figure 2: Maximally-parallel FIR architecture (A) non-pipelined (B) word-level pipelined.

If we expand each word-level adder into its individual bits we see that the carry ripple path causes the bit-level additions to occur sequentially. Assuming that the FIR of figure 2B has a 6 bit data path, the critical path is the propagation delay of a carry through six full adders. In order to reduce the critical path even further, a finer grained pipelining must be considered, where the bit-level computations are pipelined if possible. Traditionally this approach has not been exploited due to the expected increase in register complexity as well as the resulting irregularity of the layout. To analyze the power/area tradeoff, we have developed a retiming program to perform bit-level pipelining, and applied it to several examples. Some results are shown in the next section.

While the word-level pipelining shown in figure 2B can be manually derived from the maximally-parallel form of 2A, at the bit-level it is often not feasible to perform the pipelining by hand. Figure 3 shows the results of bit-level pipelining for the FIR from figure 2, where a latency of 1 has been allowed. The number of register bits required has increased from 24 to 34, while the critical path has been reduced to half that of the word-level pipelined form of figure 2B.

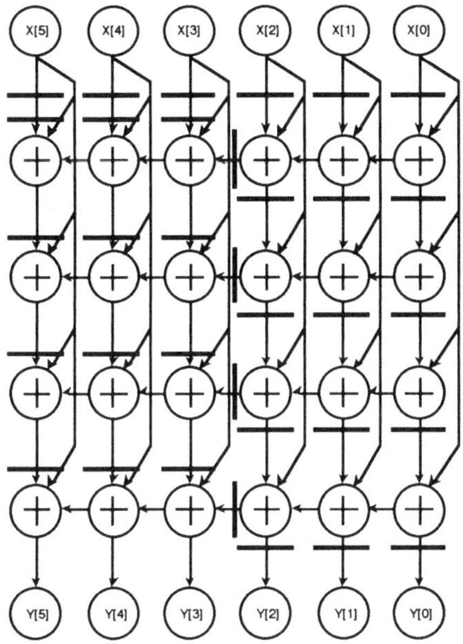

Figure 3: Bit-level pipelined FIR filter

The bit-level pipelined circuit of figure 3 will have an increased value of Π due to the extra registers. At a given data sample rate, however, the bit-level pipelined design would be able to run at a lower voltage than either design of figure 2. Interestingly, our initial designs that employ bit-level pipelining show that the lower value of V_{dd} will more than compensate for the increase in Π and result in significant power savings.

V. Design Results

CAD tools for generating maximally-parallel architectures with bit-level pipelining have been developed and integrated in a system called Hi-PASS [2][4]. The input to this system is a 'C' description of the DSP algorithm, such as that shown in figure 4 for an 11-tap word-level pipelined FIR filter. Starting from this input, Hi-PASS has been used to design two versions of the filter, one with bit-level pipelining and one with word-level pipelining [3][9]. The resulting architectures have been implemented in 2 μm CMOS using standard cells in the Lager IV system [8] and have been fabricated through MOSIS.

```
#define SIZE 11
static int out;

int sinc(int in)
{
  static int tap[(SIZE + 1)/2];
  static int z[SIZE];
  int i;
  tap[0] = - (in>>9);
  tap[1] =   (in>>8);
  tap[2] = - (in>>7);
  tap[3] =   (in>>6) + (in>>8);
  tap[4] = - ((in>>4) + (in>>8));
  tap[5] =   (in>>1) + (in>>2);

  out = z[10];
  for(i=SIZE-1;i>0;i--) {
    if(i > SIZE/2)
      z[i] = z[i-1] + tap[SIZE-1-i];
    else z[i] = z[i-1] + tap[i];
  }
  z[0] = tap[0];
}
```

Figure 4: 'C' code 11-tap FIR filter.

The word-level pipelined version has a total of 158 single bit registers and occupies a core area of 9.16 mm^2. The second design produced by Hi-PASS was generated by using a constraint of 45 MHz in the retiming algorithm to produce a bit-level pipelined circuit. This version has 207 register bits and occupies an area of 10.62 mm^2, a 16% increase in area over the word-level pipelined design.

	Word-level Pipelined	Bit-level Pipelined
Latency	0	2
Core Area (mm^2)	9.16	10.62
Register Bits	158	207
Max Clock Rate	35 MHz	55 MHz

Table 1: FIR designs synthesized by Hi-PASS.

Both designs have been fabricated using MOSIS 2 μm pwell scmos technology. Test vectors were developed to exercise the critical paths of the filters. The word-level pipelined design yielded a maximum clock rate of 35MHz at 5V, while the bit-level pipelined design functioned properly at 55MHz for the same voltage. A summary of the design results for both FIR filter designs is included as table 1.

A power comparison was performed between the bit-level and word-level pipelined FIR designs at reduced supply voltages. Both were tested at selected operating frequencies from 7 MHz to 35 MHz. Supply currents were recorded at 5 V. The supply voltage was then reduced until the chip under test failed to function properly. Supply current readings were taken over the full range of correct functionality for each test frequency.

Figure 5 presents the results of the current measurements. It can be seen that at a given operating frequency and voltage, the bit-level pipelined version generally requires slightly more supply current. The increase in average supply current for the bit-level pipelined version is expected since additional registers cause the number of node transitions per input sample Π to increase. However, at all frequencies above 7 MHz the bit-level pipelined circuit can operate at lower voltages than the word-level pipelined version.

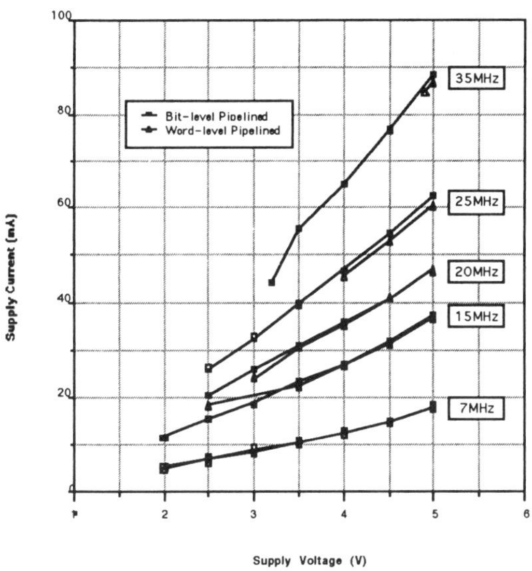

Figure 5: Supply currents with reduced supply voltages.

The power comparison between the two designs becomes clear when average power is plotted using the lowest possible voltage for each operating frequency as shown in figure 6. At 7 MHz both designs can be operated as low as 2.0 V and require essentially the same average power consumption. As the operating frequency is increased, the supply voltage of the word-level pipelined circuit must be increased much more rapidly than that of the bit-level pipelined circuit. This leads to a significant power savings for the bit-level pipelined circuit when approaching the maximum operating frequency of the word-level pipelined design. At 35 MHz the word-level pipelined circuit requires 416.5 mW while the bit-level pipelined version requires only 143.2 mW, resulting in a power savings of 66%. These measurements were taken using a single fabricated circuit for each design. Exhaustive testing with a number of different circuits of each design would take into account the statistical variation in process parameters and lead to variations in the measured power. However, the results presented here demonstrate the power savings trend to be expected for the two types of pipelining over a range of frequencies.

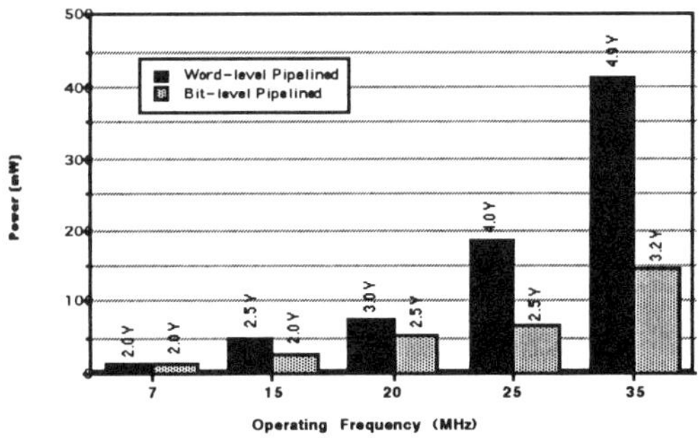

Figure 6: Power comparison for bit-level and word-level pipelined circuits.

VI. Conclusion

In designs where area is the overriding factor, time multiplexed architectures are the best implementation strategy. This is not true, however, for real-time DSP applications in which low power consumption is the major design criteria. As large transistor count circuits become ever more affordable due to shrinking feature sizes and increased levels of integration, it will become increasingly feasible to trade off complexity for low-power by using maximally-parallel, bit-level pipelined architectures at reduced supply voltages. Even with the current cost of silicon area, many applications having very low power budgets will benefit from these techniques.

A comparison of the two design examples presented in this paper clearly show that bit-level pipelining, though increasing the amount of circuitry and total area, can be used to reduce power consumption by using a reduced supply voltage. At 35 MHz, the bit-level pipelined circuit achieved a power savings of 66% while increasing the area by only 16%.

The FIR example also demonstrates that low-power designs, using maximally-parallel architectures and bit-level pipelining, can be synthesized by design automation tools such as the Hi-PASS system.

VII. Acknowledgments

The authors wish to acknowledge the support of DARPA, Texas Instruments, and the MICRO program. The authors are also grateful to Wanda Gass and Ed Deprettere for their support and technical contributions.

VIII. References

[1] A. P. Chandrakasan et al., "Low-Power CMOS Digital Design," *IEEE JSSC*, vol. 27, no. 4, pp. 473-484, 1992.

[2] P. Duncan, *Hi-PASS: A Synthesis System for High Speed DSP ASICs*, Ph.D. Prospectus, UCLA, 1992.

[3] P. Duncan et al., "Experiments with the Hi-PASS DSP Synthesis System," *International Symposium on Circuits and Systems*, San Diego 1992, pp. 168-171.

[4] P. Duncan et al., "Hi-PASS: A Computer-Aided Synthesis System for Maximally Parallel Digital Signal Processing ASICs," *International Conference on Acoustics, Speech and Signal Processing*, San Francisco 1992, pp. 605-608.

[5] C. E. Leiserson and J. B. Saxe, "Retiming Synchronous Circuitry," *Algorithmica*, vol. 6, no. 1, pp. 5-35, 1991.

[6] N. Park and A. C. Parker, "Sehwa: A Program for the Synthesis of Pipelines," *23rd Design Automation Conference*, 1986,

[7] J. Rabaey et al., "HYPER - Design Synthesis for High Performance Real Time Systems," *IEEE International Symp. on Circuits and Systems*, 1990,

[8] C. Shung et al., "An Integrated CAD System for Algorithm-Specific IC Design," *Proc. of 22nd Hawaii International Conference on System Science*, 1989,

[9] S. Swamy, *High Performance DSP Architecture Synthesis: Benchmarks and Examples*, M.S. Thesis, UCLA, 1992.

Implementation of a packet switching device as a delay-insensitive circuit

Jelio Yantchev
Programming Research Group, Oxford University, UK
Jelio.Yantchev@comlab.ox.ac.uk

Ivailo Nedelchev
Department of Electronical Engineering, University of Surrey, UK
I.Nedelchev@ee.surrey.ac.uk

Abstract

The paper describes the delay-insensitive implementation of a packet switch. It illustrates the use of Delay Insensitive Algebra for the specification and top-down design of delay-insensitive circuits. The compilation scheme has been developed that translates expressions in the algebra into networks of basic circuits; this is used to implement the specification of the final level of design refinement. The switch uses the low latency *mad postman* switching technique and the main design objective is to achieve as low as possible propagation latency through the switch.

1 Introduction

Packet switches are familiar components of contemporary computer architectures and are often used to illustrate asynchronous design techniques [3],[1]. The asynchronous packet switch design presented here is of particular interest because it implements *mad postman* switching [12] for low latency packet propagation. Mad postman switching can in principle achieve node through delay that is lower than the channel rate and is ideally suited for asynchronous implementation. Furthermore, unlike previous designs that use bundled data convention for inter-switch signaling [3],[1], the design in this paper uses a delay-insensitive 2-phase signaling convention. As a result the control path and the packet propagation (data) path of the switch are designed using the same design techniques.

Mad postman switching works by sending each incoming packet out on an output channel immediately and without waiting for a correct decision to be made first. Routing decision are made concurrently with the transmission of the head of the packet, as if in 'retrospect', but the packet moves closer to its destination on each step. The node through delay that is achieved is independent of the channel width, channel rate, and the address length. As packet addresses need not be buffered for routing purposes mad postman switches can be very small; a bit-serial switch, for example, only requires one bit buffer for each input channel.

The asynchronous implementation of packet switches is attractive because, as asynchronous circuits are allowed to operate at their own rate, packet propagation rate will be determined only by the performance of the individual switches and not limited by their global arrangement. Delay-insensitive inter-node signaling will considerably simplify the composition of large packet routing networks, as internode wires can be arbitrarily long. Increasing the lengths of the interconnecting wire will gracefully increase inter-node delays, but will not affect correct operation.

The purpose of this paper is to describe in detail the implementation of a mad postman switch as a delay-insensitive circuit. A high-level behavioral specification of the switch will be successively refined and translated to a delay-insensitive circuit of basic library components. A precise specification of the switch is presented as a recursively-defined process in a CSP-based algebraic notation [6, 5] called *Delay Insensitive Algebra* or *DIA*. The algebra captures the essential properties of delay-insensitive circuits and can be used to express not only the functionality of the device, but also obligations to be met by its environment. The high-level specification will then be decomposed into a network of concurrently active and cooperating processes.

A compilation scheme has been developed for compiling DIA expressions into circuits composed of basic library components [9]. However, rather than compile the high-level specification the approach favoured here is to first refine the specification into a concurrent network of simpler processes while trying to meet certain performance criteria and apply the compiler only at the final stage of the refinement. This produces a network of smaller and faster state machines rather than one single complex and slow state machine.

We will first describe the set of library components circuits. Then the specification in DIA, design refinement and implementation of the switch as a network of library components will be presented and the design evaluated.

2 Library of Primitive Cells for Delay-Insensitive Design

The Delay-Insensitive Algebra is used for high-level design. A delay-insensitive system is constructed from a library of primitive components. It is envisaged that the primitive components are designed once and for all, but there are methods of designing new components should the need for that ever arise.

We will now specify a small set of widely used basic delay-insensitive components.

- Merge
- Decision-Wait
- Arbiter
- Variable
- Modulo N counter

2.0.1 Specification

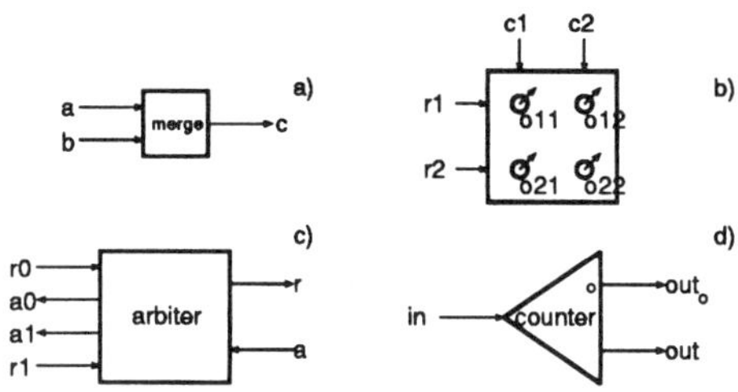

Figure 1: Merge, DW, Arbiter, Counter elements.

Merge The MERGE element is a circuit with two input wires a and b and one output wire c (figure 1.a). In response to a signal on either a or b, it will output on c:

$MERGE = [a? \rightarrow c!; MERGE \mid b? \rightarrow c!; MERGE]$

Note that this definition implies that it is unsafe for the environment of $MERGE$ to supply input on both a and b before receiving an output on c.

Decision-Wait The 2×1 Decision-Wait element [6] DW has 2 row inputs r_1, r_2 and 1 column input c. It expects a signal on some r_i and a signal on c before producing a signal on the corresponding output o_i.

$DW = [r_1? \rightarrow c?; o_1!; DW \mid r_2? \rightarrow c?; o_2!; DW]$

Note that the r_i inputs are mutually exclusive.

This definiton can be readily extended to the $n \times m$ Decision-Wait

$DW_{n,m} = [\,|_{i=1}^{n} r_i? \rightarrow [\,|_{j=1}^{m} c_j? \rightarrow o_{i,j}!; DW]]$

The circuit symbol for the Decision-Wait is shown in figure 1.b.

Two-way Arbiter A two-way arbiter is a device that gives two processes mutually-exclusive access to a single shared resource. Its circuit symbol is given in figure 1.c. A pair of request and acknowledge wires r_i, a_i connect the processes to two the arbiter. The r and a wires are used for resource request and acknowledge. The behaviour of the arbiter is defined by the following algebraic expression:

$A = [\,\Box_{i=0}^{1} r_i? \to r!; a?; a_i!; A_i]$

$A_i = r_i?; r!; a?; a_i!; A$

The request signals r_0 and r_1 can be concurrent. In this case one of them is blocked until the resource has been released by the other process.

Variable A variable (VAR) is a device which in one can store a value from a fixed set ($value_1, ..value_n$) and later read this value. Its circuit symbol is given below (figure 2).

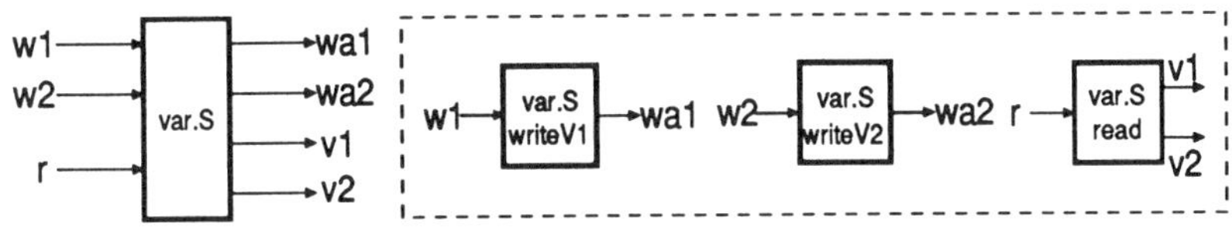

Figure 2: The Variable element.

Because we will use this circuit in different parts of the schematic representation, it is useful to split its graphical representation as it is shown with the device circumscribed with polygon line.

The specification of this device is as follows:

$VAR = [[|_{1 \leq i \leq n}\; w_{value_i}? \to wa_{value_i}!; V_{value_i}] \mid r? \to \perp]$

$V_{value_j} = [[|_{1 \leq i \leq n}\; w_{value_i}? \to wa_{value_i}!; V_{value_i}] \mid r? \to ra_{value_j}!; V_{value_j}]$

Counter modulo N A delay-insensitive counter to N is a device with one input in and two outputs $out_\bullet$ and out. It counts $N-1$ transitions on the $out_\bullet$ output and the Nth transition on out (figure 1.d). The specification in DIA is as follows:

$COUNTER = COUNT_0$

$COUNT_i = in?; out_\bullet!; COUNT_{i+1} : 0 \leq i \leq n-2$

$COUNT_{n-1} = in?; out!; COUNT_0$

3 Delay Insensitive Specification of a Mad Postman Switch

In this section we will give the DIA specification of a delay-insensitive 2×2 mad postman switching process MP with relative addressing.

A pictorial representation of MP is given in figure 3. MP is connected to its environment by four pairs of data/acknowledgement wires; the data wires encode the signals 0, 1, and e. 2-cycle signaling is used with double-rail encoding using wires 0 and 1, and a separate start/end-of-packet wire e. Signal transmission on these wires is assumed mutually exclusive. A packet

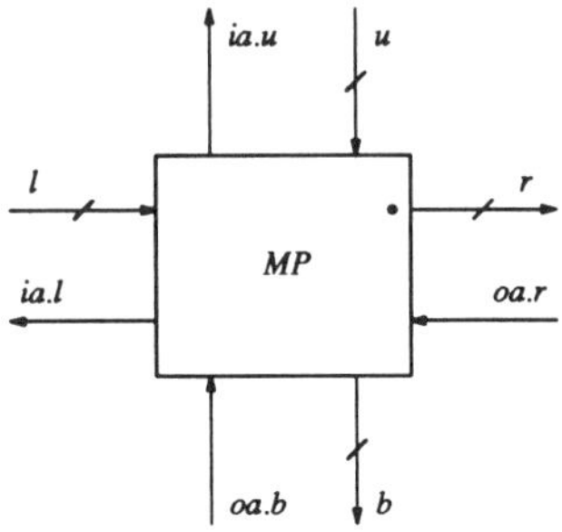

Figure 3: The MP switching device.

commences with an e signal, followed by an hl-bit, $hl \geq 1$, address flit, or *header*, consisting of 0s and 1s, followed by an arbitrary number of 0s and 1s, and completing with an e signal. This can be formally defined in regular expression syntax as $e(0|1)^{hl}(0|1)^{*}e$

The output r of MP, also marked with the bullet sign • in figure 3, is the *direct output* and b is the *complementary output.* Each incoming packets is by default first sent out on r. If the hl bits in the header of the packet are all 0, then the rest of the packet is switched out on b; otherwise it is switched out on r. The header of each packet is decremented by one before it is transmitted on r.

The specification of the MP process is given by the mutually recursive expression:

$$MP = Arb/oa.r?$$

$$Arb = oa.r?;[\Box_{in\in\{l,u\}} in?x \rightarrow \textbf{if } (x = e) \textbf{ then } ia.in!; r.e!; H_{in,0} \textbf{ else } \perp]$$

$$\begin{aligned}
0 \leq i < hl :: \quad H_{in,i} &= oa.r?; in?x; ia.in!; \\
&\quad \textbf{if } (x = 0) \textbf{ then } r!1; H_{in,i+1} \\
&\quad \textbf{elseif } (x = 1) \textbf{ then } r!0; P_{in,i+1} \\
&\quad \textbf{else } \perp \\
H_{in,hl} &= oa.r?; r!e; b!e; S_{in,b}
\end{aligned}$$

$$\begin{aligned}
0 \leq i < hl :: \quad P_{in,i} &= oa.r?; in?x; \\
&\quad \textbf{if } (x = e) \textbf{ then } \perp \\
&\quad \textbf{else } ia.in!; r!x; P_{in,i+1} \\
P_{in,hl} &= S_{in,r} \\
\\
S_{in,r} &= oa.r?; in?x; ia.in!; r!x; \\
&\quad \textbf{if } (x = e) \textbf{ then } Arb \\
&\quad \textbf{else } S_{in,r}
\end{aligned}$$

$$
\begin{aligned}
S_{in,b} \quad = \; & oa.b?; in?x; ia.in!; b!x; \\
& \textbf{if } (x = e) \textbf{ then } oa.b?; Arb \\
& \textbf{else } S_{in,b}
\end{aligned}
$$

The named states of MP can be explained as:

Arb, the initial state when the process receives the acknowledge $oa.r?$ for the last sent $r!e$ and then is waiting for a packet on either of the input channels l or u. A packet should commence with an e signal; otherwise there is an error.

$H_{in,i}$, $0 \leq i < hl$, a packet header is being switched out (decremented by 1) on the direct output r and only 0s have been received so far.

$H_{in,hl}$, a header of hl many 0s has been received, the rest of the packet must be switched to the complementary output b, and a dead flit has been released on the direct output r. The process then

1. completes the transmission on the direct output r by a transition on $r.e$ and
2. initiates transmission on the complementary output b by a transition on $b.e$.

$P_{in,i}$, a packet header is being switched out on the direct output r and is not all 0s.

$P_{in,hl}$, a non-zero header has been received and the rest of the packet must be transmitted on the direct channel as well.

$S_{in,out}$, the packet is switched out on channel *out* (either b or r) until an e is received on *in*.

4 High Level Design

In principle, the top level specification can be compiled using the compiling techniques described in [9]. As the specification is of a sequential process with no explicit concurrency it will be compiled to a single sequential state machine. Although this approach is straightforward, it is also unacceptable as it produces a complex and slow state machine.

A more promising approach and the one that is favoured here is to decompose MP into a network of simpler concurrent and interacting processes that can be implemented as smaller and faster state machines. This section describes in detail the top-down design decomposition of MP.

A *decomposition* of a process is an implementation of that process (in the sense of the refinement ($\sqsubseteq$) ordering) as a network of other processes. A *delay-insensitive decomposition* is a decomposition such that the correctness of the network does not depend on the delays in the componenet processes and the interconnecting wires.

Careful consideration of the specification of MP reveals three independent actions performed by MP

- arbitrating between concurrent requests for input on channels l and u;
- routing of each packet to either r or b, depending on the comparison of the address flit with zero;
- decrementing the address flit of each packet by one before it is output on r.

Each task can be implemented by a separate process and these can then be connected together in a pipeline, as figure 4 illustrates.

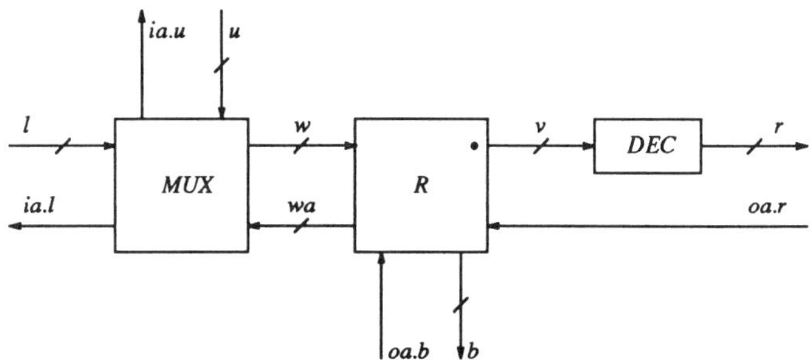

Figure 4: High level decomposition of MP.

The process MUX arbitrates between competing requests for input on l and u. It also merges at the packet level the two input streams on l and u into one stream on the internal channel w. The process R routes each incoming packet by sending it first out on v. If the address flit of the packet is all 0s, then the rest of the packet is switched out on b; otherwise it is transmitted out on v. The process DEC decrements the address flit of each packet as it is output on the direct channel and transmits the rest of the packet unchanged.

It is convenient to use two wires $wa.d$ and $wa.e$ to acknowledge forward 0s or 1s and forward $w.e$ transitions, respectively. Separating the e-acknowledgements from the data acknowledgements simplifies the implementation of the MUX process, as will become apparent later.

Each of the three processes, MUX, R, and DEC is individually simpler than MP itself, while collectively they are a correct implementation of its behaviour

$$MP \sqsubseteq (MUX \parallel R \parallel DEC)$$

In the next section we will present the circuits corresponding to each of these three processes (MUX, R, DEC) and based on the library of delay-insensitive cells.

5 Implementation of MUX

The MUX process is defined as:

$$MUX = [u?e \rightarrow w!e; wa.e?; ia.u!; D_u \ \Box \ l?e \rightarrow w!e; wa.e?; ia,l!; D_l]$$

$$D_{in} = in?x; w!x; wa?y; ia.in!; \ \textbf{if} \ (x = e) \ \textbf{then} \ MUX \ \textbf{else} \ D_{in}$$

I.e it arbitrates in its initial state between (possibly) concurent inputs on the $l.e$ and $u.e$ wires, and then copies out the 0 and 1 signals from the channel that has been selected.

The implementation of MUX as a network of primitive circuits is straightforward if we consider the following restrictions about the behaviour of its environment:

- every packet starts and ends with an e transition
- 0, 1, and e transitions on each of the l and u channels are mutually exclusive

Signals sent on the l and u channels are not mutually exclusive and we need to employ an arbitration circuit to resolve potential conflicts. If two packets arrive concurrently the circuit must block one of them while granting the other the right of way. As each packet commences and completes with an e transition we can use the opening/closing $l.e$ and $u.e$ signals as request/release inputs to an Arbiter element. As the 0 and 1 transitions are mutually exclusive, the two 0 input wires can be merged together to form the $w.0$ signal and, similarly, the two 1 input wires can be merged together to form the $w.1$ signal. Figure 5 shows the final circuit for the MUX process.

6 Implementation of R

The routing process R is the most complex component of MP as it implements all routing functionality. The DIA specification of R is as follows:

$$R = H_{0,F} \ / \ oa.r?$$

$$H_{i,b} = oa.r?; w?x; v!x; \ \textbf{if} \ (x = 1) \ \textbf{then} \ wa.d!; H_{i+1,T} \ \textbf{elsif} \ (x = eop) \ \textbf{then} \ wa.e!; H_{i+1,b} \ \textbf{else} \ wa.d!; H_{i+1,b}$$

where $b \ in \ \{F, T\}$, $i \ in \ 0..hl$

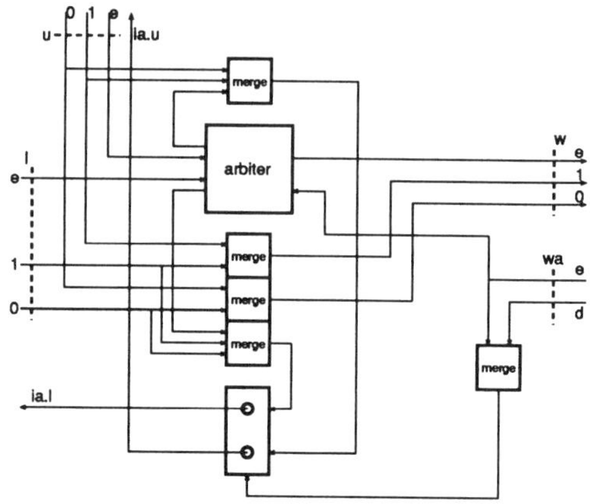

Figure 5: Compilation circuit for MUX

$$H_{hl+1,T} = \quad CP_1$$
$$H_{hl+1,F} = \quad oa.r?; v!eop; b!eop; CP_2$$

$$CP_1 = \quad oa.r?; w?x; v!x;$$
$$\textbf{if } (x = eop) \textbf{ then } wa.e!; H_{0,F}$$
$$\textbf{else } wa.d!; CP_1$$

$$CP_2 = \quad oa.b?; w?x; b!x;$$
$$\textbf{if } (x = eop) \textbf{ then } wa.e!; oa.b?; H_{0,F}$$
$$\textbf{else } wa.d!; CP_2$$

Because of the complexity of the process R and for clarity of presentation we will compile the $H_{0,F}$, CP_1, and CP_2 expressions separately and will assemble the complete circuit from the compiled parts.

6.1 Compilation techniques

The signals in a compiled circuit are divided into control and environment signals. The control signals are generated from DIA expressions during compilation process. The signals from the environment are those specified in DIA expressions as explicit inputs and outputs.

Control signals propagate through the circuit synchronising with signals from the environment. Each construct in DIA is compiled into a circuit that waits for a control signal form its predecesor, executes some actions and passes control to its successor. The actions can be

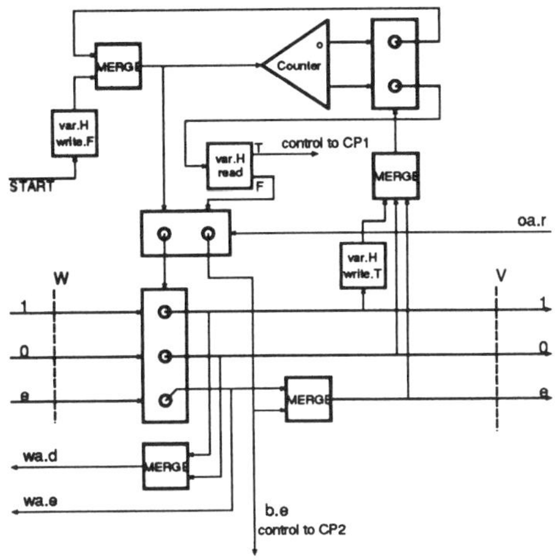

Figure 6: Compilation circuit for $H_{0,F}$.

- input/output on wire (channel),
- arbitration/choice between several inputs,
- branching of the control flow.

Outputs to the environment are produced upon synchronisation of control signal with inputs from the environment. An input from the environment is accepted exactly when the control signal reaches the point of synchronisation. An output is sent immediately after the control is passed to the operator which creates it.

For every input from the environment there is a DW element of dimension M columns and N rows, where N is the number of the control signals requesting this input and M is the number of wires comprising that channel. If the expression is constructed according to the DIA rules, all outputs to the same wire are mutually exclusive and can simply be MERGEd.

6.2 Implementation of $\mathbf{H}_{0,F}$

Figure 6 shows the compiled circuit for the process $H_{0,F}$. The first subscript i of the process H is used as a loop index variable; the number of iterations is equal to the length hl of the packet header plus 1 for the starting e. The loop is implemented using a counter of modulo $hl + 1$.

The VAR circuit records the occurrence of a 1 in the packet header. Control is passed either to the process CP_1 or to CP_2 depending on whether 1 was encountered on the channel w. The control signal $START$ that activates the circuit also initialises VAR to value F. The occurrence of a 1 in the header assigns the value T to VAR.

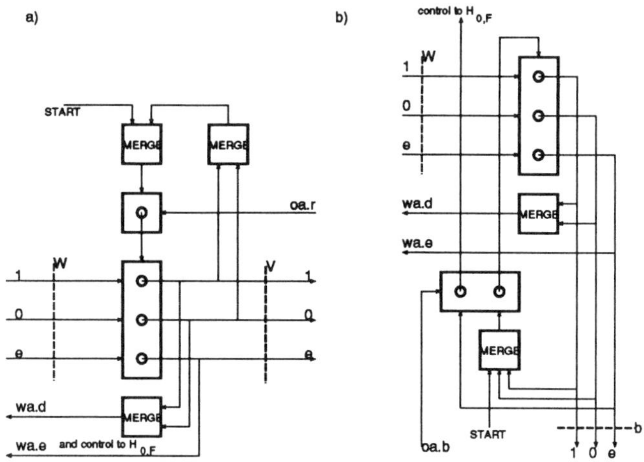

Figure 7: Compilation circuit for CP_1 and CP_2.

6.3 Implementation of CP_1 and CP_2

The compilation of these two processes is straightforward. The state machine is based only on DW elements. Figures 7.*a* and 7.*b* illustrate the compiled circuits for the processes CP_1 and CP_2 respectively.

6.4 Assembling the circuit for the process R

We can now assemble the process R from the separately compiled parts. In each circuit here is just one DW element for each input from the environment. When the circuits are assembled into one all DW elements for the same input are combined into one 3×3 DW element (for the channel w) and one 3×1 DW element (for the channel $oa.r$) and the control signals corresponding to the different states of R are connected to their columns.

Figure 8 shows the circuit for the process R.

7 Implementation of DEC

The process DEC is intended to decrement the header of each packet by one and leave the packet body unchanged. One would therefore expect that DEC will require some form of counter to count the header bits. A careful consideration of the behaviour of the environment of DEC and the process R in particular reveal that a counter is not required. This simplifies the implementation of DEC.

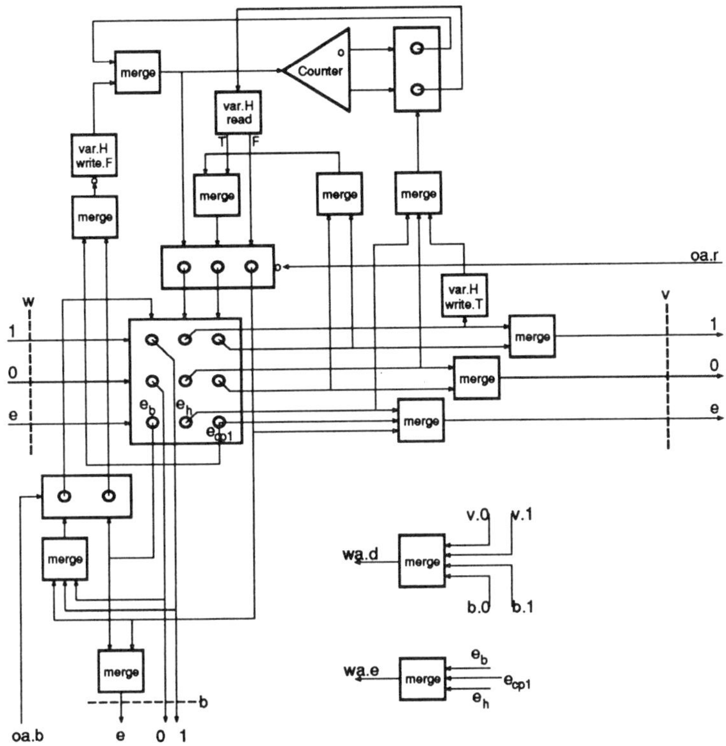

Figure 8: Compilation circuit for R.

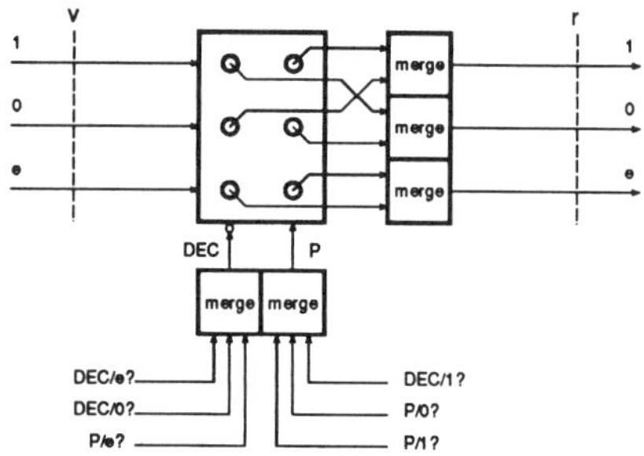

Figure 9: Implementation of DEC.

$$\begin{aligned} DEC = \;& v?x; \\ & \textbf{if } (x = 0) \textbf{ then } r!1; DEC \\ & \textbf{elseif } (x = 1) \textbf{ then } r!0; P \\ & \textbf{else } r!e; DEC \end{aligned}$$

$$\begin{aligned} P = \;& v?x \\ & \textbf{if } (x = e) \textbf{ then } r!e; DEC \\ & \textbf{else } r!x; P \end{aligned}$$

The process can be in one of two possible states

- still decrementing, and the next input is an e signal which will take it back to the initial state, or
- passing the bits on to r unchanged (not decrementing) and will continue to do so until another e is received.

Although the DEC process as specified above has the potential to corrupt the body of a non-empty packet with zero address header by decrementing the body as well, the specification of R excludes such possibility as all such packets are switched out on the complementary channel b.

The implementation of DEC derived from the above specification is shown in figure 9.

8 Conclusions

The paper describes the delay-insensitive implementation of a packet switch. It also shows how from a high-level behavioral specification via refinement the final circuit is compiled. The compilation techniques are based upon a library of basic delay insensitive cells. The switch uses the low latency *mad postman* switching technique and as we already mentioned the main design objective is to achieve as low as possible propagation latency through the switch. To evaluate the propagation latency of the implementation we need only consider the forward data path between the inputs and the direct output. Figure 10 illustrates the forward propagation path extracted from figure 5, figure 8 and figure 9.

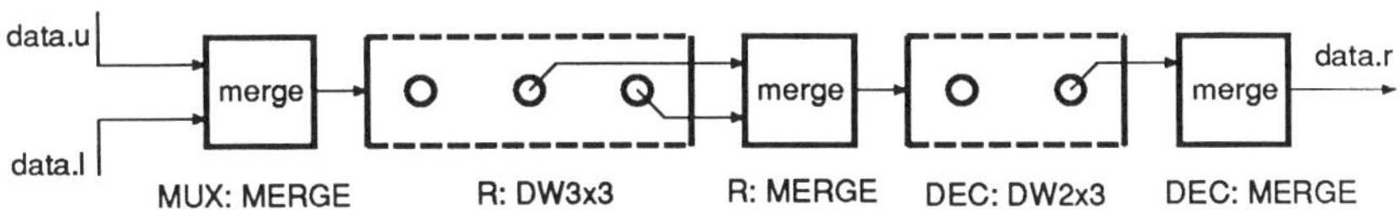

Figure 10: The forward data path.

The delay on the e signal path is similar and it comprises
$t_{p_{mp}} = t_{arb} + 2 \times t_{dw} + 2 \times t_{merge}$
where $t_{arb} = t_{and} + t_{or}$ due to the novel low-latency arbitration scheme. Otherwise if conventional arbiter is used
$t_{arb} = t_{mutex} + t_{and} + t_{or}$
where t_{mutex} might be nondeterministically long.

The simplicity of the final design is due to the virtues of mad postman switching as simple but efficient routing strategy. This routing technique can achieve node through delay that is lower than the channel rate and the latter ideally suits the asynchronous implementation.

9 Acknowledgements

We gratefuly acknowledge the support of the European Commission under ESPRIT III Basic Research Working Group 7225 (ACiD-WG) and SERC under grant *no* GR/G 18605.

References

[1] Brunvand, E., Implementing Self-Timed Systems with FPGA's, FPGAs, Abingdon EE&CS Books, 1991

[2] Dally, W.J., Seitz, C.L., 'Deadlock-Free Message Routing in Multiprocessor Interconnection Networks,' *IEEE Trans. Comput.*, Vol C-36, No.5, May 1987, pp 547-553.

[3] Dally, W.J., Seitz, C.L., 'The Torus Routing Chip,' *Journal of Distributed Computing*, Vol 1, No.3, 1986.

[4] Jo C Ebergen, Arbiters: An exercise in specifying and decomposing asynchronously communicating components, Research report CS-90-29, Computer Science Department, Univ of Waterloo, Canada, July, 1990 .

[5] Josephs, M., 'Receptive Process theory,' *Eindhoven Uni. Tech. Rep. 90/8*, 1990.

[6] Josephs, M.B., Udding, J.T., 'Delay-Insensitive Circuits: An Algebraic Approach to their Design,' Lect. Notes in Comp. Sci., Vol. 458, pp. 342-366, Springer-Verlag, 1990.

[7] Josephs, M.B., Udding, J.T., 'The Design of a Delay-Insensitive Stack,' In: Jones, G., Sheeran, M. (eds.) Designing Correct Circuits, pp. 132-152, Springer-Verlag, 1990.

[8] Josephs, M., Udding, J.T., 'An algebra for delay-insensitive circuits,' *Proceedings DIMACS/IFIP Workshop on Computer-Aided Verification*, Rutgers University, New Jersey 1990, DIMACS Series in Discrete Mathematics and Theoretical Computer Science, Vol. 3, pp 147-175, AMS-ACM 1991.

[9] I.M.Nedelchev, C.R.Jesshope, C.G.Huang, Compilation Techniques for Asynchronous Controllers, ACiD-WG/EXACT Workshop on Asynchronous Controllers and Interfacing, Leuven, Belgium, Sep 14-16, 1992

[10] Udding, J.T., *Classification and Composition of Delay-Insensitive Circuits*, PhD thesis, Dept. of Math. and C.S., Eindhoven Univ. of Technology, 1984.

[11] Udding, J.T., 'A formal model for defining and classifying delay-insensitive circuits,' *Distributed Computing*, 1(4):197–204, 1986.

[12] J.Yantchev, C.R.Jesshope, "Adaptive, low latency, deadlock-free packet routing for netowrks of processors", IEE Proc, Vol 136, Pt E No 3, MAY 1989

Electrical Design Of A 1GByte/Sec High Performance Backplane Using Low Voltage Swing CMOS (GTL)

Christopher Cheng, Leo Yuan

Sun Microsystems Computer Corporation

Abstract

A high speed backplane that can support up to ten system boards with a peak data bandwidth of more than 1Gbyte/sec has been implemented using the low voltage swing CMOS (GTL)[1]. The entire backplane is passive and all drivers and receivers are implemented in CMOS ASIC. The logic low level is 0.4V and the logic high level is 1.2V.

1 Introduction

A high performance multi-processor server has been implemented. The server can accommodate up to 20 Super SPARC™ processors in ten system boards. The main system bus is XDBus™ which is a high bandwidth, packet-switched, data consistent shared memory bus. A high data bandwidth bus such as XDBus™ is required to support the computing throughput for the 20 Super SPARC™ processors and their shared memory system. This paper describes a single board backplane which can support XDBus with ten system boards running at a peak frequency of 66 MHz with a maximum data rate exceeding 1GByte/sec. Low voltage swing CMOS (GTL) is used for the data and control signals while the clock signals are in Pseudo ECL (PECL) level.

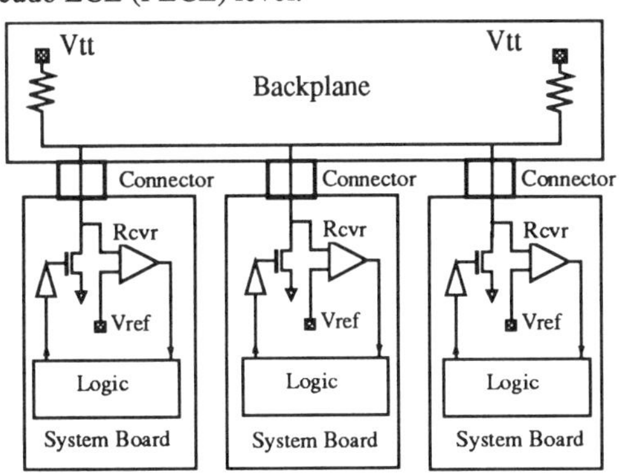

Figure 1 Bidirectional Data Bus

In the early design stage, CMOS, TTL, Pseudo ECL [2] and BTL [3] I/O levels were all evaluated. Selection was based on the system's architecture, electrical performance, cost as well as the thermal requirements. Pseudo ECL and BTL were eliminated due to their high cost. With up to 144 bidirectional I/O per chip, Pseudo ECL and BTL consume too much chip power to meet the maximum limit of 8 watts per chip. CMOS drivers were not able to drive the 10 slot 17 inch backplane or the 5 large PGA loads on a 16 inches long on board XDBus at the desire frequency. None of the existing logic levels could meet all the requirements above. Therefore, a low voltage swing open drain transceiver was implemented using standard CMOS ASIC technology.

The backplane has to be able to support both bidirectional data and unidirectional control signals used by the XDBus protocol. (Figure 1,2) GTL drivers are open-drain n-channel devices and the receivers are differential comparators. The output levels are: Vol_{max}=0.4V, Voh_{min}=Vt=1.2V and Vref=0.8V. The bidirectional signal lines are terminated at both ends while the unidirectional signal lines have termination at the furthest receiver to prevent reflection. In some heavily loaded unidirectional signals, terminators are used at both ends to maintain good signal quality.

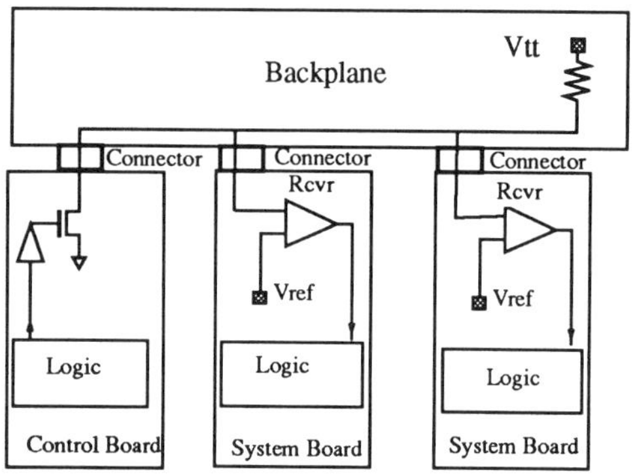

Figure 2 Unidirectional Control Signal

There are 144 data (including parity) and more than one hundred control bits divided into two physical buses in the backplane. The total length of the bus is 17 inch with 1.7 inch slot spacing between system boards. Standard four row two hundred pin HDI connectors are used to connect the backplane to the system boards.

2 GTL circuit

Figures 3,4 show a simplified driver and receiver circuit respectively.

Transistor N4 in the driver circuit provides the current to pull the terminating resistor to logic low level when Vin is low. When Vin switches to high, transistors N2 and N3 will momentarily short the gate of N4 to its drain and cause N4 to conduct. When N3 turns off, N1 will eventually pull the gate of N4 to ground. This circuit can reduce the turn-off di/dt and the underdamped overshoot caused by package inductance. Typical delay from Vin to output is about 1ns.

The receiver is a simple differential receiver. Vref is provided by the system which is set to be 2/3 of Vtt (0.8 V). Typical propagation delay is about 1.5ns.

The worst case on chip power consumption of the GTL driver is Vol x Iol = 0.4 x 32mA = 12.8 mW per driver. For 36 active signals the power consumption will be 0.46W which is substantially lower than both BTL and ECL. Low cost plastic quad flat packages can be used instead of more expensive packages required by BTL and ECL due to thermal cooling reasons. This is the reason why it is advantageous to implement GTL in conventional CMOS ASIC technology.

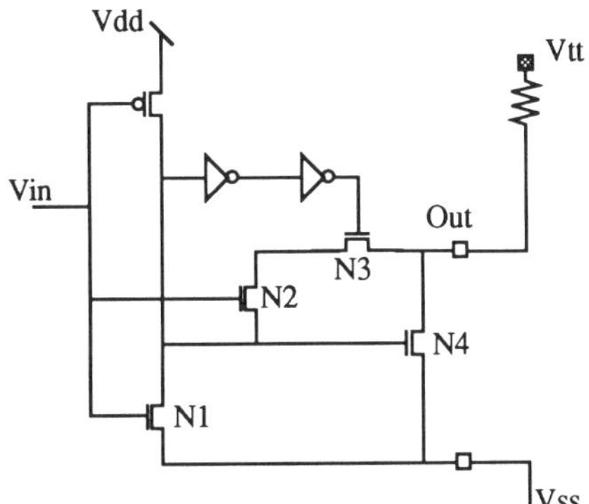

Fig. 3 GTL Driver Circuit

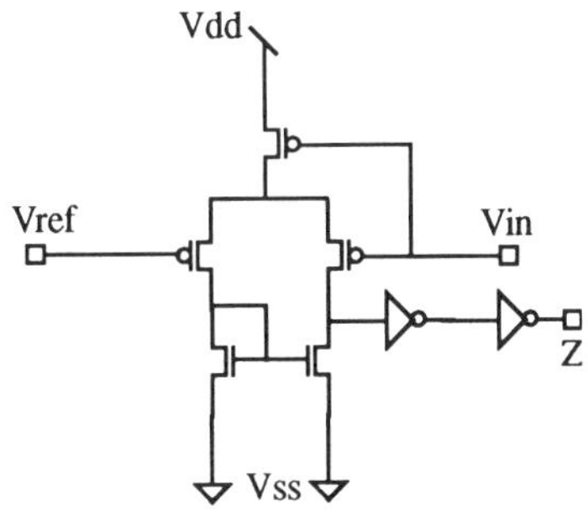

Fig. 4 GTL Receiver Circuit

3 Signal quality and propagation delay

The biggest challenge to the electrical design of the backplane is maintaining good signal quality between the drivers and receivers and meeting the cycle time requirement by minimizing the propagation delay between signals. Loading effect, simultaneous switching output (SSO), wire or glitches and other secondary effects will degrade the signals and can cause the receiver to sense an incorrect logic level. Key electrical design considerations are described below to address these problems.

A fully loaded backplane achieved a cycle time of 15.15ns (66MHz) in the laboratory and the following parameters were measured:

15.15ns (cycle time) = 2.5ns (clock to output) +6.75ns (round trip propagation delay) + 4.8ns (receiver setup) + 600ps (SSO delay) + 0.5ns (approx. clock skew)

A Loading Effect

Individual system board loading can affect the signal quality. The system board loading consists of the following components : driver/receiver pad capacitance inside the chip, chip package capacitance, PCB signal trace, connector pin inductance and through hole capacitance.

The electrical length of the loading is also important. If the combined signal trace, package and connector delay is short compared with the signal rise and fall time, the loading can be lumped together as a capacitive load. When the system boards are uniformly connected to the backplane, they become distributed capacitive loads and will affect the impedance and propagation delay of the signal line as in the following equations [4]:

$$\text{LoadedDelay} = \text{IntrinsicDelay} \times \sqrt{1 + \frac{C_{load}}{C_{line}}} \quad \ldots (1)$$

$$Z_{load} = \frac{Z_{unload}}{\sqrt{1 + \frac{C_{load}}{C_{line}}}} \quad \ldots\ldots (2)$$

Where Z_{load} is the impedance of the loaded line, Z_{unload} is the unloaded line, C_{load} is the capacitive board loading per unit length and C_{line} is the capacitance per unit length of the signal line.

As can be seen from equation (1), system loading directly affects the cycle time by increasing the signal propagation delay. Also, the variation of signal line impedance produces reflections and ringing which will decrease the input noise margin. If ringing is excessive, the system cycle time must be increased to allow the ringing to settle.

The key to minimizing the impedance variation and propagation delay is to minimize the ratio of C_{load} and C_{line}. Both the driver and receiver circuits and the ESD protection circuit have to be optimized to maintain minimum loading while providing enough drive for the signal. Typical bidirectional driver on chip loading is about 2.5pF. The bus logic is broken down into ten smaller ASIC blocks so that 100 pin plastic quad-flat packages (PQFP) with less loading can be used instead of bigger packages such as pin grid arrays (PGA). Package loading is lowered to around 1.5pF.

The off chip loading is also minimized by shortening the signal traces between ASIC and the connector; the pin capacitance of connector is also minimized. The overall system board loading to the backplane is about 8pF/system board. Measured results of the system board load is :

$$C_{system\ load} = C_{driver} + C_{package} + C_{trace} + C_{connector/hole} \quad (3)$$

$$2.5\text{pF} + 1.5\text{pF} + 1.7\text{pF} + 2.3\text{pF} = 8\text{pF}$$

Another important factor is to increase the signal line capacitance per unit length. This will lower the ratio of the two capacitance terms in equations (1) & (2). The impedance of the microstrip signal traces is set to 40 ohm instead of the normal 50 ohm used by the rest of the system. The capacitance per unit length is increased from 2.9pF per inch for the 50 ohm line to about 3.8pF per inch for the 40 ohm line. Miscrostrip lines are used to minimize the intrinsic delay, which is around 165ps per inch. Based on equation (1), the fully loaded propagation delay will be 3.7ns with the 40 ohm line compared with 4.1ns for the 50 ohm line. The backplane has to tolerate round trip propagation delay due to the wire-or glitch which will be described in the next section. The impedance is optimized to lower the round trip propagation delay instead of first incident switching. The backplane is designed such that it does not require dummy load board on empty slots.

B Wire-Or Glitch

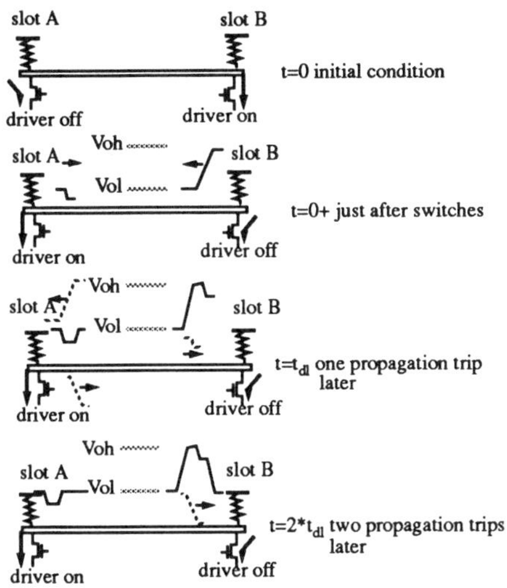

Figure 5 Wire-Or Glitch

Open drain/collector technologies such as GTL and ECL have wire-or glitches [5] on bidirectional buses when one driver on the far end location of the bus is releasing to the high impedance state while the near end driver is going from high impedance to active. (Figure 5)

When the driver in slot B is in high state, the terminators at that end will pull up to about Voh. The rising edge will propagate down to the other side and will be negatively reflected back since the driver at slot A is active and has low impedance. When the negatively reflected wave comes back, slot B will settle back to the low state. The wire-or glitch takes about 6.75ns to settle and contributes to the largest portion of the cycle time.

The control signals are arranged in a unidirectional bus and do not suffer from the wire-or glitch problem.

C Simultaneous Switching Output (SSO)

In order to drive the heavily loaded backplane, driving current is increased at the expense of higher power and ground noise. When the GTL drivers switch from tristate to active, a large current is flowing from the drivers through the package ground return to the system board ground planes. When a large number of drivers are switching in the same direction, the inductance of the package ground return will raise the on chip ground voltage level higher than the board ground. A data bit which is already in a low state will appear to have a rise in voltage, and the fall time of drivers will be increased since the on chip logic low voltage is rising. The same problem will happen between the connector and the backplane. Rise and fall time will be further increased due to simultaneous switching. The net result is an increase in output delay which has to be budgeted in the overall cycle time.

Power and ground inductance should be minimized to reduce the simultaneous switching delay. Since GTL is an open drain technology, the power is supplied from the terminators through the backplane. Terminator SIPs with internal bypass capacitors are used, and the lead inductance is minimized to decrease the simultaneous switching effect on terminator power. The on chip simultaneous switching delay is reduced by increasing the ratio of ground to signal pins. To minimize SSO effects, a ground return pin should be provided to each signal pin, however, due to the high signal count of the bus, only three signal pins to one ground pin can be achieved with the small PQFP package. The SSO delay is measured to be around 600ps with worst case switching which includes both package and connector SSO penalty. Ground bounce is measured to be about 200mV.

D Secondary Effects

Crosstalk and lossy transmission line effects also degrade signal quality. The design goal is to achieve less than 5% crosstalk between signals. The edge rate degradation due to skin effects should be minimum. Based on simulations, a minimum spacing between signal traces is used to guarantee the crosstalk is below the 5% specification. Simulation and experimental measurement, which will be discussed in later sections, verify that the skin effect is very small with GTL technology.

4 Simulations and measurements

Figure 6 shows the electrical model used to analyze the signal traces.

All signal traces are modelled as lossless transmission lines. The package and connector parasitics are modelled as lumped inductor or capacitor elements. The wire-or glitches and fully loaded backplane signals are simulated using SPICE and the results are compared with measurements. Figures 7 & 8 show the measured and simulated waveforms of the driver and receiver respectively.

The simultaneous switching outputs were measured and figures 9a and 9b show the output delay penalty and ground bounce due to SSO respectively.

The fastest rise and fall time of GTL signals is around 1ns and the maximum bus length on backplane is about 17 inches. A time domain skin effect transmission line simulator is used to predict the signal degradation, and a TDT (time domain transmission) measurement (figure 10) is used to analyze the skin effect of signal traces. Both simulation and measurement confirm that the lossy transmission line does not degrade signal significantly.

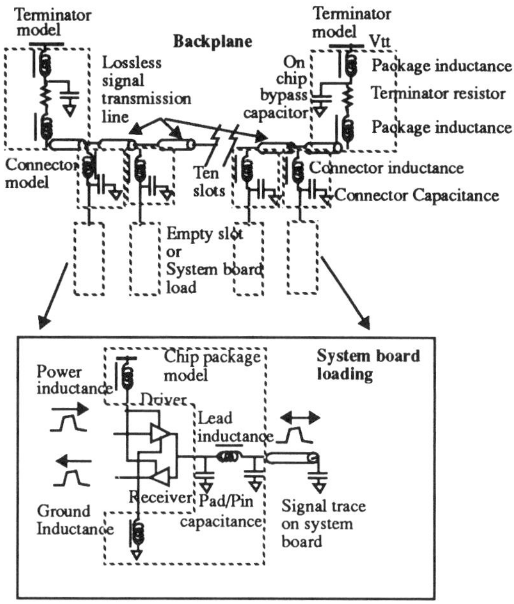

Figure 6 Electrical Model of Backplane

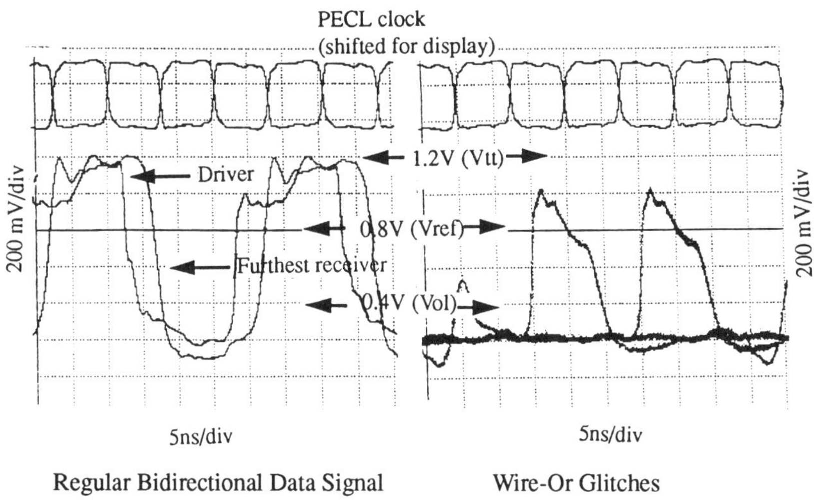

Figure 7 Measured Waveforms

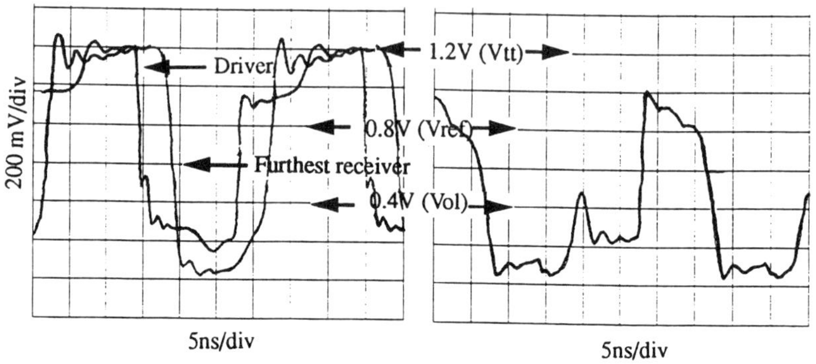

Regular Bidirectional Data Signal

Wire-Or Glitches

Figure 8 SPICE Simulated Waveforms

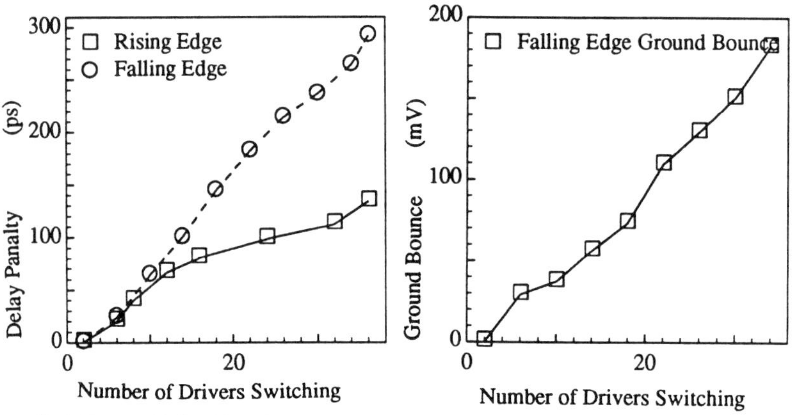

9 a) Output Delay Penalty

9 b) Ground Bounce

Figure 9 Simultaneous Switching Outputs (SSO)

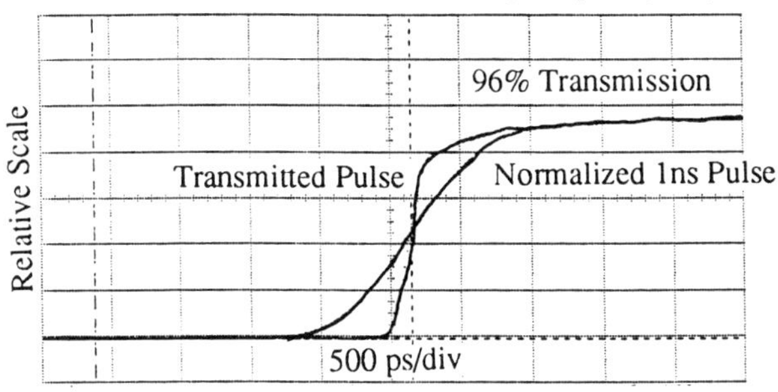

Figure 10 TDT measurement of lossy transmission line

4 Conclusion

A 1 GByte per second data throughput backplane has been implemented using low voltage swing CMOS (GTL). Electrical design trade-offs on terminator, line impedance, power and ground pin assignments have been described. Simulation and measurement are in good agreement. This backplane can be used for high end servers with ten system boards and very high sustained performance. The electrical performance of the backplane also shows the advantages offered by the GTL technology in high speed and low power consumption when compared with existing I/O technologies such as ECL, BTL and CMOS. Future improvement will be in the area of current mode drivers which can eliminate the wire-or delay which is the largest portion of the cycle time. A current mode drivers output impedance will remain high at low state so that the turn off edge will only take a single trip to settle. With advanced PCB and connector technology, the overall electrical length of the backplane can be shortened to further reduce the cycle time.

Acknowledgments

The authors would like to express their appreciation to the following persons for their help, support and technical discussions: David W. Yen, Jean Gastinel, Bill Gunning, Jeff Price

References

[1] B. Gunning *et al.*,"A CMOS Low-Voltage-Swing Transmission Line Transceiver," in *ISSCC Dig. Tech. Papers*, 1992, p. 58-59.

[2] MOTOROLA,"MECL DEVICE DATA", MC10H123, p.2-25,26

[3] National Semiconductor,"Interface Databook", DS3890, p.2-68-73.

[4] MOTOROLA,"MECL SYSTEM DESIGN HANDBOOK", p. 129-131

[5] D. Gustavson & J. Theus,"Wire-Or Logic on Transmission Lines," in *IEEE Micro*, June 1983, p. 51-55.

High-Performance Bidirectional Signalling in VLSI Systems

Larry R. Dennison Whay S. Lee William J. Dally

Artificial Intelligence Laboratory
Laboratory for Computer Science
Massachusetts Institute of Technology
October 12, 1992

Abstract

Interchip I/O bandwidth is a critical bottleneck in VLSI systems. To make the best use of this resource the conventions and circuits used for inter-chip signaling must be optimized to achieve the maximum bit rate with minimum power dissipation. This paper describes a set of I/O pads that we have developed at MIT. They operate with small signal levels to reduce power dissipation. To achieve high bandwidth, they operate simultaneously in both directions over a single wire and are isolated from supply noise. Bidirectional signalling is achieved by superimposing transmitted current waveforms and can be used with either transmission lines or lumped loads. This paper describes the theory behind our signaling method, the circuit design of the I/O transceivers, and the results of our experiments with prototype transceiver test chips.

1 Introduction

As integrated circuit densities increase, the density and operating frequency of logic gates is growing at a much greater rate than that of chip I/Os. While the number of devices per chip is increasing quadratically as critical dimensions are reduced, I/Os on the other hand are usually confined to the chip perimeter and hence scale only linearly with linear dimensions. Even then, I/O pad pitch scales at a slower rate than on-chip wire pitch. Gate switching frequency scales with improvements in device dimensions while I/O bandwidth per pin is limited by external wiring. Also, device switching energy scales as the cube of linear dimensions [6], while with standard signalling levels, I/O switching energy remains constant.

The net effect of this uneven scaling of logic and I/O is that the increase in gate switching events per second on a chip is greatly outpacing the increase in I/O switching events per second. As a result, the bandwidth of a chip's pinout has become a major bottleneck to system performance unless locality

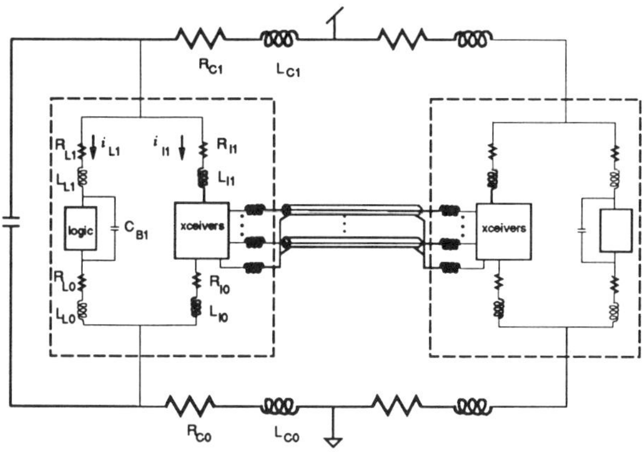

Figure 1: Transceivers on chips A and B exchange data over a limited number of signal pins connected by transmission lines. The signalling system must deal with noise due to power supply parasitics, reflections, crosstalk, and component tolerances.

can be exploited to keep communication on chip. Hence efficient signalling between chips in a VLSI system is critical.

In this paper we describe an approach to high-performance digital signalling currently under development at MIT. Section 2 introduces the problems involved in high-speed signalling and discusses the shortcomings of conventional signalling. Section 3 describes our approach to overcoming these problems: current-mode signalling and simultaneous bidirectional transmission. The circuits to realize this approach are the topic of Section 4. The limits of the performance of interconnections are discussed in Section 5. Finally, Section 6 discusses some early experiments with preliminary versions of these circuits.

2 The Problem

The digital signalling problem is illustrated in Figure 1. Two integrated circuit chips, A, and B must exchange information at high-bandwidth over a limited number of signal pins. In general A and B may each also exchange signals with other chips, not shown. The problem is to transmit signals over each wire at maximum rate while dissipating minimum power.

The big obstacle to high-speed, low-power signalling is *noise*. In contrast to analog systems where noise is often due to physical phenomena (e.g., Johnson or shot noise), in digital systems almost all noise is self induced. The major sources of noise in digital systems affecting an I/O transceiver are:

1. Self-induced power supply noise.
2. External power supply noise.
3. Transmitter and receiver offsets.
4. Reflections.
5. Crosstalk.

Self-induced power supply noise, sometimes called "ground bounce", occurs when the I/O transceivers draw a current i_{IX} from power supply X[1]. This current induces a self-induced power supply noise voltage on supply X of:

$$v_{PX} = i_{IX} \sum_{Y} R_{YX} + \frac{di_{IX}}{dt} \sum_{Y} L_{YX}. \tag{1}$$

This voltage is measured between the *root* of the power supply tree for supply X and the *leaf* of the tree where it connects to the transceivers. It is due to voltages developed across the parasitic inductances L_{YX} and resistances R_{YX} in each branch Y of the X power supply tree between the root and the transceivers.

Self-induced noise can be extreme in chips that attempt to drive fast edges with high-voltage swings. In addition to corrupting signals that are referenced to the power supplies, I/O induced supply noise can cause the system logic to malfunction because of the noise voltage developed across the parasitics on shared branches of the power supply tree.

External power supply noise is the voltage developed across shared branches of the power supply tree by other parts of the system:

$$v_{EX} = \sum_{S} \left(i_S R_{SI} + \frac{di_S}{dt} L_{SI} \right). \tag{2}$$

This voltage is developed across parasitics R_{SI} and L_{SI} through which both the transceivers and noise generator S are connected to the root of the X supply tree. For example, R_{C0}, is in a branch of the ground supply tree that is shared between the logic and transceivers on the left chip in Figure 1. Even if the I/O pads themselves are very quiet, they are connected to power supplies that may have substantial noise induced by on-chip logic and clock generators, and even other chips on the same board. While I/O drivers may have their own power supply pads to limit shared parasitics, board ground planes and board connector pins are shared.

Bypass capacitors can reduce supply noise when it is balanced, however they are not a panacea. Figure 1 shows bypass capacitors both on chip, C_{B1}, and off chip, C_{B2}. To zeroth order, these capacitors act as AC shorts between their

[1] In the figure, X is 0 for the ground supply and 1 for the V_{DD} supply.

terminal points on the power and ground supply trees. As an AC short, bypass capacitors are very effective when the power and ground currents are equal and opposite (as in the case of entirely on-chip logic or perfectly balanced differential signalling). In this balanced case, the AC current flows out of the bypass capacitor, rather than from one supply tree to the other. However, when driving single-ended lines (or differential lines whose termination points are far apart compared to a signal risetime), the AC power current is not equal and opposite that of the AC ground current. In this unbalanced case, all the bypass capacitor does is put the two power supply trees in parallel. At best, this will reduce the noise due to unbalanced currents by a factor of two. Also, even in the balanced case, power supply oscillations can result when the bypass capacitor forms a tank circuit with parasitic power supply inductances.

Transmitter and receiver offsets occur when device parameter variations cause the transmitted signal level and receiver threshold to differ from their nominal values. These offsets can be particularly large when a level is set by ratioing two devices whose parameters do not track one another – for example in a CMOS inverter receiver where the receive level is set by ratioing an N-FET against a P-FET.

Reflections, sometimes called "ringing", are due to an impedance mismatch at any point of the transmission line. The ends of the line are the most common place for a mismatch to occur. If the line has a characteristic impedance of Z_0 (for typical PC boards, $50\Omega \leq Z_0 \leq 75\Omega$) and is terminated into an impedance of R to the return plane, an incident waveform v_i will cause a waveform

$$v_r = v_i \left(\frac{R - Z_0}{R + Z_0} \right) \tag{3}$$

to be reflected back down the line. An impedance mismatch in the middle of a line (e.g., at a connector) can be tolerated only if it is small compared to the rising edge of the signal. Usually an impedance mismatch of 10% can be tolerated resulting in a reflected waveform that has 5% of the amplitude (and .25% of the power) of the incident waveform.

Rise-time control is important to minimize reflections because: (1) the impedance of termination structures may be frequency dependent, and (2) small impedance discontinuities are only *seen* by high frequencies. Thus, to minimize the high-frequency component of the transmitted signal, it is important that the signal rise time be kept at the maximum allowed by the design bit rate. A rule of thumb is that the rise time should be about half of the bit cell time (e.g., a 1ns rise time for a 500Mbit/s signal)[2].

Crosstalk is the injection of one signal into another signal via parasitic capacitance and mutual inductance. If two signals X and Y run on adjacent

[2]This ratio gives a good compromise between maximum bit rate (cell size equal to rise time) and jitter tolerance.

copper traces on a PC board for a length l, a transient on X will induce two types of crosstalk on Y: forward crosstalk which has the same waveform as Y with amplitude scaled by the coupling coefficients and l, and backward crosstalk which has a waveform of Y convolved with a pulse of length l. Forward crosstalk can be completely eliminated by adjusting the capacitive and inductive coupling coefficients to cancel each other while backward crosstalk is dealt with by having the source impedance matched to the line so that the crosstalk noise is absorbed at the source. Rise-time control is also important to minimize crosstalk. While the coupling coefficients of a stripline are independent of frequency, coupling due to package and device parasitics is frequency dependent and usually increases at high frequencies.

Not only do these five noise sources affect a signal by causing the level of a signal to be misinterpreted, they also introduce *jitter* into the system. Adding a noise source with magnitude v_N to a signal that swings through ΔV with rise time t_r introduces an uncertainty of

$$t_j = \frac{t_r v_N}{\Delta V} \tag{4}$$

in the timing of when the signal is detected crossing the receiver threshold. Power supply noise introduces additional jitter by modulating the delay of logic elements and drivers.

Conventional approaches to high-speed digital signalling in the presence of these noise sources have attempted to extend standard signalling techniques by heroic but brute-force methods involving very large numbers of power and ground pins (to minimize power supply noise) and very high current drivers (to achieve incident wave switching with standard voltage swings). The result is a system with high power dissipation.

The remainder of this paper discusses our solution to the problem of high speed signalling which (1) isolates the signal from noise using current-mode signalling, rather than suppressing noise using brute force techniques, (2) simultaneously sends signals in both directions over a single conductor to maximize the communication bandwidth per package pin, and (3) uses very low voltage levels to minimize I/O power dissipation.

3 Our Solution

Our approach to digital signalling is to isolate our transmitter signal levels and receiver threshold from power supply noise rather than to try to reduce the noise through brute force. We also extract the maximum bandwidth per pin by operating single-ended and in both directions simultaneously[3]. The key elements of our approach are:

[3] Related previous efforts include [4] and [5]

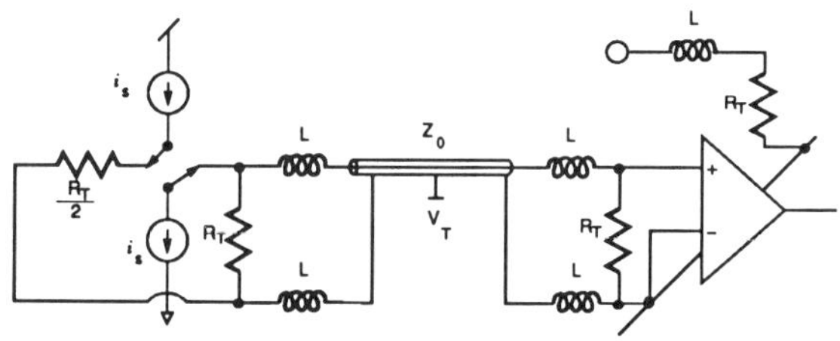

Figure 2: A high-performance current-mode signalling system. A current mode driver isolates signal levels from transmitter power. The receiver senses the voltage across the termination.

1. Connections are made point-to-point by connecting a controlled impedance transmission line between a pad on one IC, A, and a pad on another IC, B.

2. Digital signals are propagated in both directions simultaneously between A and B, doubling the available I/O bandwidth of the technology.

3. Signals are transmitted using a current-source driver with voltage sensed across the on-chip termination resistor in the receiving chip. This isolates the signals from power supply noise.

4. Signals are parallel terminated on-chip in each transceiver.

5. The signal return plane is independent of the IC power supplies. This isolates the signals from power supply noise in the transmission plane or through the terminators.

Figure 2 shows a simplified circuit of our driver and receiver. The driver operates in current-mode. That is it supplies a current of $+I_S$ or $-I_S$ to the line to signal a 1 or a 0 respectively. The figure shows the driver sinking I_S to send a zero. To transmit a 1, the two switches reverse and the driver sources I_S. This switching is performed at a controlled rate (as slow as possible given the design bit rate) to minimize $\frac{dI}{dt}$ into the transmission line. With an ideal current source, the signal level is completely isolated from transmitter power supply noise. Also the impedance of the current source is infinite, so the source impedance is entirely due to the on-chip termination resistor, R_T. With a 50Ω line (a 25Ω load), a value of I_S of 10mA gives a 250mV swing and a termination power dissipation of 2.5mW (however, without on-chip transformers, transmitter power is at least $V_{DD}I_S$, 30mW for a 3V supply). Compare this with the 360mW required to drive a 3V swing signal into the same 25Ω load.

The receiver is an ideal differential amplifier that senses the voltage across the on-chip termination resistor, $\pm\frac{1}{2}I_SR_T$. The ideal amplifier rejects all power supply noise and supplies are not used to generate the threshold. The push-pull current signalling results in a threshold equal to the termination voltage which is particularly easy to generate without offsets.

The lead inductances of the signal and V_T leads are shown in the figure. The signal lead is an issue if its distributed capacitance is such that it appears as an impedance discontinuity just before the termination. In that case, signal rise times will be limited to be at least some multiple of the lead length. Also transmitter transients result in $L\frac{di}{dt}$ drops across the signal inductance that are seen by the receiver. The edge rate must be selected to make these drops small compared to the signal level. The V_T inductance is more of a concern because it is shared by several signals. A signal transition on one lead will induce a voltage, $L\frac{di}{dt}$, that will be seen by the other receivers and injected as a backward traveling signal onto their attached transmission lines. For a given signal rise time t_r the number of signal leads N that can share a V_T lead will be limited to keep the voltage across the inductor less than some small fraction f (typically $f \leq 0.25$) of a full signal:

$$V_L = L\frac{di}{dt} = \frac{LNI_S}{t_r} < fI_SR_T, \tag{5}$$

so

$$N < \frac{ft_rR_t}{L}. \tag{6}$$

This design produces very little self-induced power supply noise. The current driver draws constant current from the power supply and injects zero net current into V_T. The upper current source always sources I_S into V_T while the lower current source always sinks I_S out of V_T. One of the two sources passes all of its current through the local $R_T/2$ resistor while the other splits its current between the line and the local terminating resistor. The low voltage swings result in current levels that are an order of magnitude lower than in a conventional system. Moreover the current mode operation completely isolates the transmitted voltage levels and the receiver threshold from external power supply noise.

The on-chip terminations, which can be made self-adjusting, minimize noise due to signal reflections and absorb crosstalk noise when it reaches its first termination. Careful layout to minimize coupling coefficients, to match coefficients to eliminate forward crosstalk, and to enforce an upper limit on parallel wire length is still required to keep crosstalk from being a problem.

Figure 3 shows how simultaneous differential signalling is achieved. Current-mode transmitters are attached to both ends of the line. Signals generated by A propagate from left-to-right while signals generated by B propagate from right-to-left. The voltages and currents in the line are the superposition of the two waveforms. At A we receive B's signal by subtracting out the voltage due to A's signal on the transmission line. Figure 4 shows how the two signals

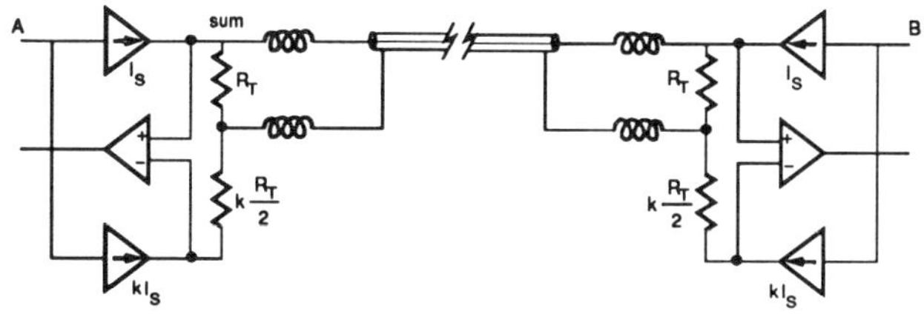

Figure 3: Simultaneous bidirectional signalling is accomplished by transmitting signals in both directions across a transmission line. At each end, the received signal is recovered by subtracting out the effects of the transmitted signal. A current source scaled by factor k is used to subtract out the transmitted signal to reduce power dissipation.

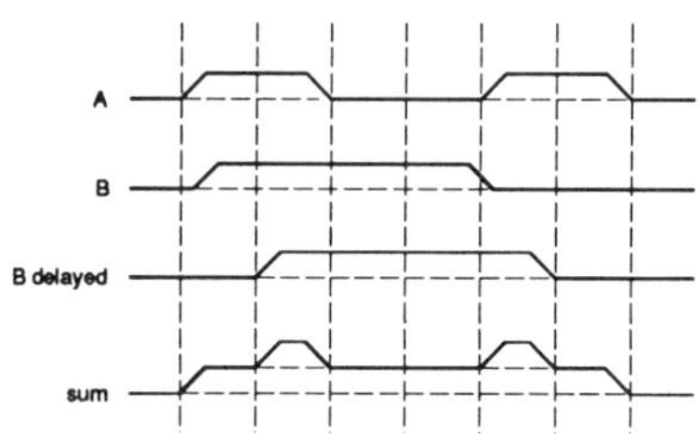

Figure 4: Waveforms for bidirectional signalling. The superposition of signal A with a delayed version of signal B gives the waveform on the left end of the transmission line.

are summed together on the transmission line.

We have avoided using differential signalling because while it provides twice the signal swing per unit pad current and gives completely balanced loads, it is less bandwidth efficient than single-ended signalling. To achieve the same bandwidth per pin with differential signalling would require operating at almost twice the signal rate as with single-ended signalling[4]. While this requires the same $\frac{di}{dt}$ to get the same voltage swing, the faster rise-times make smaller transmission line discontinuities a factor and the shorter bit-cell periods make jitter of timing signals more of a concern. Finally, the gain-bandwidth product of the receiving amplifier would have to double to achieve this rate. Hence, it is usually not possible to go twice as fast with differential signals.

We have not used series terminating drivers [4] because these drivers inherently send a voltage signal referenced to a local power supply voltage. Thus it is impossible to isolate them from power supply noise. We have attempted to duplicate the power advantages of these drivers by using pulsed signalling as described in Section 6.

4 Circuit Design

In this section we describe the circuit designs that approximate the ideal elements used to describe the concept in Section 3. The driver design described draws zero AC power and injects zero DC current into the termination plane, V_T. The receiver separates the two bidirectional waveforms by subtracting a copy of the transmitted waveform from the waveform on the line.

4.1 Driver

The driver may be constructed by using standard CMOS current sources, as shown in Figure 5. These are implemented using transistors operating in the saturation region. If the output driver is designed using cascoded current sources, output impedances greater than 100K ohm are easily obtained. The power supply rejection in this case is bounded below by -100dB.

Power supply noise also enters the line from parasitic capacitances, such as gate-source/gate-drain overlaps, back-gate, diffusion-well, and diffusion-bulk capacitances. Since all plates are connected via DC paths to some portion of the circuit, these capacitors couple high-frequency noise into the line. The effects of some of these parasitics may be reduced by careful layout techniques and the use of source-coupled wells.

Figure 6 shows an HSPICE AC analysis of the power supply rejection. The cascode current sources lead to 60dB of power supply rejection for frequencies

[4]Because fewer V_T pins are required for differential signalling the same bandwidth per pin can be achieved at slightly less than twice the single-ended signalling rate.

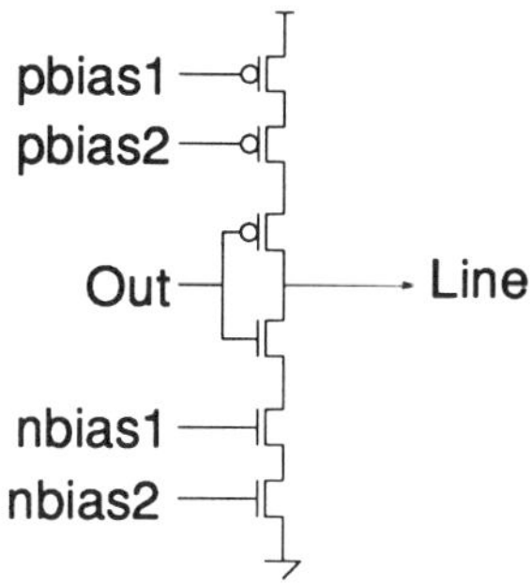

Figure 5: Cascoded Current-Source Driver

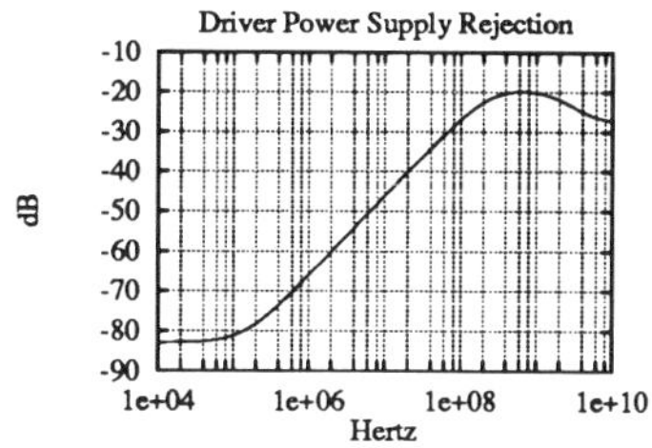

Figure 6: Driver Power Supply Rejection

below 1 MHz. Power supply noise in this range usually results from switching power supplies. Between 1MHz and 100MHz, we still achieve 20dB of rejection. Noise in this range of frequencies is typically the result of other IC's on the same board. Coupling becomes most noticeable at frequencies over 100MHz, which is typically self-induced. We thus want to limit the amount of power supply noise V_{PX} in the higher frequencies.

4.2 Current Steering

One of the problems with the cascoded current sources is that the internal nodes tend to have large parasitic capacitances. When decoupled from the line, the current sources charge these capacitances to the rails. Later, when the output switches, the drive transistor sees the parasitic capacitor, not the current source. This results in undesirable initial current overshoot.

A solution is to *steer* the current from the cascoded current sources into either the line or a dummy load. The circuit show in figure 7 is similar to that used for a differential driver except that the complementary output never leaves the chip. In the absence of a received signal, this circuit gives a constant voltage on all nodes in the current source, eliminating the initial overshoot.

The current steering driver has two other beneficial properties. First, it draws constant current from both the V_{DD} and GND supplies, irrespective of the output state. This reduces V_{PX} which in turn reduces the required number of supply pins and the noise effects on the system logic. Second, the total current supplied from an output driver into V_T is close to zero[5]. This suggests that the V_T supply could be as simple as a capacitor and a resistor divider.

As we will explain further in section 5, we need to limit the edge rate of the driver. This is done by splitting the driver into several small parallel drivers. The turn-on of each driver is then staggered via a buffer chain. This technique allows the construction of not only a slow straight edge, but also a piecewise

[5]The DC current into V_T is equal to the mismatch between the two current sources.

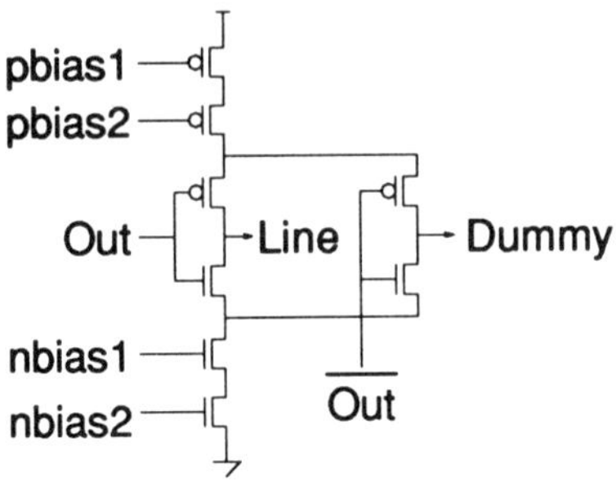

Figure 7: Cascoded Current Sources with Steering

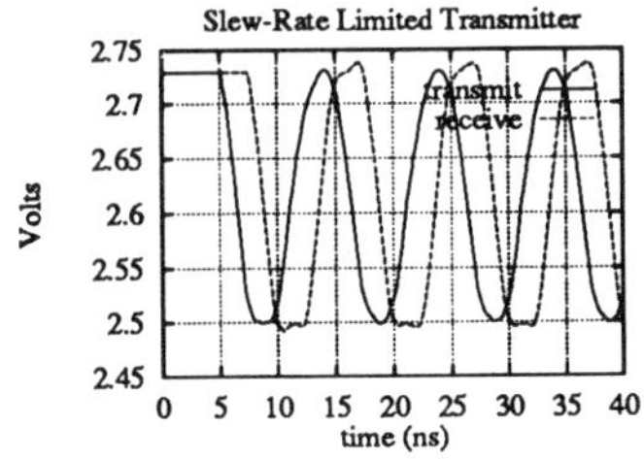

Figure 8: Transmitter Transient Response, 200Mbit

approximation to more elaborate curves. These could more closely match the desired low-pass characteristic. Figure 8 shows the transient response of a slew-rate limited driver.

4.3 Receiver

As we decreased the signalling voltage level, we kept the output voltage swing of the receiver constant (full CMOS levels). In doing so, we increased the required gain-bandwidth product of the receiver. The input voltage swing of the differential amplifier is 250mV, ignoring noise. For 3V CMOS levels, a gain of 12 (22dB) is required. To achieve a 200Mbit signalling rate, we need a receiver with a gain-bandwidth product of 1.2GHz.

The gain-bandwidth product must be measured large-signal, as the high-frequency output swing must match the low-frequency output swing to avoid inter-symbol effects. Using a good CMOS process, it is fairly easy to construct a standard transconductance amplifiers with a unity-gain frequency near 2GHz. For a standard transconductance amplifier, these gains are only for a limited output range which is typically $\pm 1V$. Outside of that range, the transistors in the output stage come out of saturation and enter the triode region. Once in triode, they supply less current and the output slew rate falls off.

The transconductance amplifier can be adjusted to give large output swings by allowing one of the two output drive transistors to enter cut-off. This produces the rail-to-rail swings shown in Figure 9. The input waveform (labeled "input") is 200mV peak-to-peak, with 1ns edges. The waveform labeled "amp output" is the output of the differential amplifier. This output is then cleaned up using an inverter whose output is labeled "output".

Figure 10 shows the result of a 500Mbit/sec input waveform. The input waveform is 200mV peak-to-peak, with .25ns edges. Notice that the amplifier output no longer goes rail-to-rail and is now slew-rate limited. The inverter following the amplifier restores full rail signalling.

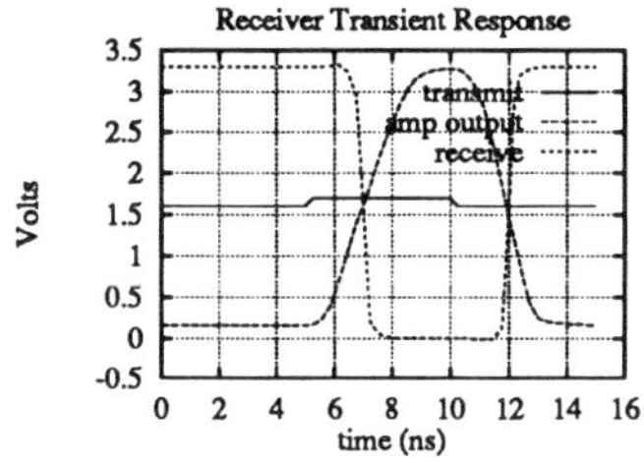

Figure 9: Receiver Transient Response, 200Mbit

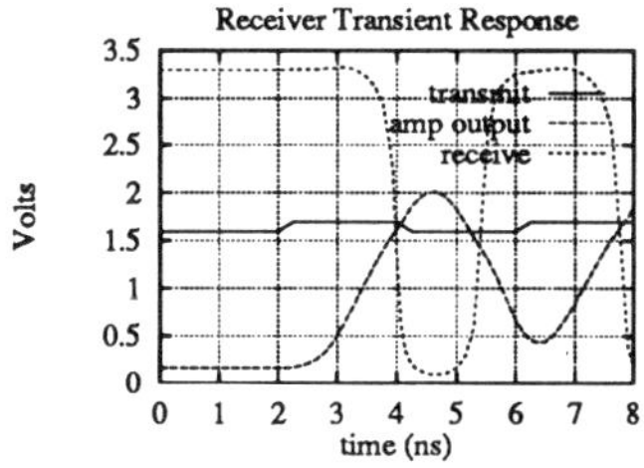

Figure 10: Receiver Transient Response, 500Mbit

5 Achievable Performance

The performance of an inter-chip signalling system is limited by the following factors:

- The bandwidth and frequency characteristics of the transmission media and its terminations.
- The gain-bandwidth product of the transmitter.
- The gain-bandwidth product of the receiver.
- The voltage noise and timing jitter present in the system.

The eye diagram 11 illustrates the effect of these factors on system performance. The transmitter produces a waveform with amplitude v_{tx}, bit cell period t_c, and rise time t_r. The rise time t_r is limited both by the bandwidth of the transmitter and by the bandwidth of the media. If the transmitter input amplitude is v_i, the transmitter gain-bandwidth product, f_{tx}, must be at least

$$f_{tx} \geq \frac{v_{tx}}{2v_i t_r} \tag{7}$$

to generate this waveform. Transmitter gain-bandwidth is unusually not the factor limiting performance because it is easy to build transmitters in 0.8μ CMOS with gain-bandwidth products in the GHz range and the transmitter is attenuating ($v_{tx} < v_i$). The transmission media is the limiting factor on the achievable transmitted waveform. Generating a waveform with significant frequency components above the flat response of the transmission system will result in excessive attenuation, excessive reflections, or both.

The received waveform is also shown in Figure 11. Here the reliable received amplitude has been reduced by a combination of voltage noise v_N and attenuation A to $v_{rx} = Av_{tx} - v_N$. The rise-time and timing jitter limit the

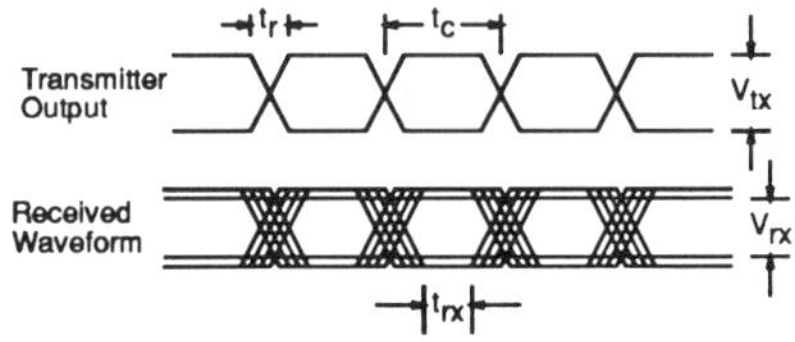

Figure 11: An eye diagram showing transmitted and received waveforms. Attenuation, noise, and jitter reduce the height and width of the *eye* that the receiver must sense.

window or *eye* within each bit cell during which v_{rx} can be sensed by the receiving amplifier to t_{rx}. If the receiver output has amplitude v_o, the receiver gain-bandwidth product, f_{rx}, must be at least

$$f_{rx} \geq \frac{v_o}{v_{rx} t_{rx}}. \tag{8}$$

At present, receiver gain-bandwidth product is a limiting factor on system performance. For very small signal swings, two stage amplifiers are required to achieve the required gain without sacrificing bandwidth. Also, the required f_{rx} can be traded off against the required f_{tx} by increasing signal level v_{tx}, although this is at the expense of increased power dissipation. However, in 0.8μ CMOS, one can build a receiver that operates at bit rates that push the capabilities of conventional package and PC-board transmission media. As circuit bandwidth improves with smaller devices, transmission media will become the single limiting factor on performance.

Interconnect Model: As described above, even with perfect circuits, performance is limited by the transmission media; bond-pad parasitics, lead inductance, pin capacitance, and transmission line imperfections all affect signal quality and bandwidth. Consider the model for the interconnection between two integrated circuit chips shown in Figure 12. The circuits are in separate packages attached to a common PC board. The PC trace is modeled as an ideal transmission line with a 50Ω characteristic impedance. Both ends of the transmission line are terminated on-chip in the characteristic impedance. We add in three parasitic elements to each end: pad capacitance, lead inductance, and pin capacitance. For these, we assign the nominal values of $5p$F, $3n$H, and $5p$F. These values are representative of high-performance ASIC packages available today. This model is representative for a short connection across a PC board where package parasitics dominate. For a longer run or one that included connectors, line attenuation and discontinuities would be included as well.

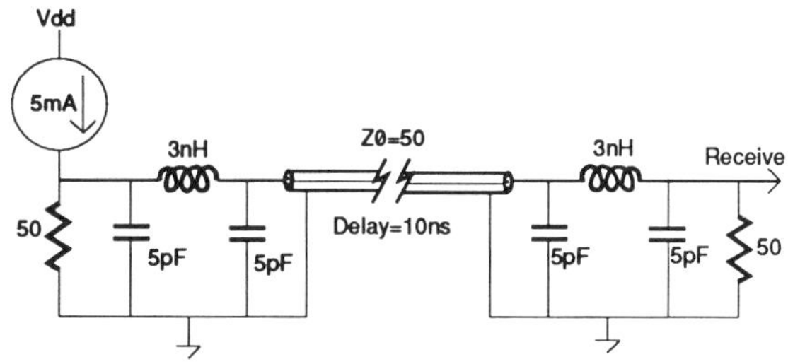

Figure 12: A circuit model of a package and PC board transmission media. Packaging parasitics at either end of the line affect the frequency response of the termination.

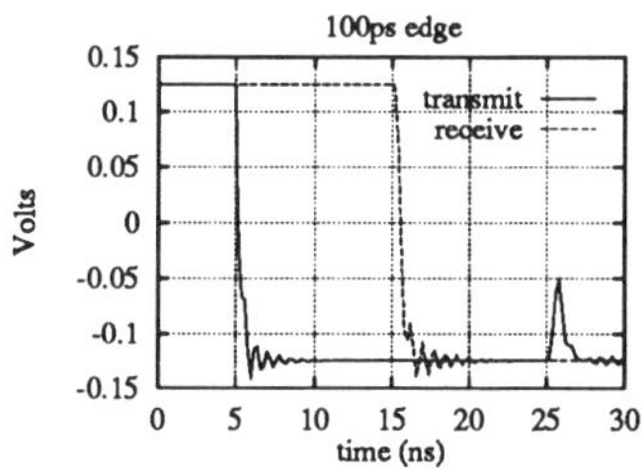

Figure 13: Line response to 100ps edge. Ringing is caused by the parasitic tank circuit and a pulse is reflected back to the transmitter because the parasitics make the termination frequency dependent.

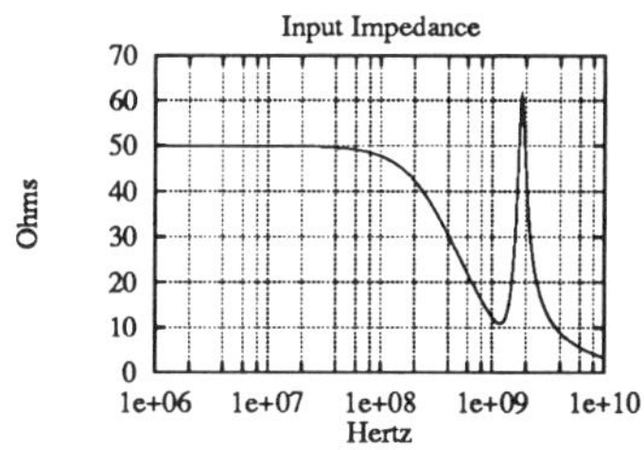

Figure 14: Effective input impedance for the model circuit. The *Pi* low-pass filter causes the impedance to roll off above 100MHz and then peak at 2GHz, the resonant frequency of the parasitic tank circuit.

Step Response: Using SPICE, we look at the step response of the interconnect model (Figure 13). We see that the initial edge at the driver pad looks clean (with only a small amount of ringing due to the parasitic tank circuit), as does the initial edge at the receiver. However, a very large spike is reflected back at the driver. In the actual circuit, we are recovering data at both ends of the interconnect. A spike of this magnitude could easily cause an error in data recovery.

This spike is mainly the result of the frequency-dependent termination. The topology of pad capacitor, lead inductor, and pin capacitor is that of a Π (low-pass) filter. The effective termination impedance is shown in Figure 14. High frequency components are reflected back toward the driver. Very little energy at those frequencies is actually transferred to the pad, explaining the clean waveform seen at the pad.

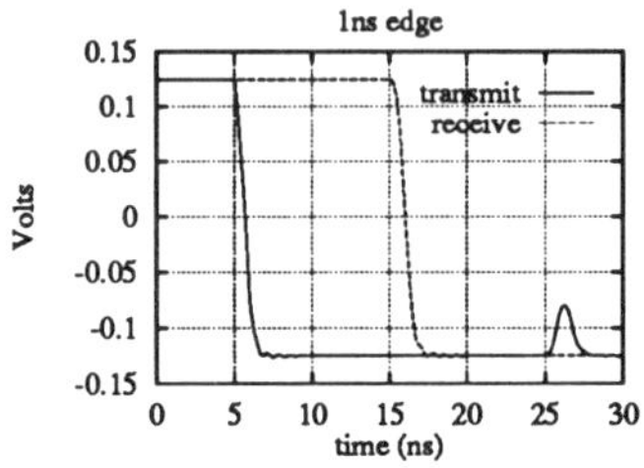

Figure 15: Line response to 1ns edge. Slowing the edge rate reduces the ringing and halves the amplitude of the reflected pulse.

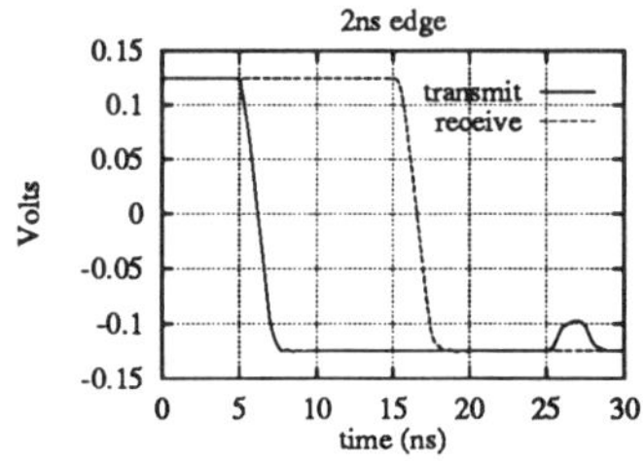

Figure 16: Line response to 2ns edge. The reflected pulse is now reduced to about 10% of signal amplitude.

Rise-Time Control: The results of the step response analysis suggest that the source waveform not have any frequency components which are incorrectly terminated. Such components may be attenuated by low-pass filtering or by shaping the drive current transition. We adopt the second approach by slowing down the transition. The SPICE runs in Figure 15 and Figure 16 show the result of edge times of 1ns and 2ns respectively. These figures show that slowing the edge rate is effective in reducing the reflection.

Performance Limits: Consider the circuits described in Section 4. The driver is able to supply very high speed edges, however, as described above, edge shaping is needed to reduce reflections. To limit the reflections to less than 10%, we restrict $t_r \geq 2$ns. Thus, based on packaging constraints, we are restricted to less than 500Mbit/sec per direction per pin.

As the line rate increased or edge rate decreased, the amplifier in the receiver became slew-rate limited. This results in a kind of inter-symbol interference. A long transmission of zeros results in difficulty in recovering an isolated one. A similar effect holds true for a long transmission of ones followed by a zero. Even at 500Mbit/sec, the isolated pulse width at 90% of final value is slightly over 1ns.

Thus, for an 0.8μ CMOS process, using the circuits of Section 4, the performance-limiting factor is the package and the interconnect, not the circuits or devices.

6 Experiments

The design presented in Section 4 has been influenced by our experience in fabricating and testing three prototype chips. Our first chip demonstrated the basic concept of bidirectional signalling. However its performance was limited by slow on-chip test circuitry. This chip was described in [5] and will not be discussed further here. The second chip, described below, overcame

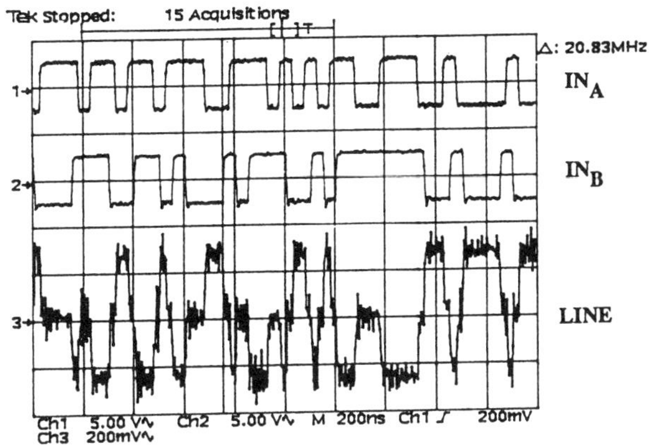

Figure 17: Transmitted Data and Transmission Line Signal Waveforms for Static Test Chip.

the problems with on-chip test circuitry. Its performance was limited by noise and the gain-bandwidth of the 2μ CMOS receiver. The third chip tested a pulsed version of the bidirectional pads. Unfortunately fabrication difficulties prevented this chip from being fully operational.

For economy and fast turnaround, all three chips were fabricated in 2μ CMOS. The first chip was fabricated through MOSIS while the second and third chips were fabricated through the Massachusetts Microelectronics Center.

6.1 Static Pad Test Chip

The system shown in Figure 3 was fabricated and tested. The driver was realized as a simple two-transistor CMOS inverter, sized to source and sink 10mA. A differential sense amp was used as the receiver. A scaled down (5:1 ratio) driver was used to produce the reference copy of the transmitted signal for subtraction, to help reduce power dissipation. A static differential sense amp was used as the receiver . The complete design was fabricated in a 2μ N-well process. Total simulated power consumption was 41.5mW per transceiver.

A pseudo-random data stream generator and a compare circuit were included on the prototype chip for the purpose of testing. Operating bidirectionally at 100Mbits/s through a 75Ω, 2m long coaxial cable, our transceiver was monitored for 24 hours without a compare mismatch. Functionality of the system is tolerant of fluctuations of about ± 200mV in V_T.

Figure 17 shows the two transmitted data waveforms along with the superimposed signal seen on the transmission line. The noise seen on the superimposed signal is rejected by the differential receiver amplifier.

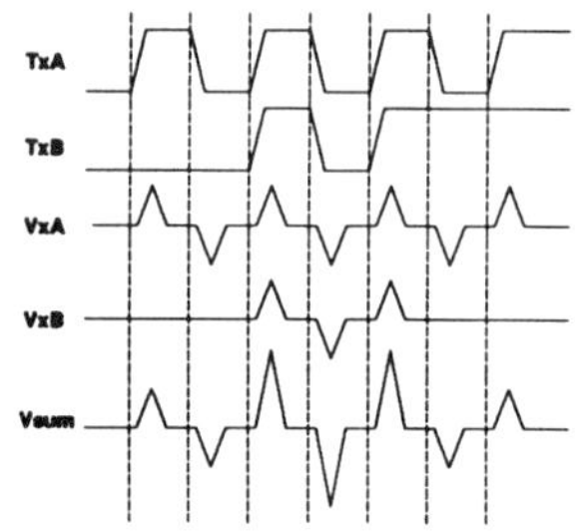

Figure 18: Timing Diagram for Pulsed Transceiver System

6.2 Pulsed Bidirectional Test Chip

Our third test chip included a pulsed bidirectional transceiver based on a technique reported by Gabara & Thompson [2]. This design pulses the current drive high or low in response to a rising or falling transition as shown in Figure 18. Thus the transmitter dissipates no static power. A combination of fabrication problems (transistor betas off by a factor of three) and design problems with the pulse shaping circuitry prevented these chips from operating as designed.

6.3 Self-Adjusting Terminations

Both of the designs described above include self-adjusting termination resistors. A finite state machine (FSM) switches a series of exponentially sized complementary MOS passgate resistors. During the reset sequence, the FSM switches passgates in and out until the termination resistance matches an external reference resistor. Using an on-chip terminator reduces board parts count, and avoids the lead inductance between the transceiver and terminator inherent with external termination.

6.4 Discussion of Experimental Results

Both of the prototypes described above suffered from a large discrepancy between the SPICE model parameters used for simulation[6] and the actual device parameters. Actual transistor transconductance was measured to be less than half the simulation value.

Despite modeling inaccuracies, the digitally-adjusted termination mechanism operated properly. By operating digitally, it eliminated many of the stability and noise problems associated with the analog-controlled termination scheme used in [5].

[6] We based our design on the data supplied in [3], as recommended by the foundry.

Our experiments with the static test chip revealed several noise problems with our initial circuits. Our CMOS inverter drivers did not deliver zero total current into V_T and hence induced a substantial amount of $L\frac{di}{dt}$ noise across the shared V_T inductor that corrupted the received signals. This driver had inferior power supply rejection and passed substantial amounts of power supply noise into the transmitted signal. Without the use of current steering, the driver also generated substantial power supply noise. Finally, this driver did not employ rise-time control.

Our experiments, particularly since they were conducted using a slow 2μ CMOS process also highlighted the importance of receiver gain-bandwidth product. In particular careful layout is required to minimize capacitance on the critical nodes of the differential amplifier.

The DIP package we used for the prototypes, contributed a large amount of lead inductance (up to 19nH [7]), which was the another major cause of noise and signal distortion. A high-performance package is essential to successful high-speed signalling.

7 Conclusion

The number of logic switching events per second per chip is growing much faster than the rate of I/O switching events per chip making I/O bandwidth a critical resource. Conventional I/O signalling conventions operate with large voltage swings, reference signal levels and receive thresholds to power supplies, and send signals in only one direction at a time. The net result is that conventional signalling techniques squander the scarce I/O pads.

We have described a high-performance bidirectional signalling method and its implementation with CMOS circuits. Our method uses small voltage swings (250mV) to minimize power dissipation and self-induced noise. We use current mode signalling to completely isolate the transmitted signal levels and receiver threshold from the noisy power supplies. Finally, we send signals simultaneously in both directions over a single point-to-point link between two chips – in effect doubling the effective I/O bandwidth.

Our circuit implementation uses a cascode current source driver to minimize power supply coupling. Current steering is used to eliminate the initial overshoot problem, reduce self-induced power supply noise, and reduce noise coupled through the V_T lead inductance. The driver rise-time is controlled to minimize reflections, crosstalk, and self-induced noise. Our receiver is implemented as a differential CMOS amplifier. In 0.8μ CMOS the amplifier achieves a gain-bandwidth product of 2GHz enabling operation at 500Mbits/s.

To gain experience with high-performance bidirectional signalling, we have implemented three test chips. The results from two of these experimental chips are reported here. Using a 2μ CMOS process, we have successfully demonstrated bidirectional signalling at rates in excess of 100Mbits/s. However, experience with these prototype chips has led us to modify our driver design to reduce its noise generation and increase its noise immunity.

The current-mode, low-voltage swing, bidirectional signalling method we describe here is an attractive alternative both to (1) the conventional full-voltage swing CMOS signalling in use in many VLSI systems and (2) the low-voltage swing power-supply referenced signalling common in ECL systems. In the first case it dramatically reduces the power required to achieve a given data rate. In both cases it greatly increases noise immunity hence permitting operation at smaller signal levels.

This signalling method is suitable for use in all types of digital systems that use point-to-point signalling. It is particularly appropriate for connections between similar VLSI components, such as between the routers in a multicomputer interconnection network [1]. In these cases, there is no need to interface to any existing signalling standard and only one chip type needs to be designed to take advantage of the benefits of bidirectional current-mode signalling.

We envision industry adopting a signalling convention similar to the one described here as clock frequencies increase and straightforward extensions of TTL compatible signalling become impractical because of power dissipation and the number of supply leads required to keep power supply noise under control. Adapter chips will be required at first to allow chips using high-speed signalling to interface to chips using existing standards.

With today's 0.8μ CMOS chips, data rates are limited both by receiver gain-bandwidth product and by the bandwidth of the transmission media. With high-performance packages and specially designed substrates, one can make the receiver circuit the rate limiting factor. In the long run, however, as circuit speeds continue to increase with shrinking device dimensions and as we gain the benefit of the high f_T bipolar devices found in many modern BiCMOS processes, the transmission media will be the rate limiting factor. Within a very few years, the package design and the characteristics of the substrate used to connect the chips will largely determine the available inter-chip bandwidth.

Acknowledgment

The research described in this paper was supported in part by the Defense Advanced Research Projects Agency under contracts N00014-88K-0738 and N00014-91J-1698, in part by a National Science Foundation Presidential Young Investigator Award, grant MIP-8657531, with matching funds from General Electric Corporation, IBM Corporation and AT&T Corporation.

References

[1] William J. Dally. Network and processor architecture for message-driven comput-

ers. In Suaya and Birtwhistle, editors, *VLSI and Parallel Computation.* Morgan Kaufmann, 1990.

[2] Thaddeus J. Gabara and David W. Thompson. High speed, low power cmos transmitter-receiver system. *IEEE Transactions on Computers*, pages 344–347, May 1988.

[3] Lance. A. Glasser and Daniel. W. Dobberpuhl. *The Design and Analysis of VLSI Circuits.* Addison-Wesley, 1985.

[4] Thomas Knight and A. Krymm. Self terminating low voltage swing cmos output driver. *IEEE Custom Integrated Circuits Conference*, pages 289–292, 1987.

[5] Kevin Lam, Larry. R. Dennison, and William. J. Dally. Simultaneous bidirectional signalling for ic systems. *ICCD : VLSI in Computers and Processors*, pages 430–433, 1990.

[6] Carver Mead and Lynn Conway. *Introduction to VLSI systems.* Addison-Wesley, 1980.

[7] Signetics. *Signetics ABT Handbook.* Signetics, 1992.

A Family of Routing and Communication Chips Based on the Mosaic

Charles L. Seitz and Wen-King Su
Computer Science 256-80
California Institute of Technology
Pasadena, CA 91125

Abstract

This paper describes the specifications, internal design, and performance characteristics of the latest series of Caltech Mesh-Routing Chips (MRCs), slack-protocol-conversion chips, and router-interface chips. These high-performance, self-timed designs were derived from the corresponding parts of the message-passing system of the Mosaic (see the paper "The Design of the Caltech Mosaic C Multicomputer" in this volume), but are packaged as separate chips to allow their use by other projects engaged in the development of highly concurrent computers.

In addition to n data bits, an MRC channel includes a bit to mark the tail of the packet, a self-timed request signal, and an opposite-going self-timed acknowledge signal. MRC channels operate in a queue (FIFO) discipline, in which the request and acknowledge signals provide both timing and flow control. The timing and electrical characteristics of these signals are critical to reliable communication between MRCs. The timing of the internal channels of an MRC is based on a transition-signaling, ripple-through FIFO whose properties are exploited to simplify the internal design the MRC. The MRC implements oblivious, dimension-order, blocking-cut-through (wormhole) routing with an internal organization based on elementary routing automata.

Because the zero-slack protocol of an MRC channel limits the communication performance over long cables, Slack chips provide a conversion to a k-slack protocol. These Slack chips can be implemented with the same FIFO that is used in the MRCs. Dialog chips provide this same slack function, but within a protocol that allows a dialog of control symbols to be exchanged on a channel.

In order to help the system designer deal with these high-speed, asynchronous channels, we also developed router-interface chips to provide synchronization and data conversion between MRC channels and microprocessor buses.

1 Introduction

Our research group has for many years been providing routing chips to other projects, but, until recently, in only one configuration similar to the Mosaic router (Figure 1). These Caltech Mesh-Routing Chips (MRCs) provide packet communication and oblivious, dimension-order, cut-through routing in a 2D-mesh network that connects a set of computing nodes. The north-, east-, west-, and south-bound channels that form the mesh fabric are referred to as the *"news"* channels. The channels to and from the node are referred to as the *"p"* channels. Dimension-order routing assures freedom from deadlock provided that the *po* channel consumes the packets routed to it, and also permits this 2D router to be implemented, as indicated in Figure 1, as a cascade of two, identical, 1D routers. The Caltech MRCs are asynchronous, and their *news* and *p* channels operate according to a zero-slack, self-timed protocol on 8-bit-parallel flow-control units (*flits*) at a rate that is typically throttled to ≈70MB/s.

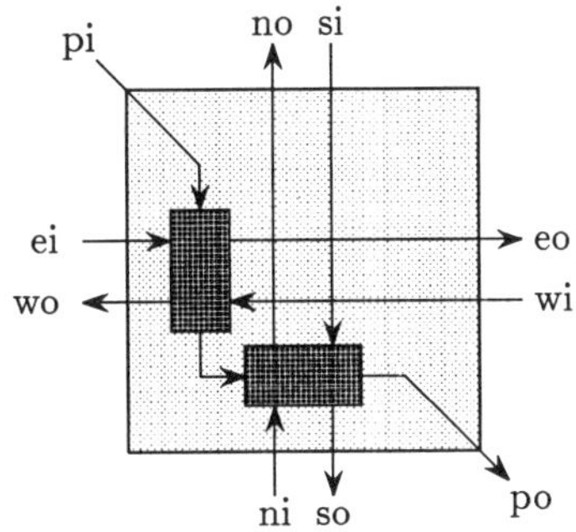

Figure 1: A Caltech Mesh-Routing Chip (MRC)

MRCs are typically used to construct the message-passing network of a multicomputer, or the switch between the processors and memories of a multiprocessor. For example, MRCs form the active, routing-mesh backplane for the Ametek S2010 multicomputer [5, 7] and the Intel Delta multicomputer. The *news* channels are connected on the backplane. The multicomputer nodes plug into the backplane, and connect to the *p* channels. Speculations on other applications of this routing technology can be found in [8].

The "Elko" router used in the production version of the Mosaic C is the latest in a series of pin-compatible but internally different MRC designs that started with the 1987, 30MB/s, "Ginsu" routers (GMRCs) used in the Ametek S2010 multicomputer and the MCC ES-kits. The 1989, 80MB/s, "Frontier" routers (FMRCs) were used in early Mosaic systems, the Intel Delta, and several other projects. In 1991–92, Intel developed their own version of the "Frontier" router, the iMRC, for the Intel Paragon multicomputer. The iMRC employs the same internal, routing-automata scheme [3] as the FMRC, and many of the same circuits, but has 16-bit rather than 8-bit flits, and supports packet broadcast. Also, its external channels operate according to a streaming (also known as a transmission-line write-ahead) protocol similar to the slack protocol.

The 2D, 8-bit version of the "Elko" MRC, designated the EMRC-2D8, was designed by the authors at the end of 1991. In addition to providing better setup- and hold-time margins than earlier routers, the EMRC is more modular and more tolerant of process variations. It was, accordingly, easy to generate EMRC-2D9 (2D with 9-bit flits) and EMRC-1D16 (1D with 16-bit flits) variants of the basic EMRC-2D8 design.

The zero-slack, self-timed protocol used in MRC channels operates correctly even when the routers are connected through cables, but the cycle time increases with the delay of the cable. In order to maintain the channel bandwidth over long cables, a series of Slack chips was developed using the same FIFO circuits that are used internally in the EMRCs. These slack chips convert between the zero-slack protocol of the router channels and a k-slack protocol that permits the sender to get k requests ahead of the acknowledges from the receiver.

The *pi/po* channels presented to the node create a challenging design problem for building interface circuits from commodity parts. On the *pi/po* side, the interface is asynchronous, self-timed, one or two bytes wide, and capable of operating at rates in excess of typical clock frequencies. On the node side, the interface must typically be synchronous and word-wide. Elko-Router Interface Chips (ERICs), based in part on the Mosaic packet interface [9], provide highly reliable synchronization, and reduce the data rate by providing a 4- or 8-byte synchronous, microprocessor-bus interface to and from the node.

This family of routing, communication, and interface chips can be used together to build the message-passing network and interfaces for highly concurrent computers. The layout modules can also be combined to provide more highly integrated solutions on single chips.

2 Specifications for the Routing-Chip Channels

Packets traverse the routing network in a queue discipline. Each n-bit data item presented on a channel by the sender is acknowledged by the receiver, and is, accordingly, referred to as a *flow-control unit*, or *flit*. A packet is composed of a sequence of flits starting with a routing *header*, and ending with a flit that is tagged as the *tail*. The "wormhole" routing used in the EMRCs is a blocking form of cut-through routing. If the required output channel is available, the head of an incoming packet is advanced immediately into this channel, and the packet is spooled through this established path until the path is broken by the tail flit. If the required output channel is already in use, the packet is blocked in place by the flow-control discipline until the required output channel becomes available.

Although routing is recognized as the function of the EMRC chips, reliable, high-bandwidth, packet communication between EMRC chips has required nearly as much design attention as the routing itself, and is the most appropriate starting point for this exposition.

2.1 Router-Channel Signals, Timing, and Performance

Each n-bit router channel is composed of $n+3$ signals: a request (r) in the direction of the channel, an acknowledge signal (a) in opposite direction, a tail bit (t), and the n bits of data ($0, \ldots, n\text{-}1$). The names of these signals are formed as illustrated in Figure 2. The bidirectional channels generally used in multicomputer networks (see section 3 of [6]) are implemented simply as a pair of unidirectional channels.

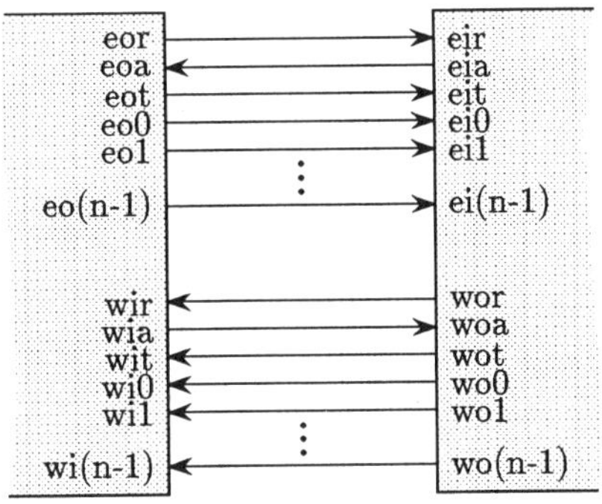

Figure 2: East- and west-bound channels connecting two routers.

The request and acknowledge signals of a channel conform to a pipeline form of 2-cycle (transition) signaling (see figures 7.16 and 7.24 in [4]), and provide both timing and flow control. The use of transition signaling assures that the rate of transitions on the request and acknowledge signals is no higher than the highest rate of transitions on the data signals. Each flit is conveyed by a transition of the request signal followed by a transition of the acknowledge signal. This is a *zero-slack* protocol: The channel output may not generate another request until the outstanding request has been acknowledged by the channel input.

At a channel output, data signals become valid at least t_{dr} before the transition of the request signal, and remain stable until after the corresponding transition of the acknowledge signal. At a channel input, data inputs are sampled coincident with the transition of the acknowledge signal ($t_{\mathrm{ad}} = 0$). The minimum values for the three parameters shown in Figure 3, $t_{\mathrm{dr}}, t_{\mathrm{ra}}$, and t_{ar}, can be observed directly on a channel that is operating at full rate. The unobservable parameter t_{ad} is, by design, determined by the difference between two very small delays within the channel input circuits, and is so small in magnitude that it can be taken to be zero. The t_{ra} interval is determined by the receiving router, and may be larger than its minimum value if the flow is blocked downstream. The t_{ar} interval is determined by the sending router, and may be larger than its minimum value if the "upstream" packet is being supplied to the sending router more slowly than the output channel can operate.

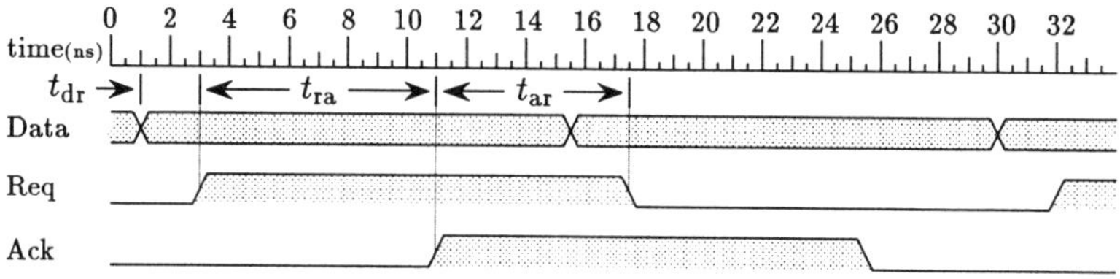

Figure 3: Router-channel timing between two routers that are physically close and operating at maximum speed.

Typical values for these key parameters at 25C and V_{dd}=5V, with EMRCs connected by 10cm PCB traces (approximately 20pF, consisting of 15pF wire and 5pF input-pin capacitances) are: $t_{dr} = 2.0$ns, $t_{ar} \geq 6.5$ns, $t_{ra} \geq 8.0$ns, and $t_{ad} = 0.0$ns. In terms of these parameters, the period is $(t_{ra} + t_{ar}) \geq 14.5$ns, corresponding to a bandwidth $\leq$ 69MB/s. The setup-time margin is $(t_{dr} + t_{ra}) \geq 10$ns. The hold-time margin is $(t_{ar} - t_{dr}) \geq 4.5$ns.

The 10ns setup-time margin was deliberately favored within the 14.5ns period. For routers that are close together, the 4.5ns hold-time margin is more than adequate. The expressions above can be augmented to allow for the the additional delay, t_p, in the wires that connect two routers that are separated by a greater distance. (These expressions can also be augmented to allow for skew.) Additional delay due to increased capacitance and flight time increases the period by $2t_p$, and increases the hold-time margin by t_p, but does not change the setup-time margin. The routers were, accordingly, designed to have a larger setup-time margin than hold-time margin.

2.2 Electrical Characteristics

Although it would be possible and desirable to use techniques such as low-voltage drivers and receivers, or simultaneous bidirectional communication [2], the routing chips that we have developed to date have been intended principally for applications in which the routers are physically close, and in which the simplicity of conventional 5V CMOS signals outweighs the benefits of these other techniques. For these ordinary CMOS inputs and outputs, typical input-pin capacitance is 5pF, and input and output pins are protected against ESD and latchup to at least ±100mA. The input switching threshold at V_{dd}=5V is 2.4V ± 0.1V.

EMRCs and Slack chips are ordinarily fabricated in the MOSIS 1.2μm scalable CMOS process ($\lambda = 0.6\mu$m (n-well), the Hewlett-Packard CMOS-34 process).

All output pins are driven by a p-channel—n-channel MOSFET pair sized to produce nearly the same transition and delay times for positive-switching and negative-switching signals. The i–v characteristics plotted in Figure 4 were obtained from electrical measurements of the pins of a typical EMRC chip at 25C and V_{dd}=5V. Although the static saturation current of the p-channel output driver to 0V is ≈35mA, whereas the static saturation current of the n-channel driver to V_{dd}=5V is ≈31mA, the devices exhibit asymmetric nonlinearities due to differences in velocity saturation between the p-channel and n-channel transistors, and due to the n-well process.

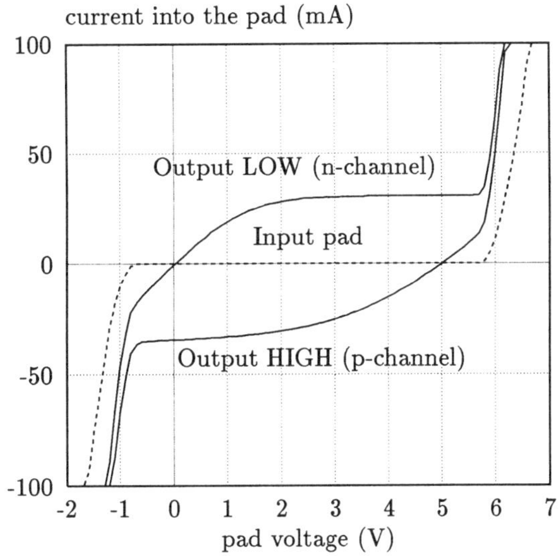

Figure 4: EMRC I/O-pad i–v characteristics.

The net result of the transistor sizing is to produce very similar (±0.2ns difference) switching and delay times into typical (20pF) lumped-capacitive loads. The source impedances are also closely matched. Load lines to the half-voltage points yield effective output impedances for source-terminated driving of transmission lines of ≈88Ω for the high output, and ≈84Ω for the low output.

Although these CMOS pad drivers are far from ideal, they are adequate as source-terminated transmission-line drivers for long PCB runs and even for cables in the 90Ω-impedance range. Channels may be connected through up to about 10m-long cables, typically shielded ribbon cables, or ordinary ribbon cables with alternate wires grounded. Because of the zero-slack protocol, performance is degraded by the flight time of the signals, but reflections from the far end of long cables return while the outputs are assured not to be switching.

Power dissipation of the routing chips is dominated by the output drivers, and can be calculated based on the load capacitance and expected number of transitions/s. For example, the average number of transitions per flit for random data with an 8-bit-router channel is six (one request, one acknowledge, and four data transitions, ignoring the occasional tail), resulting in an average energy of 1.5nJ/flit if the signals drive 20pF loads ($\frac{1}{2} \cdot 6 \cdot 20\text{pF} \times (5\text{V})^2 = 1.5\text{nJ}$). Thus, the power required to maintain a throughput of 60Mflits/s on such a router channel is 90mW.

3 The Elko Mesh-Routing Chips

The three current versions of Elko Mesh-Routing Chips — the EMRC-2D8 (2D with 8-bit flits), the EMRC-2D9 (2D with 9-bit flits), and the EMRC-1D16 (1D with 16-bit flits) — are identical in timing and performance, and similar in packet format.

3.1 Packet Format

A packet may be of any length so long as it includes the appropriate header. A header flit of the EMRC-2D8 is represented as follows:

bit:	t	7	6	5	4	3	2	1	0
	0	sign	B	MSB					LSB
				|←		Δx or Δy			→|

The sign bit is 0 for positive and 1 for negative; the Δx or Δy distance is represented in sign and magnitude, with the 6-bit magnitude represented in binary. B specifies packet broadcast, which is not implemented in the current EMRCs.

The EMRC-2D9 has a 9th data bit that can be used for parity checking or other purposes; this 9th bit is simply passed through, and does not influence the routing. The EMRC-1D16 has the same header format as the EMRC-2D8, except that the sign bit appears in bit 15 and the magnitude in bits 5–0; bits 14–6 are passed through but ignored. The Mosaic router has the same header format as the EMRC-2D8, except that there is no broadcast bit and a 7-bit magnitude appears in bits 6–0.

On the *pi*, *ei*, or *wi* channels of 2D routers, the header is composed of a Δx flit followed by a Δy flit. The sign of the Δx flit is significant only for the *pi* channel: if the magnitude is non-zero, the sign determines whether the packet is routed east (+) or west (−), and the packet will leave the router with the magnitude of the distance decremented. When a packet enters a router on *pi*, *ei*, or *wi* with the magnitude equal to zero, the Δx header flit is stripped off, and the packet is passed to the y router. For a 1D router such as the EMRC-1D16, there is no y router, and the packet is passed to the *po* channel. These EMRC-1D16 routers can, of course, be used in pairs to provide 2D routing.

For packets that enter the y router from the x router, or that enter the y router from the *ni* or *si* channels, the header is composed only of a Δy flit. The y router is identical to the x router, but connects to northbound (+) and southbound (−) rather than eastbound and westbound channels. When a packet enters the y router with the Δy magnitude equal to zero, the Δy header is stripped off, and the packet is routed to the *po* channel.

If $\Delta x = \Delta y = 0$ on the *pi* channel, the packet is routed as usual through the x and y routers back to the *po* channel. The packet is terminated and the path freed by any non-header flit in which the tail (t) bit is 1.

3.2 The EMRC Internal FIFOs

Caltech MRCs are optimized for throughput rather than latency. Throughput is maintained through the input, head-flit-decrementing and -testing, decision, merge, and output stages of the routing process by operating these stages in a pipeline. Instead of the clocked registers used in synchronous pipelines, the MRCs use asynchronous, ripple-through FIFOs (*cf* [4, 10]). The "Ginsu," "Frontier," and "Elko" routers have each used different FIFO designs that determined the internal timing and overall design style of the routing chip.

The EMRC uses the unusual, two-cycle FIFO shown in Figure 5. This FIFO is similar to the two-cycle FIFO that Ivan Sutherland advocates for micropipelines [10]. The output of a Muller C-element [4] becomes 1 if both inputs are 1, becomes 0 if both inputs are 0, and otherwise holds its value. The effect of this chain of C-elements is to control the flow of data in a FIFO discipline through the two rows of latches. A cell is "empty" if its C-element output is the same as the C-element output (or acknowledge signal) to its right. For example, the entire FIFO is empty if the input request and acknowledge, the output request and acknowledge, and all of the C-element outputs are all 0 or all 1. A cell is "full" if its C-element output differs from the C-element output (or acknowledge signal) on its right. Its data is latched in the upper row if its output is 0, and in the lower row if its output is 1. Whenever an empty cell appears to the right of a full cell, the empty cell's C-element will change state, latching the data from the full cell, and making the full cell empty. Fullness and data thus propagates to the right, and emptiness to the left.

Although the control part of this self-timed FIFO is speed-independent, correct operation requires that the propagation delay of the latches be less than the propagation delay of the C-elements and their output drivers. In fact, the propagation delay of the latches is 1ns, and of the C-elements and their driver amplifiers is 3ns, for transitions in either direction. The structure is so symmetric that it is easy to assure that fullness ripples forward at the same 3ns/cell rate that emptiness ripples backward, a property that is inessential but simplifying. The minimal period of the FIFO is thus 6ns, corresponding to a maximal throughput of 166MB/s. The FIFO is capable

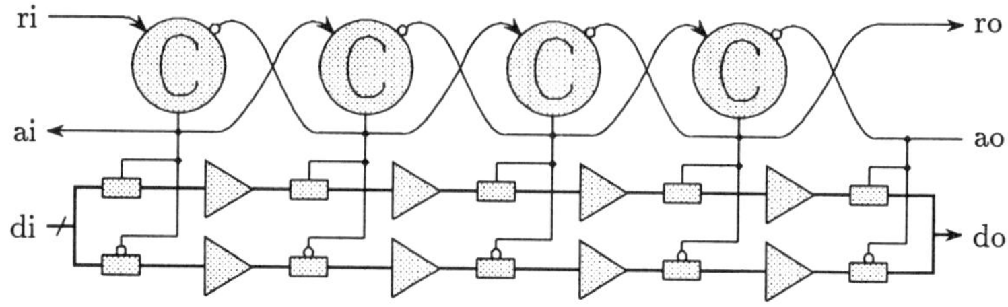

Figure 5: The EMRC, two-cycle, ripple-through FIFO.

of operating at about twice this speed, but extra delays are included in the C-element output driver for reasons described in the following paragraph. The external $\approx$70MB/s throughput rate is throttled by the pad-input and pad-output circuitry.

The difference between the 3ns C-element delay and the 1ns latch delay is responsible for a very useful property of this FIFO. On successive stages, the data will be available 2ns earlier than in the previous stage, relative to the request transition, up to a maximum of the 6ns period of the FIFO. Thus, after only three FIFO stages, the data is assured to lead the request by 6ns. The EMRC design takes advantage of this property by performing the head-decrementing operation (see Figure 6) in ordinary combinational logic (without the need for a completion signal) after a 3-stage input FIFO. The decrementer produces its result in somewhat less than 3ns, and two following FIFO stages restore the lead time of the data. A head bit is generated at the beginning of the three-stage input FIFO from reset and a unit delay from the tail bit of the previous flit, and is carried in an extra bit of the FIFO; thus, a packet is framed internally by both a head and a tail bit. The head bit is the carryin of the head decrementer. The carryout from the head decrementer indicates whether the head flit was zero, in which case the head flit will later be stripped. This zero bit is carried in still another extra bit of the FIFO.

One other useful property of this FIFO is that the lead time of the data over the request transition can be at least partially preserved when combining the data from the two rows of latches. The lead time of the data into the output stage, together with the acknowledge-switched output multiplexor, is what provides the 2ns t_{dr} interval (Figure 3) in the EMRC routers.

Although the basic cell is somewhat larger, the EMRC FIFO has numerous advantages over the FIFO used in the earlier FMRC. Correct operation of this FMRC FIFO, which is similar to the asynchronous FIFO illustrated in figure 7.24 of [4], depends on more precise determination of internal circuit delays. Also, whereas the EMRC FIFO-control output switches only once to transfer the data into the cell, the FMRC FIFO-control output switches twice to generate a narrow pulse. Only a limited number of transitions and their timing margins can be accommodated within the period of a high-speed, ripple-through FIFO. The EMRC FIFO reduces the number of required transitions to the minimum of one, maximizes timing margins, and provides two useful properties that simplify the rest of the router design.

3.3 The Internal Design of the EMRC

The internal structure of the EMRC 1D router is shown in block-diagram form in Figure 6. Just as the 2D router is composed of two 1D routers, the 1D router is composed from more elementary automata that also operate on packet streams. For this reason, we refer to the elements within the block diagram of Figure 6 as *routing automata*, and to the schema of this router as its routing-automaton formulation.

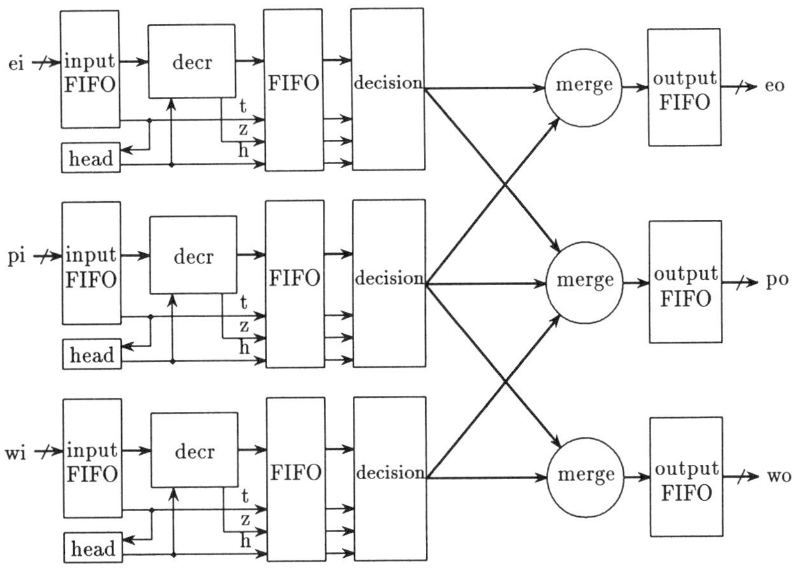

Figure 6: Internal routing-automata structure of a 1D EMRC router.

A "switching yard" follows the previously described preprocessing in the input FIFOs and decrementers. Based on the zero and sign bits of the packet header, the decision automata produce a packet-level request to the appropriate merge element. The merge elements arbitrate these packet-level requests, and set their internal multiplexors to accept flits from the source that receives the packet-level acknowledge. The packet is then spooled through this established path by flit-level requests and acknowledges until the tail flit is acknowledged, after which the decision element removes the packet-level request. The three-way merge also strips the header by acknowledging the first flit without generating a request to the output FIFO.

The merge operations are strictly fair. If a packet-level request to a two-way merge occurs when the other source is acknowledged, this request will be granted as soon as the other source removes its request. Thus, in heavy traffic, packets from the two merge sources alternate. The three-way merge is implemented with two, two-way arbitrations, such that *pi* can source one-half of the output packets, and *ei* and *wi* one-quarter each. This scheme gives *pi* some priority for injection into the network, and is also symmetrical. Giving priority to injection is based in part on the hypothesis that the multicomputer runtime system (see the paper "The Design of the Caltech Mosaic C Multicomputer" in this volume) may need to evacuate packets from a node at least as fast as they enter.

Due principally to the FIFOs in the internal routing paths, the typical $ei \rightarrow eo$, $wi \rightarrow wo$, $ni \rightarrow no$, and $si \rightarrow so$ path-formation latency for the head of a packet is ≈30ns.

4 Slack and Dialog Chips

Slack and Dialog Chips contain a pair of protocol-conversion circuits that translate between:

- the zero-slack, self-timed signaling protocol used by the EMRCs or Mosaic channels, and
- a k-slack signaling protocol suitable for conveying packets through long cables.

4.1 Slack-chip Usage

Slack or Dialog chips are used as illustrated in Figure 7. The slack-channel CMOS outputs are not suitable for driving transmission lines because the k-slack protocol does not constrain the time at which reflections may return to the outputs. Slack or Dialog chips must, accordingly, be used together with transmission-line drivers and receivers appropriate to the type of cable. Slack and dialog chips are communication devices; they are generally oblivious to the format of packets. The nominal rate of the slack channels is 60MB/s.

Slack and dialog chips share a common pinout and signal-name convention, although not all signals are used by all chips. In all cases, the request and acknowledge signals employ transition signaling. The signals and timing specifications of the *ei* and *eo* (where *e* now stands for "Elko" rather than "east") channels are the same as those of the EMRCs, and include a 9th data bit. The signals of the slack input and output channels have 3-character names constructed as *si* or *so* (where *s* now stands for "Slack" rather than "south") followed by one of (*r a c t 0–8*), where:

r is the self-timed request (transition signaling).

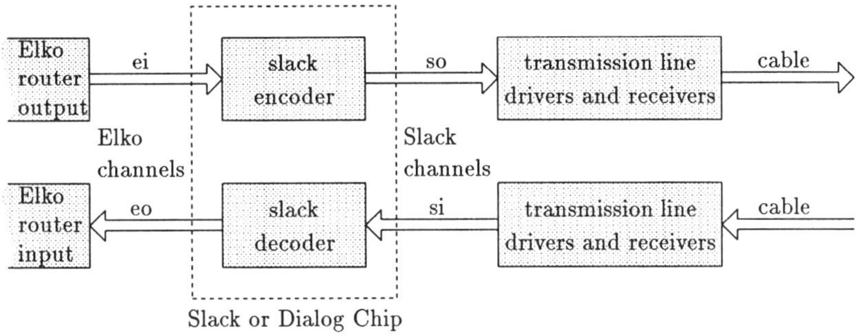

Figure 7: Slack- and dialog-chip usage.

- *a* is the self-timed acknowledge (transition signaling). This signal is not used in Dialog chips.
- *c* is a control-escape bit. This bit is not used in Slack chips.
- *t* is the tail bit, which marks the end of a packet.
- *0-8* are data bits 0-8.

Slack-chip slack channels, which include the back-going acknowledge signal but not the control-escape bit, have exactly the same signals as the zero-slack channels. The difference between the zero-slack and slack protocols is in the timing and flow control encoded in the request and acknowledge.

Dialog-chip slack channels do not include the back-going acknowledge signal; all of the signals propagate in the direction of the channel. Flow control is accomplished by the Dialog chip inserting an occasional ACK control symbol into the stream conveyed by the opposite-going channel. A control symbol is distinguished by the control-escape bit being 1. Additional control symbols are used for other purposes.

4.2 Timing specifications of the slack channels

The request and acknowledge signals of the slack channels are a variation of the transition-signaling discipline used for the zero-slack channels. The slack protocol conforms to two essential rules:

Rule 1: Each *transition* of a request signal, whether positive or negative, conveys a flit. At the slack output (*so*), data are presented a half period before the request transition, and are held at least a half period after the request transition.

The time scale in Figure 8 assumes a nominal 16ns minimal period (62.5Mflits/s maximal rate); the actual period varies somewhat from chip to chip. The use of a minimum of a half period between data and request transitions maximizes the tolerance to skew introduced by the cable between the data and request signals. At the slack input (*si*), the data are sampled at the *sir* transition ±0.2ns. Thus, the communication from an output channel to an input channel will tolerate skew between data and request signals up to 0.2ns less than the worst-case (minimum) half period.

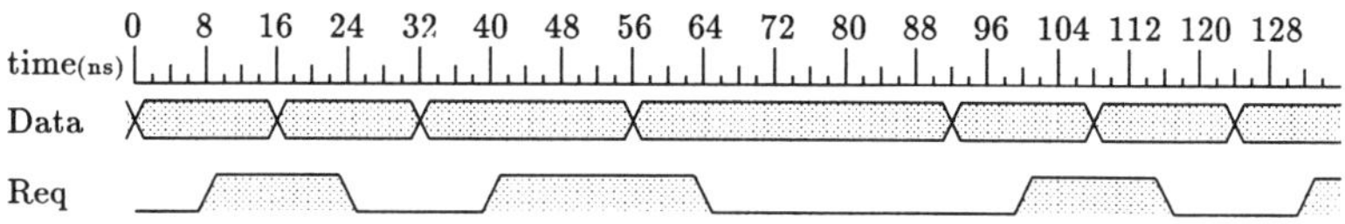

Figure 8: An example of the slack-channel protocol, as seen by the sender. The data are irrelevant a half period after a request transition, but are held at their previous value. Delay in addition to the minimum may be caused either by the data not yet being available to the encoder because of the rate at which the packet is provided to the encoder, or by the encoder not yet receiving an acknowledge indicating the freeing of slack buffers at the receiver.

Rule 2: For a slack k, corresponding to a slack decoder with k slack buffers, the encoder is permitted to present the data and request for k more flits than it has received acknowledges.

A small technicality: In the zero-slack (non-interference) protocol employed in the EMRCs, the time at which data is sampled is *referenced to the acknowledge* signal rather than to the request. The sender is, accordingly, permitted to present $k+1$ rather than k more requests than it has received acknowledges. For slack protocols in which the time at which data is sampled is *referenced to the request* signal, a slack of k corresponds to the receiver having at least k slack buffers, and the minimum value of k to support communication is one.

For Slack chips, acknowledges are also encoded as transitions. In order to limit the maximum frequency of the acknowledge signal, the receiver assures that the time between acknowledge transitions is at least the nominal period. For the Slack-20 chip, $k = 20$.

4.3 Slack-chip structure and performance

The easiest way to construct a slack chip, as illustrated in Figure 9, is to use a FIFO for the slack buffers in the slack decoder, and the FIFO control alone in the encoder as a model of the state of the decoder. The FIFO control is used here as a curious type of asynchronous, up–down, unary counter that is incremented each time a flit is sent, and decremented each time an acknowledge is received.

The slack decoder contains the k slack buffers, which are organized as a k-deep, ripple-through, asynchronous FIFO. The *sia* acknowledge is generated when a flit is passed from the FIFO to the *eo* output, and its frequency must

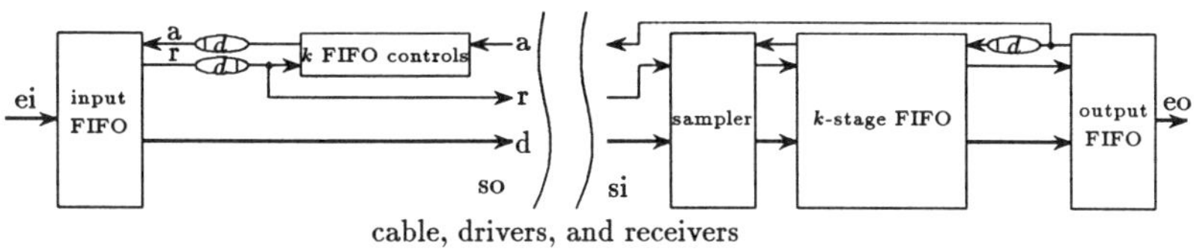

Figure 9: Structure of a slack encoder and decoder based on ripple-through FIFOs.

be limited by an internal delay. The *decoder latency,* the minimum time from the *sir* transition at the decoder input to *this flit* being presented at *eo* and being acknowledged by a transition at *sia* is ≈50ns for the Slack-20 chip.

The encoder maintains a count of the number of requests generated minus the number of acknowledges received. This count is initialized to 0 on reset, is incremented on each request generated, is decremented on each acknowledge received, and inhibits additional requests if the count is equal to k. The count is maintained implicitly in a replica of the decoder's FIFO control. It requires ≈35ns in the Slack-20 chip for the *soa* acknowledge transition to propagate through the FIFO and control circuitry to the input circuitry where *sor* and *eia* are generated. This *encoder latency* is an inevitable part of using ripple-through FIFO, and, as with the decoder latency, grows with k. The latency is as small as it is because the FIFO structures in the Slack-20 chip have been set to a higher speed than the FIFOs in the EMRC.

Starting from a condition in which all of the slack buffers are free, the encoder may generate a stream of k flits before receiving any acknowledges. For $k = 20$ and a period of 16ns, this stream lasts for 320ns. For this stream to continue without pause, the sum of the round-trip cable delay, encoder latency, and decoder latency must be not more than 320ns. The ≈85ns encoder and decoder overhead limits the round-trip cable delay for full-speed operation to ≈235ns. For cables with typical propagation velocities of $\frac{c}{1.5}$, $k = 20$, and 62.5MB/s operation, this limitation corresponds to a cable length of ≈23m. The slack protocol conveys flits correctly with larger cable delays, but not at the full bandwidth.

4.4 Dialog-Chip Protocols and Performance

We expect in the near future to replace the current use of Slack chips in Mosaic cables and in the ATOMIC LAN [1] with a more general scheme that we refer to as "dialog" protocols. For Dialog chips, acknowledges are encoded as control flits ($c = 1$) on the opposite-going channel. In order to minimize the use of channel bandwidth for acknowledges, a single ACK symbol is sent to acknowledge r flits. We expect to standardize on $k = 64$ and $r = 16$.

In an effort to reduce the magnitude and dependence on k of the

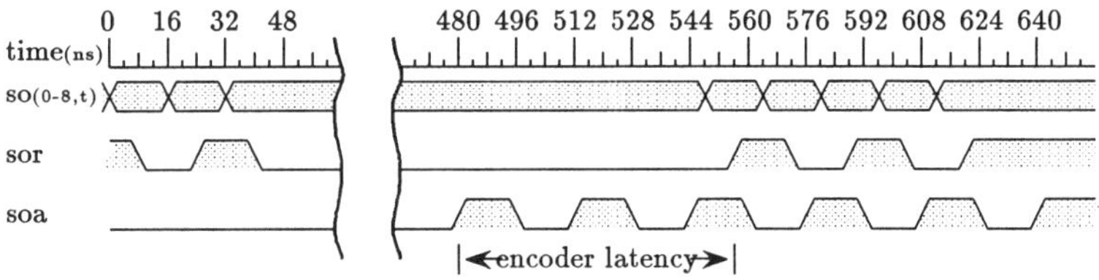

Figure 10: Example of the slack-channel protocol for an 8-flit packet with slack $k = 20$, as seen at *so,* and starting from a condition in which the receiver is blocked and all but 3 slack buffers on the receiving end are filled. The encoder may fill the 3 remaining slack buffers, but must then pause until it receives an acknowledge.

encoder and decoder latency, our design for a Dialog chip employs register-based rather than ripple-through FIFOs. The slack circuits are otherwise conceptually identical to those in the Slack chips: The decoder includes a k-deep, register-based FIFO, and the encoder includes only the FIFO control. The register-based FIFO, which includes a pair of cyclic, 1-of-k counters for the read and write pointers, also takes care of detecting each time the rth flit has left the decoder. Because r divides k, it is necessary only to detect when the read pointer advances past those FIFO cells that are multiples of r. On each such occurrence, an ACK control symbol is inserted into the opposite-going stream. On the receipt of an ACK, the dialog encoder advances its replica of the read pointer by r. Of course, the encoder's replica of the write pointer is advanced by one each time a flit is sent.

The only advantage of the dialog protocol over the ordinary slack protocol is that it is extensible. Control symbols do not appear on the *ei/eo* side of the Dialog chip, and may be inserted anywhere within a packet or between packets on the Dialog channels. Control symbols are not buffered; hence, they can be sent at any time, independent of the flow control. The minimal implementation of the Dialog chip employs only one control symbol for the flow-control acknowledge (ACK). However, between a pair of chips, there can be an arbitrary "dialog" of control symbols that are never blocked. Thus, dialog protocols allow, for example, continuous monitoring of the continuity of a bidirectional (full-duplex) channel, such that normal network blocking can be distinguished from such error conditions as a disconnected cable or an unpowered communication partner.

5 The Extended Family

There are several other members of this family currently being designed or fabricated, and that are described briefly here.

5.1 Elko-Router Interface Chips (ERICs)

The Elko-Router-Interface Chips (ERICs) provide an interface between the byte-wide, self-timed *p* channels of an EMRC – either an EMRC-2D8 or EMRC-2D9 – and either an 8-byte-wide synchronous bus (ERIC8 chip), or a 4-byte-wide synchronous bus (ERIC4 chip).

The bus interface can best be understood as a synchronous interface to a send FIFO and a receive FIFO, each (at least) 16 words long. Packets are sent by appending to the send FIFO from the bus, and are received by causing the receive FIFO to drive the bus and to advance. Between the bus interface and the self-timed channels, the ERIC provides FIFO buffering, highly reliable synchronization to the external clock, and word↔byte conversion (packing and serialization). The EMRC-2D8 and -2D9 routers strip the routing header off of each packet as it is routed. In order to maintain word alignment, the flits of an incoming packet are packed into words starting with the third byte. Thus, the packet received has the same word alignment as the packet sent, and the same word is marked with the tail bit. The ERIC chips also include a status register loaded from the third byte of each incoming packet. The status register outputs appear on pins, and may be used to control node functions.

5.2 EMRCs with broadcast

Wormhole-routing broadcast is deadlock-free only if broadcast is restricted to one direction – positive by convention – in each dimension the mesh. Broadcast can be implemented by a simple variant of the decision element of the EMRC (see Figure 6). A packet from *pi* or *ei* tagged for broadcast is sent both to *po* and *eo* if it the header is positive and non-zero. In effect, a copy is dropped off at each node until the header is decremented to zero. If the header is zero, the packet is, as usual, routed only to *po*.

5.3 Bypass chips

As noted in the paper, "The Design of the Caltech Mosaic C Multicomputer," in this volume, the bisection of a large mesh network can be augmented economically by using *bypass channels*. For example, on a network of large radix, any center-directed packets can be diverted at the $\frac{1}{3}$ and $\frac{2}{3}$ points onto the bypass channels if their destinations are outside of the central $\frac{1}{3}$ of the nodes in this dimension. The $\frac{1}{3}$ and $\frac{2}{3}$ points are optimal for maximizing average throughput under random traffic; the $\frac{1}{4}$ and $\frac{3}{4}$ points are optimal for minimizing latency. In practice, any pair of points in these regions is nearly optimal. On a 128×128 Mosaic, for example, bypass chips can be inserted between 8×8 boards 5 boards (40 nodes) from each edge.

A bypass chip requires only a very elementary router, which can be assembled from EMRC parts with no new layout, and slack or dialogs circuits to drive the cables.

5.4 A Router with Integrated Fault-Detecting and Interface Circuits

We recently developed a custom router for a commercial company that is developing a multicomputer aimed at high reliability and non-stop operation. This chip includes an EMRC-2D8 core; fault-detecting circuits on the requests and acknowedges of all of the *news* channels; synchronous, full-duplex, flit-wide channels to and from the node; and timeout and clock-detection circuits to assure consumption at *po* even when the node is unplugged or unpowered. This project has been a learning vehicle for us to understand how to achieve more nearly non-stop operation of routing networks.

Acknowledgments

The research described in this paper was sponsored by the Defense Advanced Research Projects Agency, and monitored by the Office of Naval Research. John Toole at DARPA started this effort by encouraging us to make this mature routing technology more widely available. In addition, let us acknowledge the help we have received from MOSIS in fabricating and packaging these chips, and from our operations manager, Arlene DesJardins, in packaging, testing, and distributing them.

References

[1] Cohen, D., Finn, G., Felderman, R., DeSchon, A. ATOMIC: A High-Speed, Low-Cost, Local Area Network. USC/ISI Technical Report ISI/RR-92-291, Sept-1992.

[2] Kevin Lam, Larry Dennison, and William Dally. Simultaneous Bidirectional Signaling for IC Systems. *Proceedings of the 1990 IEEE International Conference on Computer Design: VLSI in Computers and Processors*, pp. 430–433, IEEE, 1990.

[3] Charles M. Flaig and Charles L. Seitz. Inter-Computer Message Routing System With Each Computer Having Separate Routing Automaton for Each Dimension of the Network. US Patent filed 2 June 1988, issued 14 April 1992, # 5,105,424.

[4] Charles L. Seitz. System Timing. Chapter 7 in *Introduction to VLSI Systems* by Carver A. Mead & Lynn A. Conway, Addison-Wesley, 1980.

[5] Charles L. Seitz, William C. Athas, Charles M. Flaig, Alain J. Martin, Jakov Seizovic, Craig S. Steele, Wen-King Su. The Architecture and Programming of the Ametek Series 2010 Multicomputer. *Proceedings of the 1988 Hypercube Conference,* ACM Press, New York, 1988.

[6] Charles L. Seitz. Multicomputers. Chapter five in *Developments in Concurrency and Communication,* edited by C. A. R. Hoare, Addison-Wesley, 1990.

[7] Charles L. Seitz. Concurrent Computation and Programming. Chapter one in *VLSI and Parallel Computation,* edited by Roberto Suaya and Graham Birtwistle, Morgan Kaufmann Publishers, 1990.

[8] Charles L. Seitz. Let's Route Packets Instead of Wires. *Advanced Research in VLSI: Proceedings of the 1990 MIT Conference,* MIT Press, 1990.

[9] Jakov Seizovic. The Architecture and Programming of a Fine-Grain Multicomputer. Caltech Computer Science PhD thesis in preparation; publication expected June 1993.

[10] Ivan E. Sutherland. Micropipelines. (Turing Award paper) *CACM* 32:6 (720–738), June 1989.

Author Index

The MIT Press, with Peter Denning as general consulting editor, publishes computer science books in the following series:

ACL-MIT Press Series in Natural Language Processing
Aravind K. Joshi, Karen Sparck Jones, and Mark Y. Liberman, editors

ACM Doctoral Dissertation Award and Distinguished Dissertation Series

Artificial Intelligence
Patrick Winston, founding editor
J. Michael Brady, Daniel G. Bobrow, and Randall Davis, editors

Charles Babbage Institute Reprint Series for the History of Computing
Martin Campbell-Kelly, editor

Computer Systems
Herb Schwetman, editor

Explorations with Logo
E. Paul Goldenberg, editor

Foundations of Computing
Michael Garey and Albert Meyer, editors

History of Computing
I. Bernard Cohen and William Aspray, editors

Logic Programming
Ehud Shapiro, editor; Fernando Pereira, Koichi Furukawa, Jean-Louis Lassez, and David H. D. Warren, associate editors

The MIT Press Electrical Engineering and Computer Science Series

Research Monographs in Parallel and Distributed Processing
Christopher Jesshope and David Klappholz, editors

Scientific and Engineering Computation
Janusz Kowalik, editor

Technical Communication and Information Systems
Ed Barrett, editor